AF540292

Chemistry and Technology of the Cosmetics and Toiletries Industry

Chemistry and Technology of the Cosmetics and Toiletries Industry

Second edition

Edited by

D.F. WILLIAMS

Environment and Quality Assurance Manager
Givaudan-Roure UK
Surrey
UK

and

W.H. SCHMITT

Vice President, R&D
Chesebrough Ponds, Inc
Connecticut
USA

BLACKIE ACADEMIC & PROFESSIONAL

An Imprint of Chapman & Hall

London · Weinheim · New York · Tokyo · Melbourne · Madras

First edition 1992
Second edition 1996

A catalogue record for this book is available from the British Library.

Library of Congress Catalog Card Number: 96-83684

Williams & Schmitt: Chemistry and Technology of the Cosmetics and Toiletries Industry, Second edition

First Indian Reprint 2009

ISBN 978-81-8489-289-5

This edition is manufactured in India for sale only in India, Pakistan, Bangladesh, Nepal and Sri Lanka and any other country as authorized by the publisher.

This edition is published by Springer (India) Private Limited, (a part of Springer Science+Business Media), Registered Office: 3rd Floor, Gandharva Mahavidyalaya, 212, Deen Dayal Upadhyaya Marg, New Delhi – 110 002, India.

Printed in India by Rakmo Press (P) Ltd., New Delhi.

Contents

Contributors

Mr J.F.L. Chester Conform Associates, 80 Broadacres, Carlton, Goole, East Yorkshire DN14 9NF, UK

Mr J. Cunningham Vice-President Technical Services, Cosmolab, 1100 Garrett Road, Lewisberg, Tennessee 37091, USA

Mr A. Dallimore Perfumery Director, Phoenix Fragrances Ltd, Unit 6, Fleming Close, Park Farm Industrial Estate, Wellingborough, Northants NN8 6UF, UK

Dr W.E. Dupuy Personnel and Total Quality Director, Rimmel International, Carlton Road, Ashford, Kent, UK

Mr R. Giovanniello Vice President–Technical Director, Westwood Chemical Corporation, 46 Tower Drive, Middletown, NY 10940, USA

Mr. J.L. Knowlton Vice President, Research and Development, Justin–Avon, 326 Fifth Street, Wynberg, Johannesburg, Republic of South Africa

E.G. Murphy Esq. Bryan Cave LLP, 700 Thirteenth Street, N.W. Washington, DC 20005–3960, USA

Dr M. Pader President, Consumer Products Development Resources Inc., 1358 Sussex Road, Teaneck 07666, New Jersey, USA

Mr W.H. Schmitt Senior Vice President, Research and Development, Chesebrough Ponds USA Co., Trumbull Corporate Park, Merritt Boulevard, Trumbull, Connecticut 06611, USA

Mr J.J. Shipp Technical Director, Chester Laboratories Ltd, Unit 11, First Avenue, Deeside Industrial Park, Clwyd CH5 2NU, UK

Mr E. Spiess Sales Manager, Akzo Nobel Chemicals GmbH, Postfach 100132, D-52301, Düren, Germany

Mr M.J. Willcox Technical Director, The Standard Soap Co. Ltd, Ashby de la Zouch, Leicestershire LE6 2HG, UK

Mr D.F. Williams Environment and Quality Assurance Manager, Givaudan-Roure UK, Godstone Road, Whyteleafe, Surrey, UK

Dr P.J. Wilson Adams, Wilson & Associates Ltd, 8 Queens Park Road, Chester CH4 7AD, UK

Preface

This second edition has been designed to monitor the progress in development over the past few years and to build on the information given in the first edition.

It has been extensively revised and updated. My thanks go to all who have contributed to this work.

D.F.W. May 1996

Preface to the first edition

This book is the result of a group of development scientists feeling that there was an urgent need for a reference work that would assist chemists in understanding the science involved in the development of new products. The approach is to inform in a way that allows and encourages the reader to develop his or her own creativity in working with marketing colleagues on the introduction of new products.

Organised on a product category basis, emphasis is placed on formulation, selection of raw materials, and the technology of producing the products discussed. Performance considerations, safety, product liability and all aspects of quality are covered. Regulations governing the production and sale of cosmetic products internationally are described, and sources for updated information provided.

Throughout the book, reference is made to consumer pressure and environmental issues—concerns which the development scientist and his or her marketing counterpart ignore at their own, and their employer's peril. In recent years, many cosmetic fragrances and toiletry products have been converted from aerosols to mechanically pressurised products or sprays, and these are described along with foam products such as hair conditioning mousses.

The information set out in the following pages has been acquired by the hard work and enthusiasm of a number of international authors who hold senior positions within the industry and it has been my privilege to work with William H. Schmitt in bringing together their accumulated knowledge. Information presented is given in good faith and the greatest possible care has been taken to ensure that it is correct. No warranty can, however, be given on fitness for a particular use or freedom from patent infringement. All the information given is for consideration, study and verification. It is hoped that the book will inspire newcomers to progress within, and advance, the industry that has provided me with a livelihood and much enjoyment over many years.

D.F.W. 1992

1 Raw materials

E. SPIESS

1.1 Introduction

A wide range of chemical and natural materials is used in the formulation of cosmetic and toiletry preparations. It is outside the scope of this book to give a complete overview, therefore only those materials of primary importance will be stressed. This chapter deals with the basic raw materials for the production of hair-, skin- and oral-care products. It does not include special additives such as herbal extracts, nor raw materials, such as pigments, for decorative cosmetics. An overview of herbal extracts is given by Nowak [1].

The materials described are listed by application rather than by chemical classification, which is difficult to follow. This means that some products are described twice, for example the alkyl sulfates as basic surfactants and as oil-in-water emulsifiers, but makes it easier to provide an overview of a specific field and allows the direct comparison of possible ingredients. For each material, the chemical pathway of production and special precautions regarding stability and compatibility with other ingredients are also described.

1.2 Basic surfactants

The purpose of a shampoo or shower bath is to clean the skin and hair. Customers also expect a dense and luxurious lather, although this effect is not essential for the cleansing. To fulfil these requirements, the so-called basic or primary surfactants, which are the backbone of the cleansing products, are necessary.

1.2.1 *Alkyl ether sulfates*

$$\text{R}-(OCH_2CH_2)_xSO_4M$$

Alkyl ether sulfates are the most widely used surfactants for cosmetics and toiletries due to their well-balanced properties. They are excellent foamers,

= alkyl C chain = R

although the foam structure is relatively coarse. Foaming is not affected by hard water and this, together with their low critical micelle concentration, makes alkyl ether sulphates ideal for use in foam-bath preparations, shampoos and shower baths. Therefore Gohlke and Berghausen called it an 'ideal Tensid' [2].

The alkyl moiety usually consists of a mixture of C_{12}–C_{16} chains. The degree of ethoxylation is between 1 and 4 ($x = 1$–4). M is either sodium, ammonium or magnesium.

In former times, chlorosulfonic acid was used in a batch process. Today, however, the most common process for the production of alkyl ether sulfates is the continuous SO_3-sulfonation of ethoxylated fatty alcohols in a thin-film reactor, followed by a neutralization step. The typical products, called lauryl ether sulfates or sodium laureth sulfate (CTFA—Cosmetic Toiletry and Fragrance Association—nomenclature), are based on alcohol mixtures. Well-accepted and approved for use in cosmetics and toiletries are the products based on 'natural' alcohols derived from the coconut palm (C_{12} approx. 70%; C_{14} approx. 25%; C_{16} approx. 5%), the 'semi-natural' alcohols from the Ziegler process (C_{12}/C_{14} approx. 50/50%) and the partly branched SHOP® alcohols (C_{12}:C_{13}:C_{14}:C_{15} approx. 20:30:30:20)[3].

The properties of alkyl ether sulfates are largely determined by the number of ethylene glycol units in the starting ethoxylated alcohols. Figure 1.1 shows the influence of the degree of ethoxylation on skin compatibility as measured by the Zein number (an *in vitro* measurement of the irritancy of surfactants). The presence of 2 to 3 ethylene glycol units leads to ideal behavior regarding foam and detergency. Such alkyl ether sulphates are easily thickened with salts and show good water solubility, even at low temperatures. The skin and eye compatibility is acceptable for most applications but is usually

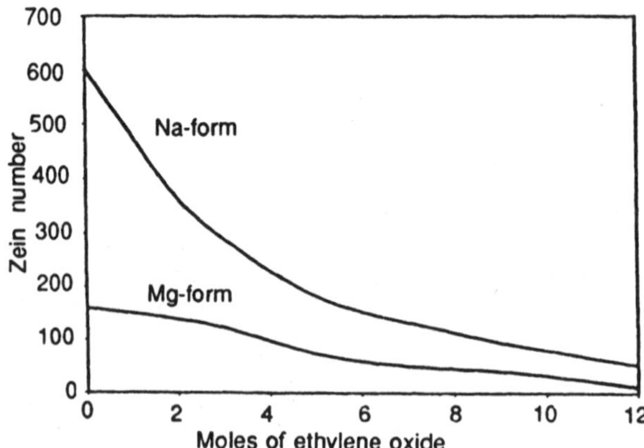

Figure 1.1 Zein number as a function of the POE-chain (moles of ethylene oxide) in laureth-x-sulfates, sodium and magnesium forms [4].

improved by combination with mild secondary surfactants. Higher degrees of ethoxylation, up to 8 or 12, can significantly reduce irritation. Combinations of these highly ethoxylated ether sulfates, especially the magnesium form with normal ether sulfates, are recommended for baby shampoos.

Alkyl ether sulfates have been found to be a source of relative high levels of 1,4-dioxane. Up to 600 ppm of this impurity has been found in shampoos. The impurity at this level presents little or no health risk, but improvements in the sulfonation process have resulted in lowering of the dioxane level below 20 ppm (calculated on active matter). This level of dioxane is now standard for ether sulfates.

Water-based and salt-free alkyl ether sulfates are commercially available in two forms: (i) low concentrate (27–30% active matter), a low viscosity clear solution; and (ii) high concentrate (70% active matter), a translucent paste with pseudoplastic flowing behavior. In low concentrations the ether sulfates form spherical micelles which are transformed into rod-like micelles when the concentration increases (Figure 1.2, phase I). At around 50% active matter these rods are tightly packed and form a hexagonal structure, a stiff gel (Figure 1.2, phase II). At 70% active matter a sandwich-like structure, the liquid crystalline lamellar phase, appears. The layers can glide on each other and the product is easily pumpable (Figure 1.2, phase III). Advantages of the 70% form are: (i) reduced transport costs, (ii) increased storage capacity; and (iii) microbiological safety without the need for preserving agents. The concentrate must be stored above 20°C and needs special equipment for dilution because the gel phase between 30 and 65% must be passed through. Tech-

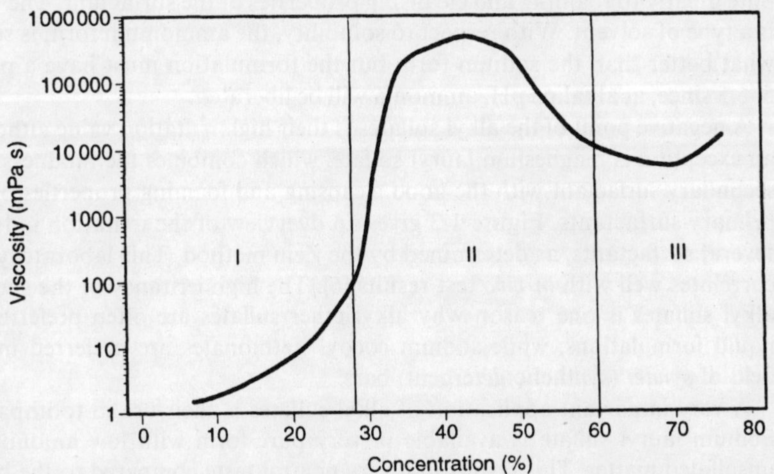

Figure 1.2 The different liquid crystalline phases of ether sulfates. I : L-phase, spherical to rod-like micelles. II : M_1-phase, hexagonal structure. III : G-phase, lamellar (sandwich) structure.

nology is available for both batch and continuous dilution. The principle is always the same: to increase the speed of dilution a large surface has to be formed using colloid mills or a spray process [5].

1.2.2 *Alkyl sulfates*

Alkyl sulfates were the first synthetic surfactant to be used in large quantities in cosmetics and toiletries and, in some countries including the USA, they are still the most popular. The alkyl sulfates used for cleansing products are usually lauryl derivatives (alkyl = C_{12} to C_{16}). As in the case of the ether sulfates, the starting alcohols are seldom pure. Sodium, ammonium, mono- and triethanolamine, magnesium and other salt forms are present in concentrations of 30–40% active product diluted in water.

Alkyl sulfates are excellent foamers, producing a rich and creamy foam, but are unstable in hard water. This is not a problem for shampoo or shower applications, but means that they cannot be used as the main component in foam baths. Instead, they must be combined with surfactants such as olefin sulfonates or sulfosuccinates, which are stable in hard water. The sodium forms of fatty alcohol sulfates are not very soluble at low temperatures, and formulations with high amounts of sodium lauryl sulfate often cloud up at temperatures below 5°C. Triethanolamine (TEA) salts show much better cold-temperature stability but relatively higher amounts must be used since TEA represents nearly one third of the molecular weight and does not contribute greatly to foaming and cleansing properties of the surfactant. The TEA is a type of solvent. With respect to solubility, the ammonium form is somewhat better than the sodium form but the formulation must have a pH of 5–6.5 since, at alkaline pH, ammonia will be liberated.

A negative point of the alkyl sulfates is their high irritation value although an exception is magnesium lauryl sulfate, which combines the mildness of a secondary surfactant with the good cleansing and foaming properties of the primary surfactants. Figure 1.3 gives an overview of the irritation index of several surfactants, as determined by the Zein method. This laboratory test correlates well with *in vivo* test results [6]. The high irritancy of the sodium alkyl sulfates is one reason why alkyl ether sulfates are often preferred in liquid formulations, while sodium cocoyl isethionates are preferred in the field of *syndet* (*syn*thetic *det*ergent) bars.

A very important application of alkyl sulfates is their use in toothpastes. Sodium lauryl sulfate is available in very pure form with low amounts of unsulfated matter. This results in a very neutral taste compared to the bitter taste of other surfactants.

Chemically, both alkyl sulfates and alkyl ether sulfates are esters of sulfuric

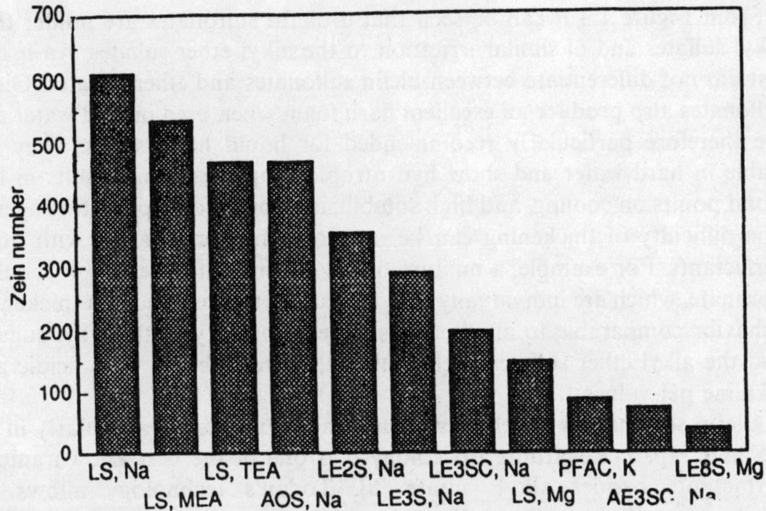

Figure 1.3 Zein number of various surfactants. LS = lauryl sulfate; LExS = laureth sulphate (x = 2, 3, 8); AOS = olefin sulfonate; PFAC = protein fatty acid condensate; LE3SC = laureth-3 sulfosuccinate; AE3SC = lauramido PEG3 sulfosuccinate.

acid and are therefore unstable at acidic pH values. In this case, the hydrolysis is an autocatalytic process where the velocity of hydrolysis increases with decreasing pH. This is mainly a problem regarding storage at elevated temperatures, but should also be taken into account during the development of formulations with low pH values [3].

1.2.3 *α-Olefin sulfonates*

CH=CH —SO_4Na

—SO_3Na
OH

α-Olefin sulfonates are a mixture of alkene sulfonates and hydroxy alkane sulfonates prepared by sulfonation of α-olefins followed by alkaline hydrolysis [7]. Contrary to most other surfactants, where the C_{12} alkyl chains show the highest surface activity, olefin sulfonates show maximum activity when C_{14} and C_{16} olefins are used. In the first step of production, which is the sulfonation in a thin-film reactor, alkene sulfonic acid and sultones are formed. These sultones are then hydrolysed into alkene sulfonates and hydroxy alkane sulfonates. The mixture usually contains approx. 60–65% of alkene sulfonates, 35–40% of hydroxy alkane sulfonates and up to 10% of disulfonates. The olefin sulfonates are commercially available as solutions with approx. 40% active matter.

From Figure 1.3 it can be seen that α-olefin sulfonates are milder than alkyl sulfates and of similar irritation to the alkyl ether sulfates. An *in vivo* test did not differentiate between olefin sulfonates and ether sulfates. Olefin sulfonates also produce an excellent flash foam when used in cold water and are therefore particularly recommended for liquid hand soaps. They are stable in hard water and show hydrotropic properties. This results in low cloud points on cooling, and high solubilizing power for superfatting agents. The difficulty of thickening can be overcome by combination with other surfactants. For example, a mixture of 60% olefin sulfonate and 40% sulfosuccinate, which are individually very difficult to thicken, shows a thickening behavior comparable to alkyl ether sulfates. Contrary to the alkyl sulfates and the alkyl ether sulfates, olefin sulfonates are stable at both acidic and alkaline pH values.

Olefin sulfonate is a well-established basic surfactant particularly in the US and Japan. Regarding toxicology it is one of the best known anionic surfactants besides alkyl sulfate [8]. Today's technology allows the production of light colored sulfonates without bleaching.

1.2.4 *Other basic surfactants*

Linear alkyl benzene sulfonate and alkane sulfonate are very powerful surfactants with good detergency and foaming properties. However, due to their strong defatting action, they leave a harsh and dry feel on skin and hair. For very cheap formulations, small amounts can be used to reduce the costs of raw materials. Their main use is in dish-washing liquids and all-purpose cleaners.

1.3 Mild anionic surfactants

Compared to the basic surfactants, a much wider group of mild anionic surfactants is used in cosmetics and toiletries. The purpose of mild or secondary surfactants is to improve skin and eye compatibility of the formulation. On the other hand, mild surfactants usually show reduced foaming and cleansing compared with basic surfactants. Fortunately there are many synergisms for combinations of surfactants which are known to compensate or partially compensate for this behavior.

1.3.1 *Sulfosuccinates*

$$\mathrm{ROOC{-}\underset{\displaystyle SO_3Na}{\underset{|}{C}H}{-}CH_2{-}COONa}$$

R = e.g. lauryl, laureth, lauramido PEG-3, oleamido MEA

Sulfosuccinates represent a diverse group of derivatives of sulfosuccinic acid. Both mono- and diesters are known. The diesters, particularly the sodium diethylhexyl sulfosuccinate, are very effective wetting agents for industrial use, while the monoesters are mild surfactants with good foaming properties, and are widely used in cosmetics and toiletries. Sulfosuccinates are prepared in a two-step process. The first step is the esterification of maleic anhydride with an alcohol. This can be, for example, a fatty alcohol, an ethoxylated fatty alcohol, a fatty acid monoethanolamide or an ethoxylated monoethanol-amide. The second step is the sulfonation of the resulting half-ester with sodium sulfite.

The most popular derivative used in Europe is disodium laureth sulfo-succinate, where laureth refers to an ethoxylated lauryl/myristyl alcohol with approx. 2–3 moles ethylene oxide. In the USA, sulfosuccinates based on amides, particularly the disodium oleamido sulfosuccinate, are more important.

The sulfosuccinates have good skin compatibilities compared with the basic surfactants (Figure 1.3). In general, the amide-based sulfosuccinates are better than the alcohol-based types. Sulfosuccinates are good foamers and are relatively cheap. A disadvantage is their instability at both low and high pH values. The highest stability is obtained at pH values between 6 and 6.5. The ethoxylated forms are stable in hard water and are useful lime soap dispersants. They are mainly used in liquid formulations. The non-ethoxylated forms are good co-surfactants for syndet bars. Sulfosuccinates are relatively poor solubilizers for perfumes.

1.3.2 *Cocoyl isethionates*

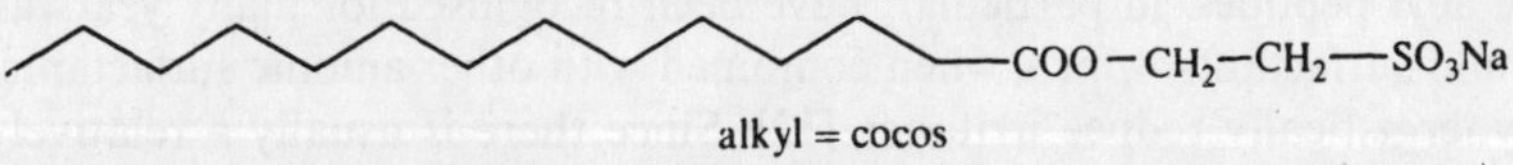

$$\text{alkyl}{-}COO-CH_2-CH_2-SO_3Na$$

alkyl = cocos

The cocoyl isethionates can be prepared in two ways: (i) by the direct esterification of sodium 2-hydroxyethansulfonate with coconut fatty acid; and (ii) by condensation with fatty acid chloride, which leaves small amounts of sodium chloride in the product.

Cocoyl isethionate is mainly used as a surfactant in syndet bars. In normal soaps it improves skin compatibility and acts as a lime soap dispersant. The extraordinary mildness of this surfactant has been proven by more than 30 years experience [9–11]. For syndet bar production, plasticizers such as stearic acids, fatty alcohols or waxes are used. Cocoyl isethionate is relatively insoluble in these plasticizers, but has to be dispersed. Consequently, cocoyl isethionates are usually offered as extremely fine powders with a particle size below 40 μm. Ready-made premixes in noodle-form, which are combinations of platicizers with cocoyl isethionate and other surfactants, are also available to the cosmetic formulator.

At room temperature cocoyl isethionate shows only limited solubility in water. This permits its use in syndet bars but can lead to crystallization in liquid preparations that are not properly formulated. Due to its excellent skin compatibility and emolliency properties, the use of sodium cocoyl isethionate in baby products and in facial-wash formulations has been reported. Since it is an ester, cocoyl isethionate shows maximum stability in the pH range 5–7.

1.3.3 *Acyl amides*

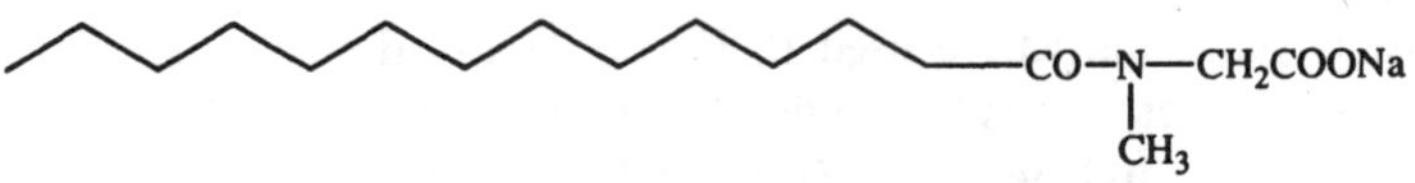

$$\text{(alkyl chain)}\!-\!CO\!-\!\underset{\displaystyle CH_3}{\underset{|}{N}}\!-\!CH_2COONa$$

Sodium acyl sarcosinate

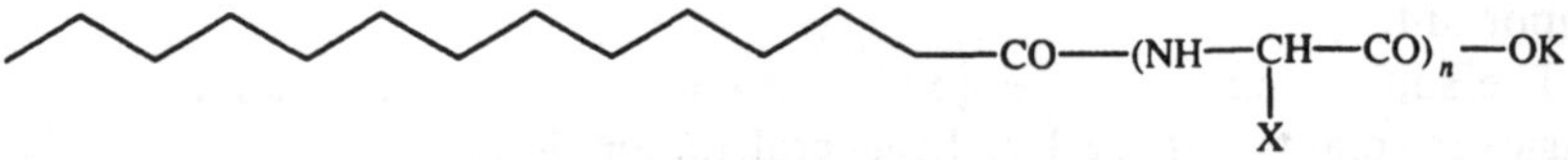

$$\text{(alkyl chain)}\!-\!CO\!-\!(NH\!-\!\underset{\displaystyle X}{\underset{|}{CH}}\!-\!CO)_n\!-\!OK$$

Potassium acyl hydrolysed animal protein

The two most important acyl amides used in cosmetics and toiletries are the sarcosinates and the protein fatty acid condensates. They are prepared by reaction of fatty acid chloride with *N*-methylglycine or with oligopeptides. The acyl amides can be considered as modified soaps with good water solubility. Due to their enlarged polar ends, they are stable in hard water. The fact that they are amides means that they are more stable than esters. Their high price and specific odor is the reason for their limited use.

The acyl peptides, in particular, have been recognised for many years as very mild surfactants, which, when combined with other anionic surfactants, can synergistically reduce irritancy [12]. Since there is usually a relatively high amount of unreacted fatty acid in the products, the potassium and TEA salts are most commonly used.

1.3.4 *Alkyl ether carboxylates*

$$\text{(alkyl chain)}\!-\!(OCH_2\!-\!CH_2)_x\!-\!OCH_2\!-\!COOM$$

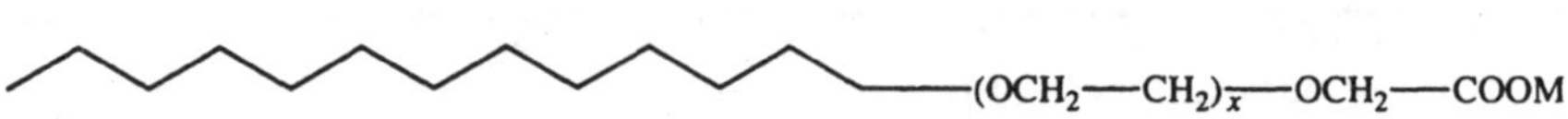

M = H or alkali

Alkyl ether carboxylates are prepared by reaction of chloroacetic acid with fatty alcohol ethoxylates. R typically represents $C_{12/14}$ alkyl groups. The structure is similar to the fatty alcohol ether sulfates, but the acidity of the carboxyl group is lower than that of the sulfate group. Consequently, ether carboxylates are milder surfactants. However, they also show reduced foaming and cleansing. In principle, the influence of the degree of ethoxylation on skin compatibility is similar to that for alkyl ether sulfates as shown in Figure 1.1.

Unlike alkyl ether sulfates, alkyl ether carboxylates have no ester linkages and are stable at low pH-values. They are available as 100% active material in the acidic form, or as neutralized solutions [13].

1.3.5 *Magnesium surfactants*

It is often stated in literature that magnesium surfactants have significantly better skin compatibility than sodium, ammonium or amine neutralized anionic surfactants. Figure 1.3 shows that there is indeed a large difference, especially in the fatty alcohol sulfates which can be considered as ether sulfates with zero EO groups. *In vivo* tests have also proven this effect. In a flex wash test on ten volunteers, nine strong and one weak reactions were found in the case of the sodium lauryl sulfate, while just one weak reaction was found in the case of a magnesium lauryl sulfate.

The advantage of the magnesium surfactants compared with other mild surfactants is that their applicational properties are practically unchanged compared with the sodium forms. From this point of view they belong to the group of basic surfactants showing high foaming and excellent cleansing properties [4, 11]. In case of the ether sulfate, easy thickening with electrolytes is still possible. The solubility for oils and perfumes is much greater when magnesium surfactants, rather than sodium surfactants are used.

The preservation of formulations based on magnesium surfactants is somewhat problematic, since divalent ions such as magnesium increase the resistance of microbes [14, 15]. Consequently, where magnesium surfactants are used in formulations, the preservation system should be carefully checked.

1.3.6 *Alkyl phosphates*

$—PO_4M_2$ M = H or alkali

While di- and trialkyl phosphates are known as emulsifiers, the pure monoalkyl phosphates are good surfactants that show high foaming combined with mildness to the skin [16]. Due to the fact that dialkyl phosphates are defoamers, a high amount of monoalkyl phosphate is necessary. In Japan, particularly, this product group is used for mild shampoos and shower-bath preparations. Both the alkyl phosphates based on fatty alcohols, and the alkyl phosphates based on low ethoxylated fatty alcohols are described in the Japanese Comprehensive Licensing Standards of Cosmetics.

1.4 Amphoteric surfactants

Amphoteric surfactants are surfactants where the charge changes as a function of the pH value of the formulation in which they are used. They are generally regarded as mild surfactants but this is not a simple matter and may not always be true. Amphoteric surfactants build complexes in combination with anionic surfactants and these complexes are milder than the individual surfactants. This is shown in Figure 1.4 for the combination of lauryl sulfate and cocamidopropyl betaine [17].

1.4.1 *Alkyl betaines*

CH_3

$-N^{+}-CH_2-COO^{-}$

CH_3

This product group derives its name from the betaine, trimethylglycine, which is found in sugar beets. The betaines are zwitterionic, with positive and negative charges on the molecules.

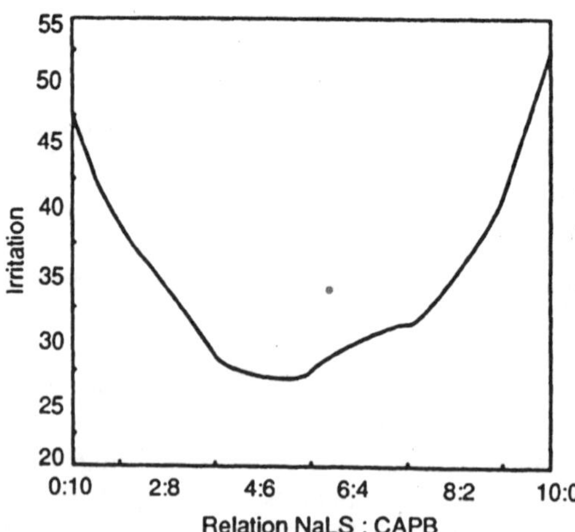

Figure 1.4 Irritation of mixtures of sodium lauryl sulfate and cocamidopropyl betaine [17].

Alkyl betaines are prepared by condensation of an alkyl dimethyl amine with sodium chloroacetate. Commercial betaines are usually 30% active products containing approx. 6% sodium chloride, which is a by-product of the reaction. Depending on the pH, the alkyl betaines can be cationic or anionic surfactants. At pH values between 5 and 7, which are typical for shampoos or shower baths, they form ionic complexes in combination with anionic surfactants such as ether sulfates or fatty alcohol sulfates. These complexes are poorly soluble in water but the presence of excess anionic surfactant or betaine aids solubilization. The improvement of the skin compatibility of an anionic formulation due to the addition of betaine is partly based on a reduction in the overall activity. A side-effect is the increase in the size of the micelles and, consequently, an increase in the viscosity. This principle is often used for thickening of a formulation, especially in case of formulas based on non-ether sulfates.

Betaines can improve the foaming of a formulation, particularly the structure of the foam, which becomes finer and more creamy. On the other hand, a 1:1 complex of anionic surfactant and betaine shows poor foaming and poor solubility in water. In conditioning shampoos, betaines support the effect of polymeric quaternaries, which impart good manageability and body to the hair.

1.4.2 *Alkylamido betaines*

$$\text{(alkyl chain)}-CO-NH-CH_2CH_2CH_2-\overset{CH_3}{\underset{CH_3}{\overset{|}{\underset{|}{N^+}}}}-CH_2-COO^-$$

Alkylamido betaines are prepared by condensation of fatty acids with dimethylamino propyl amine (DMAPA), followed by reaction with sodium chloroacetate. The amido betaines are considered to be milder than the alkyl betaines. This, together with their lower price, is the main reason why amido betaines are much more often used in cosmetic formulations. Today, efforts are being made to improve the quality and to reduce the level of by-products such as free amine and free chloroacetic acid. The application properties of alkylamido betaines are generally similar to those of the alkyl betaines. The coco-type is by far the most important.

1.4.3 *Acylamphoglycinates and acylamphopropionates*

$$RCO-NH-CH_2CH_2-\underset{CH_2CH_2OH}{\underset{|}{N}}-CH_2COONa \qquad \text{Acylamphoglycinate}$$

$$RCO-NH-CH_2CH_2-\underset{CH_2CH_2OCH_2COONa}{\underset{|}{N}}-CH_2COONa \qquad \text{Acylamphocarboxyglycinate}$$

$RCO-NH-CH_2CH_2-N(CH_2CH_2OH)-CH_2CH_2COONa$ Acylamphopropionate

$RCO-NH-CH_2CH_2-N(CH_2CH_2OCH_2CH_2COONa)-CH_2CH_2COONa$ Acylamphocarboxypropionate

The most important products of this group are the cocoderivatives—cocoamphoglycinate, cocoamphocarboxyglycinate, cocoamphopropionate and cocoamphocarboxypropionate. In former times these were called imidazoline derivatives but chemical studies have shown that the imidazoline structure hydrolyses completely during the production process. In the first production step, aminoethyl ethanolamine (AEEA) is reacted with fatty acid to produce an amide. During this stage the imidazoline ring is formed. The next step is reaction with sodium chloroacetate. This leads to the production of glycinate or, in case of reaction of two moles sodium chloroacetate, the carboxyglycinate [18]. As in the case of the betaines, salt (NaCl) is a by-product of the reaction. When acrylic acid is used in the second step, salt-free products, which are the propionates, can be obtained.

The glycinates and propionates have good skin compatibilities and show some conditioning properties in shampoo application. They are excellent foamers but the foam is not very stable in hard water. When combined with anionics, they synergistically reduce the irritancy [19]. In the US the glycinates and propionates are widely used, while in Europe the betaines are more often selected. The glycinates and propionates are generally regarded as milder than the betaines.

1.4.4 *Amine oxides*

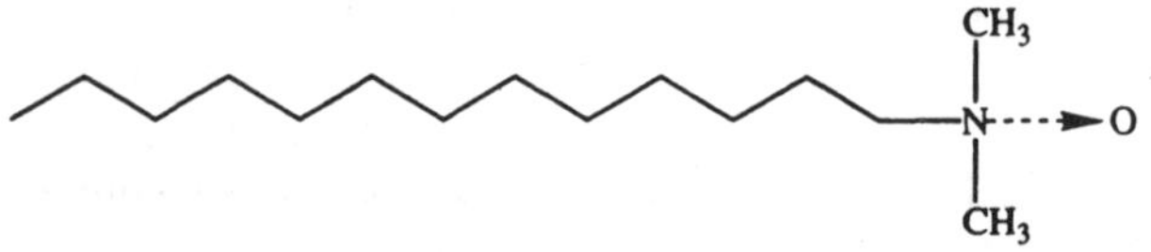

Alkyl chain–$N(CH_3)_2 \rightarrow O$ Alkyl dimethyl aminoxide

Unlike betaines, amine oxides are never anionic. However, they do show cationic or non-ionic behavior depending on the pH, and therefore behave quite similarly [20]. Amine oxides are produced by oxidation of tertiary alkyl amines (e.g. dimethyl or ethoxylated amines) with hydrogen peroxide. They are colorless liquids and free from salt (NaCl). As such, they are excellent foamers. In combination with anionics, small amounts can act as foam boosters and can improve the foam structure. Like betaines, they are good thickeners for anionic surfactants. In addition, amine oxides are good conditioning agents in hair rinses. They are commercially available as 30–50% aqueous solutions.

1.5 Non-ionic surfactants

1.5.1 *Ethoxylated products*

$(OCH_2CH_2)_xOH$

Fatty alcohol ethoxylate

$CO—(OCH_2CH_2)_xOH$

Fatty acid ethoxylate

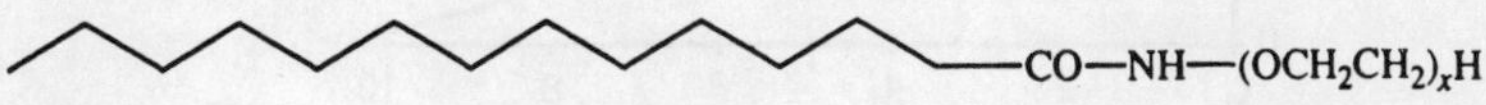

$CO—NH—(OCH_2CH_2)_xH$

Fatty amide ethoxylate

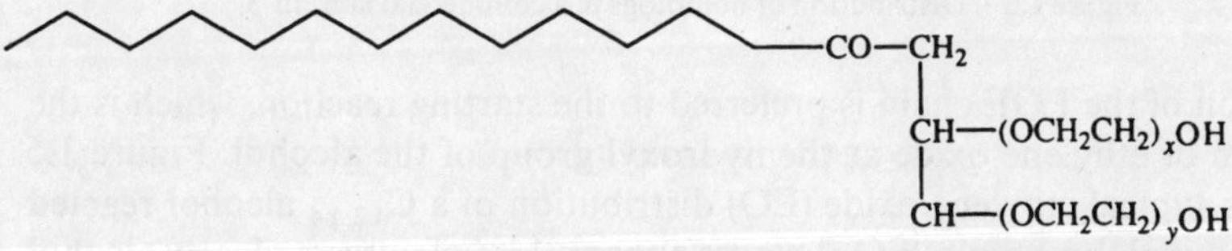

$CO—CH_2$

$CH—(OCH_2CH_2)_xOH$

$CH—(OCH_2CH_2)_yOH$

Monoglyceride ethoxylate

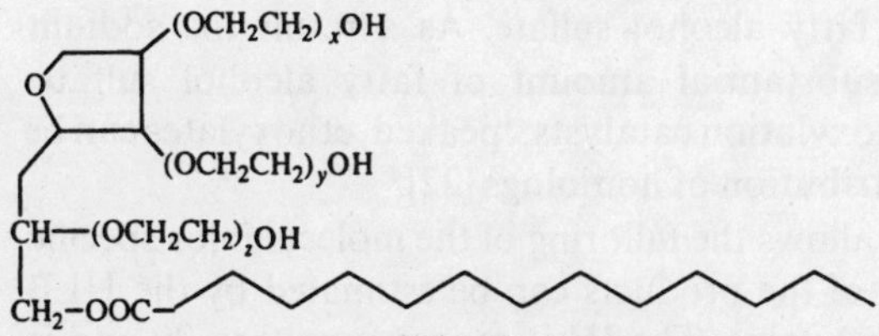

$(OCH_2CH_2)_xOH$

O

$(OCH_2CH_2)_yOH$

$CH—(OCH_2CH_2)_zOH$

CH_2–OOC–

Sorbitan ester ethoxylate

Due to their poor foaming properties, non-ionic surfactants are rarely used as shampoo surfactants, although they are often added as solubilizers for perfumes. An exception is polysorbate 20, an ester of lauric acid and sorbitan, ethoxylated with approx. 20 moles of ethylene oxide. Due to its extraordinary mildness, this is used as the main surfactant in 'non-sting' baby-shampoo formulations [21]. For this application a relatively low foaming is acceptable.

Ethoxylation allows the production of a wide range of products. Variables are the hydrophobic part of the molecule, and the number of ethylene oxide units added. Pure ethers are obtained when fatty alcohols are used as starting molecules. The ethoxylation is carried out using ethylene oxide under pressure and heat. Alkalis are usually added as catalysts. The resulting product contains a broad range of ethoxylated alcohols differing in the polyoxyethylene (POE) chain length. The nominal degree of ethoxylation represents an average.

Ethoxylated alcohols do not follow the Poisson distribution. Typically they contain a relatively high amount of unreacted fatty alcohol because the

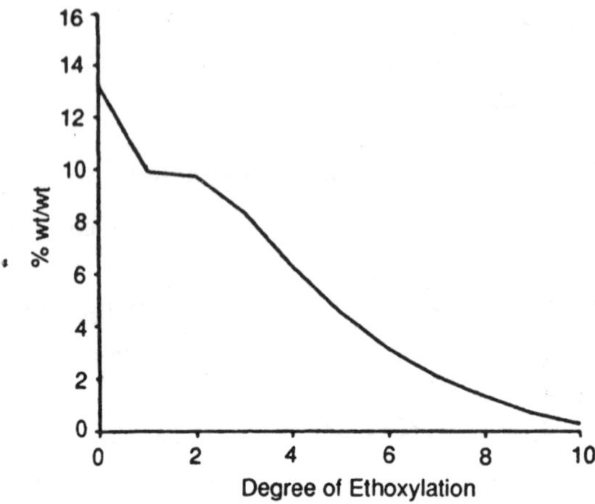

Figure 1.5 Distribution of homologs in a commercial laureth-3.

extension of the POE chain is preferred to the starting reaction, which is the addition of ethylene oxide at the hydroxyl group of the alcohol. Figure 1.5 shows a typical ethylene oxide (EO) distribution of a $C_{12/14}$ alcohol reacted with nominal three moles EO. There is approx. 13% of unreacted fatty alcohol in the product (degree of ethoxylation = 0). During the following sulfatation step, the free alcohol forms fatty alcohol sulfate. As a result the sodium laureth sulfate contains a substantial amount of fatty alcohol sulfate. Using a new generation of ethoxylation catalysts, 'peaked' ethoxylates can be obtained with a narrower distribution of homologs [22].

The degree of ethoxylation allows the tailoring of the molecules for specific requirements. The properties of the products can be estimated by the HLB (hydrophilic lipophilic balance) value. The HLB ranges from 0 to 20 and is calculated from the following expression:

$$\text{HLB} = \frac{\%\ \text{wt/wt ethylene oxide}}{5}$$

Table 1.1 outlines the main properties and applications of ethoxylates as a function of the HLB value.

Table 1.1 Properties and applications of ethoxylates as a function of the HLB value [23]

HLB value	Solubility in water	Application
0–2	Insoluble	Anti-foam additives
3–6	Poorly dispersable	W/O-emulsifiers
7–9	Dispersable	Wetting agents
8–18	Dispersable	O/W-emulsifiers
13–15	Soluble	Surfactants, detergents
15–18	Soluble	Solubilizers

Other non-ionic surfactants are ethoxylated glycerol esters, ethoxylated castor oil derivatives, ethoxylated fatty amides, and ethoxylated fatty acids. Nonylphenol ethoxylates are also very effective emulsifiers and solubilizers but, due to ecological reasons, are now rarely used in Europe [22].

1.5.2 *Alkyl polyglycosides*

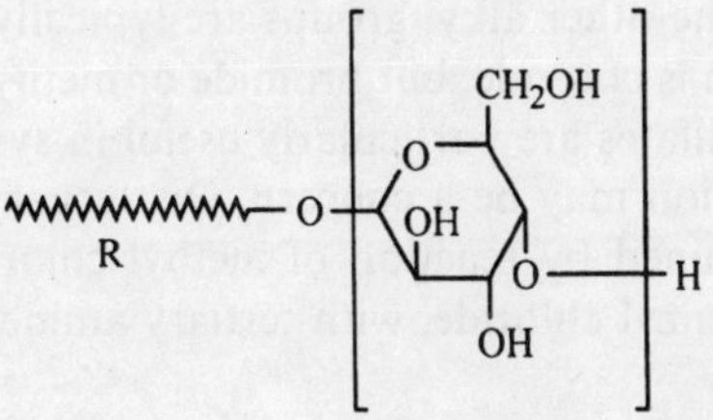

R=octyl/decyl or lauryl/myristyl

x = 1-5

A relatively new range of commercially available surfactants is the alkyl polyglycosides (APG). First described by Emil Fischer approx. 100 years ago they have been produced on a large scale since 1992. Based on carbohydrates, APGs are mixtures of different isomers, (i) stereoisomers, (ii) binding isomers and (iii) ring isomers. This together with the C-chain distribution of the alkyl group and the variation in the degree of glucosidation makes them of extremely complex composition [24].

Surface active APGs are prepared by the glycosylation of starch or monomer glucose with fatty alcohols. This can be done directly or by transglycosylation using butylpolyglucosides as intermediates in a two-step process. These processes are proton catalysed and carried out at 120–140°C under pressure and with excess fatty alcohol. In a separate step the excess fatty alcohol is removed by vacuum distillation and recycled [25].

The optimum surface activity is obtained with an alkyl chain C_8–C_{16} and a mean value degree of polymerization between 1.1 and 3. Although not as good as anionic surfactants these show surprisingly good foaming properties compared with ethoxylates [26]. This combined with their excellent skin compatibility makes them useful secondary surfactants. Since the raw materials used are from renewable sources and they are considered neutral with regard to the Earth's CO_2 balance these surfactants can be expected to play an important role in the future.

1.6 Cationic surfactants

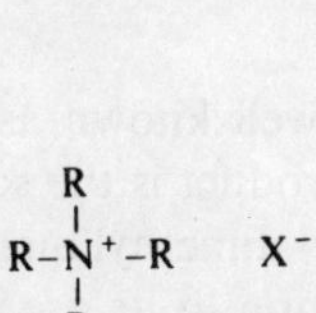

$$R_4N^+ \; X^-$$

Unlike the anionic or amphoteric surfactants, cationics are seldom used for cleansing applications [27]. Due to their conditioning properties, cationic surfactants show a high substantivity to surfaces such as skin and hair. They can act as emollients for the skin, or as conditioning agents on the hair. The main effects are to stop 'tangling', to improve wet- and dry-combing and, as a result of antistatic properties, to prevent fly-away of the hair.

A quaternary ammonium salt can generally be considered as ammonium chloride (NH_4Cl) where the four hydrogen atoms are substituted by alkyl groups. At least one of them is a hydrophobic entity with a carbon-chain length between 12 and 22. The other alkyl groups are typically methyl. The most widely used counter-ion is chloride, but bromide or methyl sulfates can also be found. The methyl sulfates are particularly useful in systems such as aerosols, where metal corrosion may be a concern. Quaternary ammonium compounds are usually obtained by reaction of methyl chloride, dimethyl sulfate or, in special cases benzyl chloride, with tertiary amines.

1.6.1 *Monoalkyl quaternaries*

$$\begin{array}{c} CH_3 \\ | \\ R-N^+-CH_3 \quad Cl^- \\ | \\ CH_3 \end{array}$$

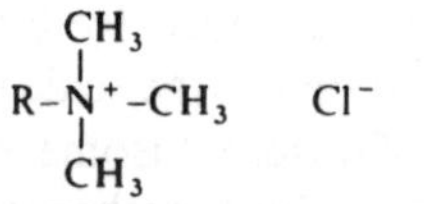

R = cetyl

The monoalkyl quaternaries are still of major importance in conditioning. The most important is cetyl trimethyl ammonium chloride (CTMAC) which shows a light to moderate conditioning intensity combined with excellent water solubility. The substantivity and conditioning properties are improved as the length of the carbon chain increases. Behenyl quaternaries are now offered as products with good conditioning properties, but are more difficult to incorporate compared with CTMAC.

1.6.2 *Dialkyl quaternaries*

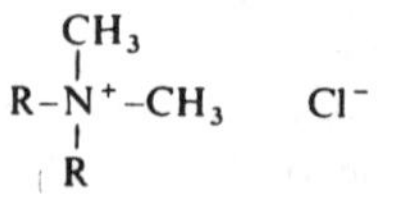

$$\begin{array}{c} CH_3 \\ | \\ R-N^+-CH_3 \quad Cl^- \\ | \\ R \end{array}$$

Quaternium-18
R = cetyl/stearyl

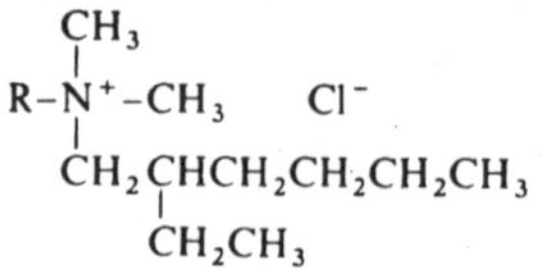

$$\begin{array}{l} CH_3 \\ | \\ R-N^+-CH_3 \quad Cl^- \\ | \\ CH_2CHCH_2CH_2CH_2CH_3 \\ \quad\;\; | \\ \quad\;\; CH_2CH_3 \end{array}$$

Hydrogenated tallow octyl dimonium chloride
R = cetyl/stearyl

The dialkyl quaternaries are well known as relatively strong conditioning agents. The most important product is the so-called Quaternium-18 (CTFA nomenclature). This is a dialkyl dimethyl quaternary based on hydrogenated tallow (DHTDMAC). In addition to its use in cosmetic formulations, it has also become the most widely used quaternary for fabric softeners. It exhibits strong conditioning properties, good detangling and combing, and adds manageability to unruly or damaged hair. Quaternium-18 is not soluble in water but forms stable dispersions. It is usually offered in form of a paste (75% active in water/isopropanol), or as powder or granular forms.

Quaternium-18 represents a mixture of dicetyl-, distearyl- and cetyl/stearyl dimethyl quaternaries in a statistical distribution. A relatively new quaternary available to the cosmetic industry is the so-called hydrogenated tallow octyl dimonium chloride (CTFA adopted name). This new quaternary has strong conditioning properties, close to Quaternium-18, but is clearly soluble in water. In addition, it forms stable dispersions in mineral oil and isopropyl myristate. This is untypical for quaternaries [28].

1.6.3 *Trialkyl quaternaries*

$$R-\overset{\displaystyle R}{\underset{\displaystyle R}{\overset{|}{\underset{|}{N^+}}}}-CH_3 \quad Cl^-$$

Tricetylmonium chloride
R = cetyl/stearyl

An increase in the number of alkyl chains leads to considerable improvement in properties (Figure 1.6). A relatively new product is tricetyl methyl ammonium chloride (TCMAC), which shows superior detangling, static control, and excellent combing properties on both wet and dry hair. It is a suitable product when really intensive conditioning is required, for example in split-end fluids and in hair cures.

It is interesting to note that the TCMAC forms stable dispersions in solutions of anionic surfactants. In this case, the two-phase titration for the determination of active matter using anionic surfactants as titrating agents

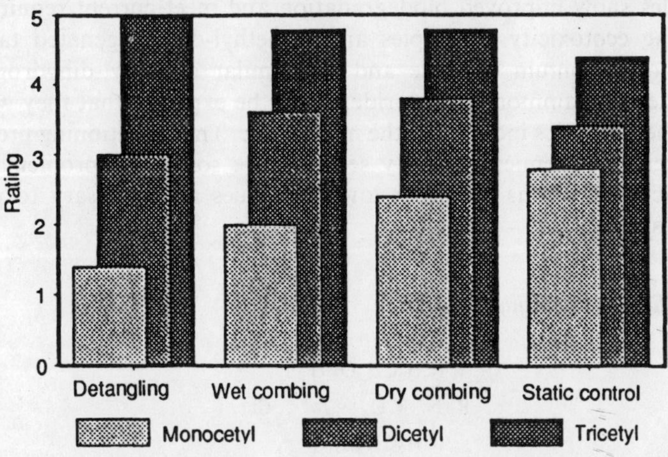

Figure 1.6 Conditioning properties of a cetyltrimonium chloride, dicetyldimonium chloride and a tricetylmonium chloride in a hair-rinse formulation [28].

is not possible. The reason could be steric hindrance, which prevents the complexation.

1.6.4 *Benzyl quaternaries*

Benzyl quaternaries are obtained by reaction of alkyl dimethyl amines with benzyl chloride. Based on short-chain amines, quaternaries with good anti-microbial properties are obtained. Products with a carbon-chain length between 12 and 14 are particularly effective. As the length of the carbon chain increases, the microbial activity is reduced. Stearyl benzyl ammonium chloride (stearyl alkonium chloride in the CTFA nomenclature) is a very effective conditioner. It was the first to be used in cosmetics and is still widely used, particularly in the USA.

1.6.5 *Ester quaternaries*

$$R_1\text{–CO–OCH}_2\text{CH}_2\text{–}\overset{\displaystyle R_2}{\underset{\displaystyle CH_2CH_2O\text{–OC–}R_1}{N^+}}\text{–CH}_3 \quad Cl^-$$

$R_1 = C_{15}/C_{17}$

$R_2 = CH_2CH_2OH$ or CH_3

Fabric softeners provide by far the biggest market for cationic surfactants. Under pressure of the environmental authorities the European detergent industry gave up the use of the DHTDMAC which was for years the standard quaternary for this application (see section 1.6.2) [22]. It has been replaced by so-called 'ester quats' which have at least one ester group between the alkyl functions and the cationic nitrogen. Due to this 'breaking point' the molecules show improved biodegradation and meet current requirements regarding ecotoxicity. Examples are *N*-methyl-dihydrogenated tallowyl-triethanolammonium chloride and the similar dimethyl-dihydrogenated tallowylethanolammonium chloride. It can be expected that they will find use in the cosmetics industry in the near future. The conditioning properties are excellent. Chemical stability seems to be somewhat problematic for cosmetic applications as rather low pH values are necessary to prevent hydrolysis [29].

1.6.6 *Ethoxylated quaternaries*

$$R\text{–}\overset{\displaystyle (CH_2CH_2O)_xH}{\underset{\displaystyle (CH_2CH_2O)_yH}{N^+}}\text{–CH}_3 \qquad Cl^-$$

Ethoxylated quaternaries (ethoquats) are obtained by reaction of methyl chlorides with ethoxylated amines. They are more hydrophilic than normal quaternaries, but show relatively good conditioning properties regarding both wet-combing and, particularly, static control. As in the case of betaines and aminoxides, the ethoquats are compatible with anionic surfactants. Due to the presence of hydrophilic groups, complexes of ethoquats and anionic surfactants are quite soluble in anionic surfactants. These combinations are ideal bases for conditioning shampoos, especially if combined with polymeric quaternaries.

1.7 Shampoo and bath additives

1.7.1 *Thickeners*

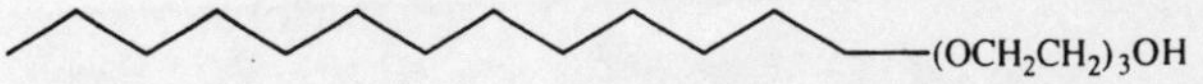

$-(OCH_2CH_2)_3OH$ Laureth - 3

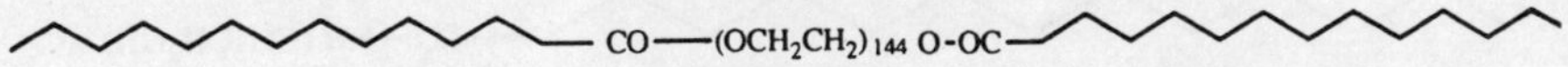

$CO-(OCH_2CH_2)_{144}O-OC-$

PEG-6000 distearate

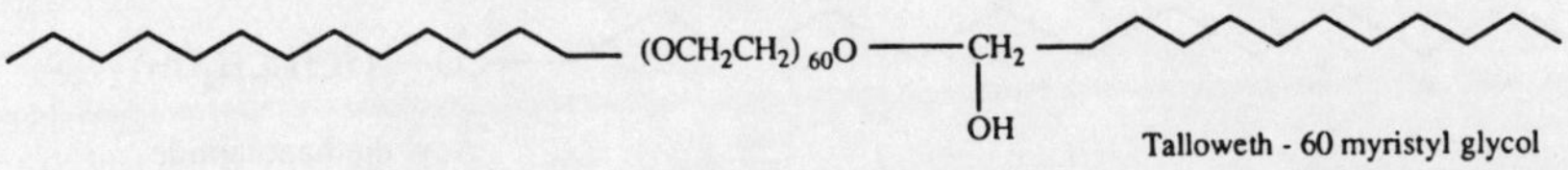

$(OCH_2CH_2)_{60}O-CH_2-$ (OH)

Talloweth - 60 myristyl glycol

A high viscosity is often very important both for product stability and for handling of a cosmetic product. For shampoos, shower and foam bath products, viscosities between 400 and 4000 mPa s are typical. Pearlescent products should have a minimum viscosity of 2000 mPa s to avoid precipitation. In the case of ether sulphate as the main surfactant, the thickening can easily be achieved by addition of electrolytes (chlorides of sodium, ammonium or magnesium, for example) to the formulation. The mechanism is by an increase in the size of the micelles. The thickening effect of alkanol amides is similar and will be described in more detail in section 1.7.2. Due to possible contamination with nitrosamines, low ethoxylated fatty alcohols such as the laureth-3 are recommended as replacements. These products are also good thickeners but, unlike the alkanol amides, they increase the cloud point of the formulations.

A different principle of thickening is obtained by the use of special, high molecular weight thickeners such as PEG-6000 distearate, talloweth-60 myristyl glycol or PEG-120 methyl glucose dioleate. The structure of talloweth-60 myristyl glycol with its hydrophobic ends is very similar to that of PEG-6000 distearate. An advantage is that it is an ether product which remains stable against hydrolysis at higher temperatures or extreme pH values.

A side-effect of all these thickeners is that they modify the flow properties, leading to increased Newtonian flow. This contrasts with salt- or polymer-thickened systems, which show typically pseudoplastic flow behavior.

Polymer thickeners like natural gums, cellulose derivatives (carboxymethyl cellulose, hydroxyethyl cellulose) and carbomers (CTFA name) are used more often in emulsions than surfactant-based formulations. An excellent review of polymers and thickeners is given by Lockhead and Fron [30].

Another type of rheological modifier is the inorganic bentonites, which can be used to obtain a yield point. This gives stability to shampoos carrying particles in suspension, e.g. pearlescent formulations or formulations containing zinc pyrithione, preventing sedimentation and separation. The systems are often thixotropic and will flow on shaking. In the absence of shear stress, such systems behave as solids.

1.7.2 *Foam stabilizers*

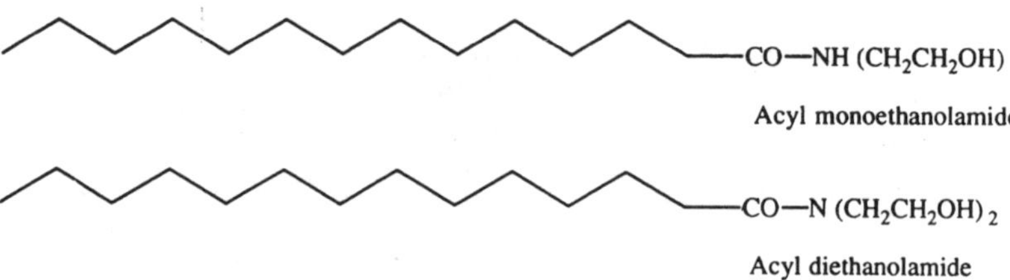

In the presence of oily soils such as sebum, the stability of shampoo foam can be drastically reduced. The so-called foam boosters act as stabilizers and also modify the foam structure to give a richer, dense foam with small bubbles. The alkanol amides are well known for this behavior. The most important types are the monoethanolamides, which are obtained by amidation of fatty acids with monoethanolamine. Diethanolamides are usually obtained by amidation of fatty acid methylester, or triglycerides such as coconut oil, with diethanolamine. The latter are liquid products with a typical glycerol content.

The monoethanolamides are the most effective foam boosters but are difficult to incorporate due to their high melting points (approx. 80°C). The diethanolamide based on coconut oil is the most popular one, although the thickening effect is reduced, due to the glycerol. The price is relatively low and the product is easy to handle compared with pure amides based on methyl ester. Today, the diethanolamides are under discussion due to possible formation of carcinogenic nitrosamines. Consequently, EC regulations allow only alkanolamides with free diethanolamine below 4%.

Other well-known foam boosters are the amine oxides and betaines described previously. Protein hydrolysates and cellulose derivatives such as CMCs are recommended as foam stabilizers.

1.7.3 *Pearlescent agents*

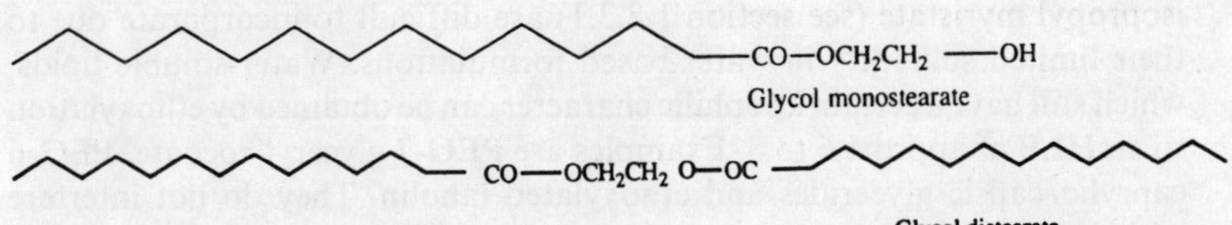

Ethylene glycol mono- and distearates (EGMS, EGDS) are most often used as pearlescent agents in surfactant formulations. They have to be incorporated at high temperatures (approx. 70–75°C), therefore ready-made liquid pearlescent bases are now very popular. A wide range of bases with different appearances, from turbid to real pearlescent, is offered on the market. Very effective opacity without pearlescent effect can be achieved with polystyrol dispersions. These are already highly effective at very low concentrations. Direct contact with perfume oils can result in coagulation of the polystyrol, and must be avoided.

1.7.4 *Conditioning agents*

Shampooing with pure anionic surfactants can leave hair difficult to comb while wet, and prone to 'fly-away' when combed after drying. Improvement of the wet-combability, and reduction of static charge build-up can be achieved by addition of conditioning agents. These are especially effective if the formulation also contains amphoteric surfactants such as betaines or amine oxides. Cationic surfactants used in hair rinses are normally incompatible with anionic surfactants and cannot be used in shampoo formulations. This problem can be overcome by the use of quaternized polymers. An example is Polyquaternium 10, a quaternized hydroxyethyl cellulose, which is compatible with most of the anionic surfactants and can therefore also be used in clear formulations. It shows excellent conditioning properties and imparts manageability and body to hair. Due to its very high substantivity to hair, very low concentrations (below 0.5%) are sufficient. Use of high concentrations may lead to over-conditioning and build-up on the hair. Used in shower- or foam-bath formulations, polyquaternium 10 can improve skin-feel after use. Other important polyquats are Polyquaternium 7, Polyquaternium 23, Polyquaternium 8, and Polyquaternium 11 (CTFA nomenclature) [30].

Small amounts of fatty components such as fatty alcohols or monoglycerides can support the conditioning effect of shampoos. Silicones can be very effective conditioners but are difficult to incorporate and may act as antifoaming agents.

1.7.5 *Emollients*

To overcome possible harsh effects of shower and foam baths on the skin, special refatting agents (emollients) are often added. Real lipids like isopropyl myristate (see section 1.8.3.1) are difficult to incorporate due to their limited solubility in water-based formulations. Water-soluble lipids, which still have a certain lipophilic character can be obtained by ethoxylation to an HLB of approx. 6 to 8. Examples are PEG-7 glyceryl cocoate, PEG-6 caprylic/capric glycerides and ethoxylated lanolin. They do not interfere with the foaming of the surfactants.

1.7.6 *Sequestering agents*

Sequestering agents like ethylenediamine tetraacetate (EDTA) are used to prevent the formation and deposition of Ca and Mg soaps and to clarify formulations when hard water is used for the production of shampoo and bath preparations. As a positive side-effect they can also give support to the preservation system as shown in Table 1.2.

Table 1.2 Influence of sequestering agents on the performance of preservatives in a solution of lauryl ether sulfate

	No additive			0.05% EDTA		
	pHB-Me 0.4%	pHB-Me 0.2%	DBDCB 0.05%	pHB-Me 0.4%	pHB-Me 0.2%	DBDCB 0.05%
Start	+++	+++	+++	+++	+++	+++
1 day	++	++	+++	++	++	–
3 day	++	++	+++	–	+	–
7 day	–	++	–	–	–	–

Scaling: +++, strong contamination to –, no bacteria (semi-quantitative determination). pHB-Me = pHB methyl ester; DBDCB = dibromo-2,4-dicyanobutane.

The outer membrane of Gram-negative organisms like *Pseudomonas* is very complex and is stabilized by Ca and Mg ions [14]. Strong chelating agents will destabilize this membrane making it more permeable for the preserving agents [15].

1.8 Oil components

1.8.1 *Mineral oil*

Mineral oils for cosmetic use are high-boiling fractions obtained from crude oil distribution that are purified and refined by treatment with sulfuric acid. Two types are of importance: (i) liquid paraffin (viscosity 110 to 230mPas); and (ii) light liquid paraffin (viscosity 25 to 80mPas). Mineral oils are colorless, clear and odorless liquids, insoluble in alcohol or water. They are excellent cosmetic emollients because they are inert and do not penetrate into the skin. They therefore have excellent skin compatibility and little or no comedogenic potential. Since they are not considered 'natural', mineral oils have been attacked repeatedly but are still the most widely used oil component in skin-care formulations. Nevertheless, 3.7% of the human *strateum corneum* lipids are n-alkanes [31].

Due to their fatty character, mineral oils form a film on the skin which increases hydration by blocking the normal evaporation of water. Combinations with synthetic esters such as isopropyl stearate are recommended in order to open the film, and to guarantee the right balance for water evaporation.

1.8.2 *Natural oils*

1.8.2.1 *Triglycerides*

CH_2—O—OC—alkyl

CH—O—OC—alkyl

CH_2—O—OC—alkyl

alkyl = C_7 to C_{21}

Natural oils from vegetable sources are mainly glycerol esters based on mixtures of $C_{8/22}$ fatty acids. The most important are the saturated lauric (C_{12}), myristic (C_{14}), palmitic (C_{16}) and stearic (C_{18}) acids and the unsaturated oleic (C_{18}) and linoleic (C_{18}) acids. These fatty acids are derived from natural sources and typically have even numbers of carbon atoms in their chains.

Important characteristic values for triglycerides are the saponification number and the iodine number. The latter indicates the amount of unsaturated fatty acids present. Although, due to the natural source, better skin compatibility compared with synthetic materials is expected, triglycerides are relatively problematic to use in cosmetic emulsions. In addition to the difficulty of emulsification, the products can become rancid due to the presence of unsaturated fatty acids. Consequently, antioxidizing agents usually have to be added.

It is important to use only high-quality triglycerides. Polycyclic aromatic hydrocarbons (PAHs) can be a problem, particularly in the case of coconut oil. This is due to contamination from flue gases when the copra, from which the coconut oil is derived, is dried. Treatment with activated carbon can remove these impurities.

Triglycerides are relatively fatty and spread very little on the skin. They generally show moderate comedogenicity, exceptions being sunflower oil and safflower oil where no comedogenic effects have been found [32]. Other examples of natural triglycerides are peanut oil, soy oil, olive oil, wheat germ oil and avocado oil.

A special triglyceride is castor oil, which is obtained from the seed of *Ricinus communis.* This oil is based on ricinoleic acid, an unsaturated C_{18} hydroxy fatty acid. Contrary to other triglycerides, it is easily soluble in alcohol. Castor oil is used in lipsticks and also in alcoholic preparations.

1.8.2.2 *Jojoba oil* Unlike the oils described in the previous section, jojoba oil is not a triglyceride. It is a liquid, unsaturated wax from the esters of long carbon-chain fatty acids (C_{20} to C_{22}) and long carbon-chain unsaturated alcohols (C_{20}, C_{22}). Its source is the jojoba plant, which grows in California and Mexico. Jojoba oil has a very fatty character and shows excellent cosmetic properties.

1.8.3 *Synthetic oils*

Synthetic oils are esters, usually obtained by direct reaction of fatty acids with alcohols. Common fatty acids are caprylic, caprinic, lauric, myristic, palmitic, stearic, oleic, linoleic and behenic acid, and also adipic acid, a dicarboxylic acid Alcohols can be, for example, isopropyl, butyl, ethylhexyl, myristyl or oleyl alcohol, as well as polyvalent alcohols such as ethylene glycol, propylene glycol and glycerol. A wide range of combinations to produce different synthetic esters is possible, and the list of synthetic oils available to the cosmetic industry is long.

In addition to products obtained via normal esterification using chemical catalysts, esters obtained by enzymatic catalysis are now also available [33]. Possibly advantages can be found in the case of oleic acid derivatives, because esterification can be obtained under milder conditions. Consequently, a light color and weak odor can be expected.

1.8.3.1 *Isopropyl esters*

$$R—CO—OCH(CH_3)—CH_3$$

RCO = myristic, palmitic, stearic

The most important synthetic ester for cosmetics is isopropyl myristate. It has a very dry character and shows good spreading on the skin. It can open films of mineral oil or other fatty oils in order to let water evaporate, and it is an excellent solvent for perfumes, ultraviolet filters, etc.

A disadvantage of isopropyl myristate is its high comedogenicity [32]. Nowadays therefore, isopropyl palmitate and isopropyl stearate, which show similar properties but reduced comedogenic effects, are increasingly used.

Due to their good solvent properties, the isopropyl esters can cause difficulties when used in formulations that are filled into polystyrene or polyethylene containers.

1.8.3.2 *Ethylhexyl esters*

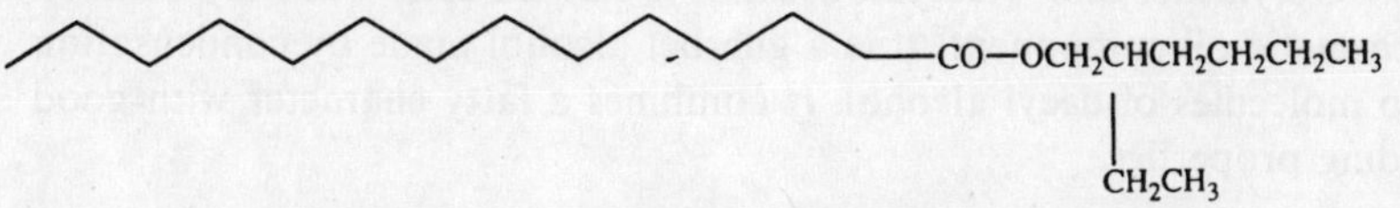

$—CO—OCH_2CHCH_2CH_2CH_2CH_3$

CH_2CH_3

Other alternatives to isopropyl myristate are the ethylhexyl esters of palmitic acid or stearic acid. These have become important products for the cosmetic industry. Compared with isopropyl esters, the molecular weight is higher, but spreading properties are similar.

1.8.3.3 *Oleic acid esters*

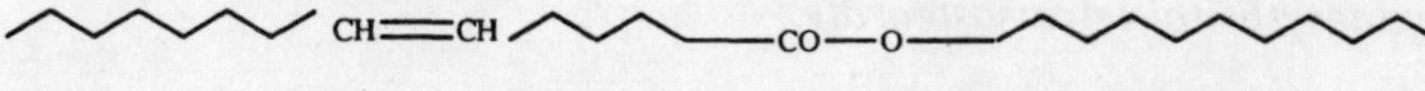

CH═CH —CO—O—

Decyl oleate

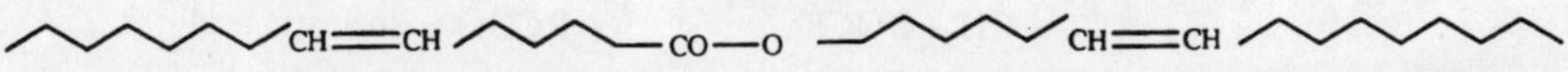

CH═CH —CO—O CH═CH

Oleyl oleate

The two most important oleates are decyl oleate and oleyl oleate. The character of these esters is relatively fatty and, as with unsaturated esters, they can become rancid. The advantage of the oleates is that, even at high molecular weight, the esters are liquid.

Interesting alternatives to the oleates are the isostearates. Esters of branched isostearic acid are also liquid and present no oxidation problem. The limiting factor for use is the price, which is much higher than that of the oleates.

1.8.3.4 *Caprylic/capric acid esters* The most important caprylic/capric esters are the triglyceride and the propylene glycol diester. The glycerol ester is mainly used as a replacement for natural triglycerides, with the advantage that it is stable against oxidation. Like the natural triglycerides, the products are relatively difficult to emulsify. Due to the typical odor of short chain fatty acids, caprylic/capric esters must be produced in very pure form. Following long-term storage, hydrolysis, which leads to changes in the odor of cosmetic products, can take place.

1.8.3.5 N-*butyl stearate* *N*-butyl stearate is an excellent skin-compatible ester with very good emollience. The character is not very fatty compared with the isopropyl esters.

1.8.3.6 *Isocetyl stearate* Isocetyl stearate is a saturated liquid ester of high molecular weight and good emollience. Water can evaporate easily through the ester and even small amounts can perforate a hydrophobic film of mineral oil.

1.8.3.7 *Octyldodecanol* Octyldodecanol is not an ester but may be used as a cosmetic oil component. It is a guerbet alcohol made by condensation of two molecules of decyl alcohol. It combines a fatty character with good spreading properties.

1.8.3.8 *Diisopropyl adipate*

$$(CH_3)_2CHO{-}OC(CH_2)_4CO{-}OCH(CH_3)_2$$

Diisopropyl adipate is an ester of a dicarbonic acid. Contrary to most other oils, it is soluble in alcohol and is therefore used, for example, as a refatting agent in aftershave formulations.

1.8.3.9 *Pentaerythritol tetraisostearate*

$$\begin{array}{c} CH_2{-}OOCR \\ | \\ RCOO{-}CH_2{-}C{-}CH_2{-}OOCR \\ | \\ CH_2{-}OOCR \end{array} \qquad RCOO = \text{isostearic acid}$$

Pentaerythritol tetraisostearate is a high molecular weight, highly branched ester of pentaerythritol and isostearic acid. It has a very fatty character and shows no comedogenic effects. Consequently, it is an ideal ester for skin-protection formulations such as baby products and, due to its water-repellent properties, water-resistant suntan preparations. The liquid pentaerythritol tetraisostearate has a melting point well below 0°C.

1.9 Waxes

The word wax has two different meanings. From a chemical point of view it means an ester of a fatty acid and a fatty alcohol. Jojoba oil is therefore a liquid wax. In this section, however, the word wax is considered from a physical point of view, and means compounds having a high melting point (approx. 50–100°C).

1.9.1 *Natural waxes*

The most important wax, and a classical component for creams, is beeswax, which is the construction material of honeycombs [34]. Chemically, it consists of mixed esters of long-chain alcohols (C_{26-32}) plus fatty acids and hydroxy fatty acids of chain length of 16–26. At room temperature the beeswax is very hard. The melting point is around 61 to 66°C. Untreated beeswax is dark-yellow and is usually treated to improve the color.

Beeswax is a very good consistency regulator in creams and ointments and is also used in stick formulations. The addition of approx. 6 per cent of borax, calculated on the beeswax, results in partial saponification and provides the beeswax with some emulsifying properties. This principle was used in the classic cold-cream formulations. Nowadays, European law restricts the use of borax in cosmetic products.

Two other natural waxes that are harder and therefore mainly used in stick formulations are carnauba wax (melting point approx. 85°C) and candelilla wax (melting point approx. 70°C). Both are extracted from South American plants.

1.9.2 *Synthetic waxes*

Due to the varying qualities of the natural products, especially yearly and source variations of beeswax, synthetic substitutes are of interest. These are often mixtures of very different chemicals, and result from empirical research. An example of a beeswax replacement is the mixture of glyceryl hydroxystearate (and) cetyl palmitate (and) microcrystalline wax (and) trihydroxy stearine. A single defined product with a structure close to beeswax is hydroxyoctacosanyl hydroxystearate, an ester of a C_{26}-β-hydroxyalcohol and hydroxystearic acid. This product, which shows very similar behavior compared with beeswax, is easy to emulsify and can be used as a general consistency regulator for cosmetic emulsions [35].

Another synthetic wax is synthetic spermaceti. The natural type is no longer available because it is obtained from whales, but was formerly an excellent wax compound for emulsions. Replacements are cetyl palmitate, which is the main component of the natural spermaceti, or special mixtures like the cetyl esters wax NF, which comes very close to the natural product.

1.10 Silicone oils

$$[\overset{\displaystyle R}{\underset{\displaystyle R}{\overset{|}{\underset{|}{Si}}}}-O]_n \qquad R = \text{methyl or phenyl}$$

The high molecular weight organo polysiloxanes (dimethicone) are hydrophobic oils with good skin protection and non-sticky skin-feel. They show very high spreading and are used in small amounts in stearate creams to avoid the soap-up effect. To prevent eye-sting, non-volatile oils should not be used in products that are applied to the eye area. Due to their water-repellent properties, silicone oils are important for waterproof sun products. They are usually classified by viscosity.

Cyclic methyl polysiloxanes (cyclomethicone) are particularly volatile. They are used in hair-care products to improve the gloss of the hair and the compatibility of the product.

1.11 Cream bases

1.11.1 *Fatty alcohols*

Fatty alcohols are important raw materials for surfactants, emollients and emulsifiers. Pure fatty alcohols, mainly cetyl alcohol and stearyl alcohol, are also used *per se* as consistency regulators and co-emulsifiers in, for example, creams, lotions and hair rinses. The so-called natural fatty alcohols are obtained by hydration of fatty acid methyl esters. Similar, linear primary alcohols can be obtained from the Ziegler process. The branched types obtained from oxo-synthesis are important as raw material feedstock, but are not used as alcohols *per se*.

Mixtures of cetyl and stearyl alcohols combined with hydrophilic emulsifiers are known as the 'emulsifying waxes' of the British Pharmacopoeia. In such combinations the fatty alcohols are self-emulsifying. Common combinations are:

(i) cetearyl alcohol plus sodium cetearyl sulfate
(ii) cetearyl alcohol plus PEG-1000 monocetyl ether
(iii) cetearyl alcohol plus alkyl trimethyl ammonium bromide

as examples for (i) anionic; (ii) non-ionic; and (iii) cationic cream bases. Type (iii) combinations are mainly used in hair-rinse formulations. The fatty alcohol gives texture and body to the formulation and acts as consistency regulator.

1.11.2 *Polyol esters*

The most important polyol ester is the so-called glycerol monostearate (GMS). Generally, it is not a pure product but a mixture of mono- and diesters of stearic and palmitic acids. The main distinguishing feature is the content of monoester. Products with approx. 40% monoester are obtained by direct esterification of stearic acid and glycerol. Products with approx. 60% monoester are produced by glycerolysis of triglycerides with glycerol. The 90% material can be obtained by molecular distillation but is seldom used in cosmetic formulations because the mono/diesters provide the best applicational properties.

GMS is poorly soluble in water. However, as with the fatty alcohols, when combined with more hydrophilic emulsifiers, it is an excellent co-emulsifier and becomes self-emulsifying. Typical combinations are:

- GMS plus potassium stearate ('GMS self-emulsifying')
- GMS plus sodium lauryl sulfate
- GMS plus ethoxylated fatty alcohols ('GMS self-emulsifying, acid stable')

Other polyol esters used as co-emulsifiers are the sorbitan esters, especially the monostearate. These are obtained by dehydration of sorbitol to sorbitan followed by esterification with fatty acid. Such products are typically offered as pure products, and function as water-in-oil (W/O) emulsifiers. In oil-in-water formulations they are most often combined with ethoxylated products of the same family.

1.11.3 *Fatty acids*

Fatty acids, like the fatty alcohols, are important raw materials for cosmetic ingredients such as surfactants, emulsifiers and emollients. Fatty acids are obtained by saponification of naturally-derived triglycerides. As such, they are seldom used in emulsions. They act as consistency regulators but tend to crystallize in the formulation.

Fatty acids especially stearic acids, are mainly used in the form of soaps. As the pH increases, they become truly anionic and act as oil-in-water (O/W) emulsifiers, forming a group of so-called stearate creams.

Commercially available stearic acids are often mixtures of C_{16} and C_{18} fatty acids, mainly 1:1. The preferred cosmetic quality with low odor is a triple-pressed stearine with an iodine value below 2. A special application of the fatty acids is their use in alcohol stick formulations, where stearic acid is dissolved in ethanol and then neutralized with sodium hydroxide. These hard sticks are used as deodorant bases.

1.12 Oil-in-water (O/W) emulsifiers

1.12.1 *Anionic O/W emulsifiers*

Anionic O/W emulsifiers can be used to obtain very stable emulsions because they can build an electrical double layer around the droplets, which prevents the droplets from coalescing. On the other hand, anionic emulsifiers are sensitive to low pH and electrolytes.

The most important anionic O/W emulsifier is soap, which forms the stearate creams. Neutralized with potassium hydroxide, the stearate is alkaline. The use of triethanolamine as a neutralizing agent is more common. A disadvantage of stearic acid is the typical 'soap-up' effect, a type of foaming that occurs when the cream is rubbed into the skin. To prevent this, silicon oil is frequently added to the formulations.

Other very effective anionic emulsifiers are alkyl sulfates, particularly sodium cetearyl sulfate. In principle, all anionic surfactants can be used. However, due to the fact that the soaps are relatively alkaline and fatty alcohol sulfates are irritant to the skin, the mild surfactants are of more interest. These include sodium cocoyl isethionate as an extremely mild emulsifier, and phosphoric acid esters, like potassium cetyl phosphate which are very effective at low concentrations.

1.12.2 *Cationic O/W emulsifiers*

Cationic emulsifiers have generally been avoided in skin-care products because they are often more irritant compared with anionic emulsifiers. They are important in hair-care formulations, where they act as conditioners and anti-static agents. Their advantage, particularly distearylammonium chloride, is that they produce emulsified products with excellent cushion feel on skin and mitigate the heavy feel imparted by glycerin.

1.12.3 *Non-ionic O/W emulsifiers*

A wealth of non-ionic O/W emulsifiers is available. These are mainly PEG derivatives such as ethoxylates or PEG esters. Typical O/W emulsifiers are:

- ethoxylated fatty alcohol
- PEG esters of fatty acids
- ethoxylated sorbitan esters
- ethoxylated monoglycerides
- ethoxylated castor-oil derivatives.

The degree of ethoxylation determines the primary properties of that part of the molecule that results in the HLB classification. Good O/W emulsifiers can be found in the HLB range between 8 and 18. However, the type of hydrophobic moiety is also very important for the stability of the emulsions. Contrary to anionic emulsifiers, the non-ionics are unaffected by changes in pH and, in the case of pure ethers, they can also be used at extreme pH values.

Following the trend towards more 'natural' cosmetics, trials have been made in order to replace the EO chain by polyglycerol or sugars [36] but up till now it has not been possible to reach high HLB values.

1.12.4 *O/W stabilizers*

Very efficient and widely used stabilizers for O/W emulsions are polymers, especially the carbomers (CTFA-name) [30]. These are polyacrylate resins which have to be neutralized in order to form gels, usually by the use of triethanolamine (TEA). Amounts up to 0.5% are recommended for cosmetic emulsions. Too much can leave an unpleasant feel on the skin. The principle of stabilization is the thickening of the outer phase. The use of preneutralized copolymers, e.g. acrylamide/sodium acrylate copolymer can make incorporation and handling easier.

1.13 Water-in-oil (W/O) emulsifiers

From the dermatological point of view, W/O emulsions are preferable to O/W emulsions. The natural lipid film on the skin is also a W/O emulsion. W/O emulsions can improve the hydration of the skin and are an ideal base for lipid-soluble active ingredients. A problematic point is that they usually leave a 'fatty-feeling' on the skin, and this is not particularly appreciated by customers.

Another problem can be 'oiling-off' or separation of the oil phase on storage. Viscosity and stability are extremely process sensitive.

1.13.1 *Single W/O emulsifiers*

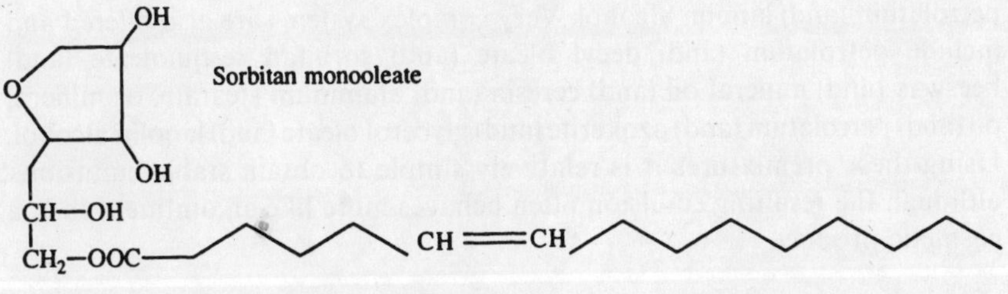

Sorbitan monooleate

$$R\text{-}CH(OH)CH_2OCH(R)CH_2OCH(R)CH_2\text{-}(OCH_2CH_2)_{22}OCH_3$$

Methoxy PEG-22 dodecyl copolymer ($R = C_{10}H_{21}$)

A limited number of W/O emulsifiers are available. This is because ionic emulsifiers will not work in the case of W/O emulsions and, since very low HLB is required, variations using ethylene oxide are not possible. W/O emulsifiers have HLB values between 3 and 6. However, the type of the alkyl chain is also important. In practice, oleyl derivatives have shown good effects. The most widely used W/O emulsifiers are sorbitan monooleate, sorbitan sequioleate and glycerol monooleate, and with increasing importance, polyglycerol esters.

In addition to oleate, ricinoleates and isostearates have also found use.

The methoxy PEG-22 dodecyl glycol copolymer is a high molecular weight W/O emulsifier, which is especially effective in formulations containing mineral oils [35]. It is a saturated ether and is therefore stable against oxidation and hydrolysis.

1.13.2 *Lanolin derivatives*

Lanolin or wool wax obtained from sheep wool is a mixture of different esters from higher alcohols, mainly cholesterol, with higher fatty acids. The alcohol fraction contains linear and iso-C_{16-30} alcohols, while the fatty acids are mainly linear and hydroxy fatty acids with carbon-chain lengths between 10 and 29. Lanolin can bind 200–300% of water in the form of a W/O emulsion. Since lanolin can cause allergies, lanolin alcohols extracted after saponification of the wool wax are now mainly used. Due to the high amount of cholesterol, they are better emulsifiers than lanolin itself.

1.13.3 *Absorption bases*

Since it is relatively difficult to obtain stable W/O systems, pre-mixtures, which are able to bind high amounts of water are available. Relatively simple absorption bases are, for example, mineral oil (and) lanolin alcohol or petrolatum (and) lanolin alcohol. Very complex systems are also offered and include petrolatum (and) decyl oleate (and) sorbitan sesquioleate (and) beeswax (and) mineral oil (and) ceresin (and) aluminum stearate, or mineral oil (and) petrolatum (and) ozokerite (and) glycerol oleate (and) lanolin alcohol. Using these premixtures it is relatively simple to obtain stable emulsions, although the resulting emulsion often behaves more like an ointment than a cosmetic product.

1.13.4 *W/O stabilizers*

A relatively simple way of obtaining highly stable W/O emulsions is to use block copolymers such as PEG-45 dodecyl glycol copolymer [35]. These products, with molecular weights of 2000 to 4000, can improve the stability of W/O emulsions by protection of the droplets from coalescence. The hydrophilic middle part is responsible for a good anchorage in the water phase, while the highly branched ends give a steric hindering effect. The high molecular weight imparts excellent skin compatibility. Contrary to other stabilizers, for example metal soaps, which increase the viscosity of the outer emulsion phase, these copolymers typically reduce the viscosity. It is possible to obtain light, highly stable W/O lotions that are good bases for sun preparations and which give a skin-feel very close to emulsions.

1.14 Humectants

Humectants protect emulsions or toothpaste from 'drying-up'. The most important humectants are glycerol, obtained from saponification of triglycerides, and sorbitol [$C_6H_8(OH)_6$], a hexa-alcohol. Sorbitol is a hygroscopic powder but is most often used in the form of a 70% aqueous solution. Glycerol is typically used at 99% with a density of 1.26 or at 85% with a density of 1.22. Both are non-toxic and have a sweet taste, which makes them ideal for the use in toothpastes.

Other humectants for emulsions are lactates, 1,3-butylene glycol and 1,2-propylene glycol. The latter should be of high quality as it has occasionally been implied as a skin irritant [23]. A positive side-effect of humectants is a reduction of the freezing point. Humectants are also important in moisturizing products.

1.15 Aerosol propellants

A few years ago only one group of aerosol propellants, the fluorocarbons, was used. Due to the well-founded suspicion that these chemicals damage the ozone layer, they are now virtually obsolete. The gap has been filled by the increased use of pump systems, and by hydrocarbons and dimethyl ether.

1.15.1 *Hydrocarbons*

The hydrocarbon propellants are propane (freezing point − 42.1), *n*-butane (freezing point − 0.5) and isobutane (freezing point − 11.7). Commercial butane is always a mixture of *n*-butane and isobutane. Mixtures with propane in different ratios are usually used to adjust the pressure. Hydrocarbons are cheap, stable and of low toxicity, but are highly flammable.

1.15.2 *Dimethyl ether*

$$CH_3{-}O{-}CH_3$$

Dimethyl ether (DME) has been known as a propellant for over 50 years. It is a gas (freezing point − 24.8) with a vapor pressure of 4.2 bar at 20°C. It shows very low toxicity [37] and does not damage the ozone layer. Like hydrocarbons it is inflammable, but unlike hydrocarbons it is miscible with water. This can reduce or prevent the risk of flammability of DME aerosol sprays, since 6% water can be solved in DME, or 34% DME can be solved in water. Alcohol/DME mixtures are good solvents for PVP/PVA resins for hair sprays, whereas hydrocarbon propellants are difficult to use in this field.

References

1. Nowak, G.A. (1985) *Cosmetic Preparations Volume 1, Process Technology of Cosmetics, Microbiology—GMP—Preservation, Data on Skin, Special Active Agents and Adjuvants.* Verlag für chem Industrie H. Zidkowsky KG, Augsburg.
2. Gohlke, F.J. and Bergerhausen, H. (1967) Alkylethersulfat, ein ideales Tensid. *Seifen-Öle-Fette-Wachse* **93** 521–526.
3. Adam, W.E. and Neumann, K. (1978) Alkylethersulfate–ein Überblick aus anwendungstechnischer Sicht. *Fette Seifen Anstrichmittel* **80** 392–401.
4. Seibert, K. and Boltersdorf, D. (1988) Magnesium surfactants. *Second World Surfactant Congress,* Paris, CESIO.
5. Dilution of concentrated ether sulfates. *Technical Information,* Verhoeven N.V., Braamstraat 229, B-2120 Schoten, 1992.
6. Götte, E. (1964) Chemistry and physics of surface active substances. *Proceedings of the Fourth International Congress on Surface Active Substances.* Volume 3, p. 83.
7. Mori, A. and Okumura, O. (1985) AOS. *Happi* (1) 51–52.
8. Ter Haar, G. (1978) AOS-The safety of alpha olefin sulfonates. *Happi* (3) 38–39.
9. Markland, W.R. (1976) The acyl isethionate surfactants. *Norda Briefs* no. 476.
10. Liebert, A.N. (1993) Final report on the safety assessment of sodium cocoyl isethionate. *J. Am. College Toxicol.* **12** 459.
11. Plate, H. (1995) Magnesium surfactants—Cleansing at its best and mildest. *Parfümerie und Kosmetik* **76** 28–32.
12. Schuster, G. and Modde, H. (1965) Anwendungstechnisches Verhalten der Kombination von Alkylarylsulfonat und Lauryläthersulfat mit Eißfettsäurekondensaten. *Fette Seifen Anstrichmittel* **67** 347–350.
13. Paassen, N.A.I. van (1984) Alkylethercarboxylate: Hautfreundliche Rohstoffe für kosmetische Anwendungen. *Kosmetik-Jahrbuch 1994.* Volume 8. Verlag H. Ziolkowsky, Augsburg p. 121.
14. Cowen, R.A. (1974) Relative merits of in use and laboratory methods for the evaluation of antimicrobial products. *J. Soc. Cosmet. Chem.* **25** 307–323.
15. Schmitt, H.G. (1987) Über die Wirkung antimikrobieller Substanzen. *Parfümerie und Kosmetik* **69** 5.
16. Imokawa, G. (1980) Comparative study on the mechanism of irritation by sulfate and phosphate types of anionic surfactants. *J. Soc. Cosmet. Chem.* **31** 45–46.
17. Zeidler, U. and Reese, G. (1983) *In-vitro*-Test zur Hautverträglichkeit von Tensiden. *Ärztliche Kosmetologie* **13** 39–45.
18. Ellis, P.R., Derian, P.-J. and Vokov, R. (1994) Amphoteric surfactants—The next generation. *Euro Cosmetics* **2** (July) 14–16.
19. Hodgson, J.E., Martin, C.G. and Nicholson, S.H. (1992) The use of Cocoampho(di)acetates as an aid in the *in-vitro* safety assessment of mild surfactants. *Parfümerie und Kosmetik* **73** 248.
20. Roerig, H. and Stephan, R. (1991) Amine oxides and their applications. *La rivista Italiana delle sostanze grasse* **68** 317–321.
21. Markland, W.R. (1977) Low eye irritation shampoo systems. *Norda Briefs* no. 479.
22. Ruback, W. (1994) Surfactant trends in Europe. *Chimicaoggi/chemistry today* **(5/6)** 16–18.
23. Schrader, K.H. (1989) *Grundlagen und Rezepturen der Kosmetika.* 2nd edn. Hüthig Buch Verlag, Heidelberg, 241 pp.
24. Balzer, D. (1991) Alkylpolyglucosides, their physico-chemical properties and their uses. *Tenside Surf. Detergents* **28** 419–427.
25. Biermann, M., Schmid, K. and Schulz, P. (1993) Alkylpolyglucoside–Technologie und Eigenschaften. *Starch/Stärke* **45** 281.
26. Busch, P., Hensen, H. and Tesmann, H. (1993) Alkylpolyglucoside–eine neue Tensidgeneration für die Kosmetik. *Tenside Surf. Detergents* **30** 116–121.
27. Wilkinson, J.B. and Moore, R.J. (1987) *Harry's Cosmeticology.* 7th edn. Longman Scientific & Technical, Harlow, 442 pp.
28. Spiess, E. (1991) The influence of chemical structure on performance in hair care preparations. *Parfümerie und Kosmetik* **72** 370–376.

29. Sebold, U. (1993) Fabric softeners worldwide. *3rd World Conference & Exhibition on Detergents*, Montreux, American Oil Chemist Society.
30. Lochhead R.Y. and Fron W.R. (1993) Encyclopedia of polymers and thickeners. *Cosmetics & Toiletries* **108** 95–135.
31. Friberg, S. (1994) Mineral oil: A natural compound? *Soap/Chemical Specialities* (**2**) 40–43.
32. Fulton, J.E. (1989) Comedogenicity and irritancy of commonly used ingredients in skin care products. *J. Soc. Cosmet Chem.* **40** 321–333.
33. McCrae, A., Roehl, E.-L. and Brand, H.M. (1990) Bio-ester. *Seifen-Öle-Fette-Wachse* **116** 201–205.
34. Brand-Garny, E.E. and Sprenger, J. (1988) Bienenwachs—Neue Aspekte eines klassischen Kosmetik-Rohstoffs. *Seifen-Öle-Fette-Wachse* **114** 547.
35. Spiess, E. (1989) Stabilizing W/O emulsions. *Proceedings of the 9th Latin American Iberian Congress of Cosmetic Chemists.* Volume 2, IFSCC, Santiago del Chile. pp. 157–170.
36. Desai, N.B. (1990) Esters of sucrose and glucose as cosmetic materials. *Cosmetics & Toiletries* **105** 99.
37. Debets, F. (1990) Health and environmental safety of DME as an aerosol propellant. *Aerosol Report* **29** 16–22.

2 Hair-care products

J.J. SHIPP

2.1 Introduction

Before discussing product formulation, some general remarks about product development are appropriate.

1. Make sure that the brief is clear. There is little point in developing a wonderful product if it is not the one that was requested. The brief should include, as a minimum:

   (i) Required performance parameters in as much detail as possible
   (ii) Benchmark products, unless the development is in a completely new area
   (iii) Guidelines on cost
   (iv) Proposed claims, and how these might be justified
   (v) The required timing

2. Do not produce a Rolls Royce when a Mini would suffice or vice versa. On the one hand, the accounts department will be extremely unhappy. On the other, the product being launched will fail to deliver its promised performance.
3. Do not use new raw materials if ones from existing stocks will do the job just as well; avoiding unnecessary proliferation of the raw material stocklist will benefit purchasing, stock control, space and cashflow.
4. Specifications often appear virtually identical, while in practice similar products from different suppliers do not perform identically. This can be very important for detergents and emulsifiers. Alternative sources of supply can be investigated as a separate issue.
5. Ensure reproducibility on scale-up to full production, making use of suitable plant and equipment. Dialogue with experienced process plant operators can be invaluable.
6. Keep formulations as simple as possible; there should be a sound reason for the inclusion of each ingredient.
7. Try to develop products which can be made as cheaply as possible, i.e. with minimum energy requirements (heating, stirring) and minimum time in the tank. Large-scale processing equipment is very expensive;

careful formulation can help to optimise the use of tank space and tank time.

2.2 Hair: structure and chemistry

Hair consists of three main layers:

(i) the central core or medulla, which is not always present and whose function is not entirely clear;

(ii) the cortex, which contributes the bulk of the hair shaft and consists of elongated keratinised cells, the whole structure having a fibrous nature. Each fibre is in turn made up of bundles of small fibres until, at molecular level, polypeptide chains are found to be twisted together to form a helix, an arrangement often favoured by nature;

(iii) the cuticle, in the form of thin overlapping scales that cover the cortex like tiles on a roof. The overlap is such that the hair is smoothest from root to tip. Techniques such as back-combing, where the comb or brush is used 'against the grain' of the cuticle can cause considerable mechanical damage. Cuticular cells also consist mostly of keratinised proteins but, in this case, in the form of thin plates; the main difference between cortex and cuticle is more a question of geometry than chemistry [1, 2].

There are, of course, other constituents of hair, e.g. water, lipids and minerals, and some work on how these are affected by the application of toiletries has been carried out [3, 4]. However, the main concern in this chapter is the interaction between hair-care products and the main protein structure of the cuticle, and of the cortex where this is exposed because of damage, chemical treatment, or weathering [5]. A great deal of finer structure has been revealed by electron microscopy and this has some bearing on these interactions [6]. The conditions under which products are applied (e.g. the use of heat) may also affect the integrity of the hair's structure [7, 8].

The cosmetic chemist is always aiming to achieve a result which looks good, and time should be given to examining the mechanical and optical properties of the hair, and how these affect our perception of 'a good-looking result' [9–11].

Hair does not grow continuously, but passes between the anagen (or growing) phase and the telagen (or resting) phase via the relatively brief transitional catagen phase. The relative time spent in each phase can vary considerably depending on the site on the body, age, sex, general health, etc. Each individual hair has its own sequence, which is unaffected by that of its neighbours. Some accurate measurements have been made of hair growth rates and the factor affecting them [12].

The size and shape of a hair is determined by the follicle from which it grows. Most Caucasian and Oriental hair grows fairly straight follicles and is approximately circular in cross-section, although varying in diameter, Oriental hair typically being of larger diameter than European hair. Negroid hair differs most, however. It grows from curved cuticles that are also somewhat flattened in cross-section, producing hair with an elliptical cross-section and a pronounced curl. This type of hair often requires different treatment to European hair and a separate chapter in this book (see chapter 6) covers its special needs.

The most important structures present in the scalp dermis and epidermis, from the point of view of the hair-care formulator, are the sebaceous glands which secrete sebum into the hair follicle. Sebum acts as a lubricant for the emerging hair and as a protectant against invasion of the follicle by hostile species (chemical, physical or microbiological). Greasy hair, caused by overproduction of sebum, is a subject which has attracted some attention and various ingredients have been examined in an attempt to overcome this problem [13, 14].

2.2.1 *Structure of hair keratin*

Keratin is a protein, a copolymer of a variety of amino acids, about 25 types in all. Keratin differs from most other proteins in that quite a high proportion (around 20%) of the amino acid units contain sulphur, notably in the form of the disulphide bond present in cystine. The basic amino acid unit may be represented as

$$\text{R}-\text{CH}\begin{matrix}\diagup\text{COOH}\\ \diagdown\text{NH}_2\end{matrix}$$

where R can either be a simple aliphatic group, or may be more complex, containing benzene rings, double bonds, heterocyclic rings, fused rings or additional carboxylic acid or amino groups. The side chains are relatively small, accounting for about one half of the weight of the protein molecule. There is an excess of carboxylic acid groups over amino groups in the side chains, leading to an overall negative charge on the surface of the hair. In that sense, the behaviour of the hair may be likened to that of a cation-exchange resin.

There are three main types of bond present in hair keratin:

1. Hydrogen bonds—these may occur between numerous sites (e.g. NH and C=O) in the same or adjacent protein chains. Individually they are weak, but collectively they contribute significantly to the overall strength of the hair. Water causes the hair to swell as water molecules 'insert' themselves into these hydrogen bonds, converting them into chains of bonds with consequent loss of strength. The breaking of hydrogen bonds and their subsequent reformation explains the temporary setting effect achieved by water treatment.

2. Salt linkages—these will occur between acidic and basic groups present in the side chains and, as with hydrogen bonds, may link groups together in the same or (more usually) adjacent polypeptide chains.
3. Disulphide bonds—cystine is a 'double' amino acid, containing COOH and NH_2 groups at both ends of the molecule, with a disulphide bond in between. It can enter into the structure of two adjacent polypeptide chains forming a disulphide bond between them:

```
-CO-CH-NH-
    |
    CH2
    |
    S
    |
    S
    |
    CH2
    |
-HN-CH-CO-
```

These bonds are extremely strong and can only be broken under severe conditions (e.g. powerful reducing agents at high pH as in most permanent-waving lotions).

Cross-linking mechanisms ensure that the keratin helices are held together in a fairly rigid structure. The order in which the side chains occur is very variable; steric hindrance is minimised by most of the side chains adopting an outward-facing position, such that reactive groups on the side chains are more exposed to the possibility of chemical reaction with hair-care products.

2.3 Shampoos

Shampoos may be made in various physical forms, liquids, creams or pastes, aerosol and dry. The majority are liquids, either clear or pearlised. The principal constituents of most liquid shampoos can be classified as:

- Primary detergents
- Secondary detergents
- Thickeners
- Foam stabilisers and boosters
- Perfumes
- Preservatives
- Diluents (usually water)
- Conditioning agents
- Other additives (functional or otherwise)
- Pearlisers/opacifiers
- Colours

Many ingredients are multifunctional and therefore do not clearly fall into any one category [15].

2.3.1 *Detergents*

There is no clear distinction between a primary and a secondary detergent. Primary shampoo detergents are usually anionic and inexpensive, with sodium laureth sulphate being easily the most widely used (particularly in Europe). A whole mass of different detergent materials are available including:

- Alkyl sulphates
- Alkyl ether sulphates
- α-Olefin sulphonates
- Paraffin sulphonates
- Isethionates
- Sarcosinates
- Taurides
- Acyl lactylates
- Sulphosuccinates
- Carboxylates
- Protein condensates
- Betaines
- Glycinates
- Amine oxides
- Alkyl polyglycosides

to name but a few.

Within these groups, there are many variants, depending on chain length, chain-length distribution, degree of ethoxylation, chain branching and numerous other variables. The properties of the different detergents have been reviewed in great detail elsewhere [15–25]. Here, therefore, only specific comments about the most commonly used materials will be made. In the case of sodium lauryl ether sulphate (CTFA (Cosmetic, Toiletry and Fragrance Association) dictionary name sodium laureth sulphate) the general formula is $R–(OCH_2CH_2)_n–OSO_3^-Na^+$ where R is the alkyl chain of variable length, predominantly C_{12} (lauryl) and the average degree of ethoxylation n is equal to 2 or 3. The various methods of preparation for this surfactant are referred to in chapter 1 and, depending on feedstock and the method used, many variations are possible. Consequently, the final product, although called by the same name, can be a complex and variable mixture. In addition, sulphation and neutralisation can leave varying minor amounts of free fatty alcohols and ethers, which have an important effect on the thickening behaviour of shampoos [26]. The higher ethoxylates behave as non-ionic surfactants in their own right although they are present in insignificant amounts in commercial grades of sodium lauryl ether sulphate. Today sulphation is nearly always carried out directly by SO_3 gas in a continuous sulphation plant, giving better control and improved yields compared with the older method using chlorosulphonic acid. Hydrolysis during neutralisation of the half-ester formed at the sulphation stage can give rise to free fatty ethoxylated alcohol and sulphuric acid, the latter being subsequently neutralised to form sodium sulphate:

$$R\text{-}(OCH_2CH_2)_n\text{-}OH + SO_3 \rightarrow R\text{-}(OCH_2CH_2)_n\text{-}OSO_3H$$

$$R\text{-}(OCH_2CH_2)_n\text{-}OSO_3H + H_2O \rightarrow R\text{-}(OCH_2CH_2)_n\text{-}OH + H_2SO_4$$

$$H_2SO_4 + 2NaOH \rightarrow Na_2SO_4 + 2H_2O$$

Temperature and pressure control, plus the avoidance of high local concentrations of the reactants, are paramount in achieving a product with the desired characteristics, i.e. near water white, low odour, active matter consisting of a well-defined and carefully controlled blend of the various species possible, low chloride, low sulphate, and controlled (generally quite low) free fatty alcohol and alcohol ethoxylates. Lauryl sulphates and ether sulphates are obtainable as solutions in the 25–30% range or as 'high-active' concentrates, usually in the 60–70% active matter range. These products are hazy semi-gels with a curious rheology due to the presence of liquid crystals, which gives them an anisotropic structure. They can also be quite difficult to handle since, on dilution with water, they pass through a gel phase. The structure of these fascinating surfactants is explained admirably in a booklet published by Albright and Wilson [27]. The potential problems associated with changing the source of key raw materials part of the way through a development are numerous.

In Europe, lauryl ether sulphates (especially the sodium salt) are the most commonly used primary surfactants, with lauryl sulphates occupying second place. Some comparisons of the properties of these materials are shown in Table 2.1. Judicious blending can optimise these various properties and other ingredients can be used to modify them. A common example is the use of coconut fatty acid diethanolamide to stabilise the foam and improve the coarse texture of the foam obtained with ether sulphates. In addition, the diethanolamide will help prevent excessive degreasing of the hair (superfatting effect). A very simple and cheap range of shampoos for dry, normal, and greasy hair could therefore be formulated as shown in Table 2.2. Crude, but reasonably effective in their primary function of washing the hair, products of this type are available on the market in the 'cheap and cheerful' sector. The rationale behind such formulation is a progressive increase in 'active matter' (i.e. overall detergent concentration) from dry to greasy variants, accompanied

Table 2.1 Properties of lauryl ether sulphates and lauryl sulphates

	Sodium lauryl sulphate (SLS)	Sodium laureth (2) sulphate (SLES-2)	Sodium laureth (3) sulphate (SLES-3)
Flash foam	Moderate	Good	Good
Foam texture	Dense, creamy	Coarse and open	Coarse and open
Detergency	←————	increasing	————
Solubility	————	increasing	————→
Irritancy	←————	increasing	————

Table 2.2 Typical shampoo formulations for dry, normal and greasy hair

	Dry hair	Normal hair	Greasy hair
Sodium lauryl sulphate (SLS), 30%	—	5	10
Sodium laureth sulphate (SLES), 27%	27	25	23
Coconut diethanolamide	3	3	3
Preservative	*q.s.*	*q.s.*	*q.s.*
Perfume	*q.s.*	*q.s.*	*q.s*
Colour	*q.s.*	*q.s.*	*q.s*
Thickener (usually salt)	*q.s.*	*q.s.*	*q.s.*
Water	to 100	to 100	to 100

by an increase in the proportion of the stronger detergent (SLS) in the mix on the basis that greasy hair will require a shampoo with more powerful soil-removal properties.

Another useful comparison that can be made between these surfactants is the effect of changing the cation. Sodium lauryl sulphate for instance, has only limited solubility; changing the cation to ammonium or monoethanolaminium, or triethanolaminium has a dramatic effect on the physical properties of the surfactant as shown in Table 2.3.

Triethanolamine lauryl sulphate (TLS) is notable for being particularly difficult to thicken and having poor colour stability. Diethanolamine lauryl sulphate, once widely used in the US is rapidly declining in use as a result of fears concerning nitrosamine formation from diethanolamine. Monoethanolamine lauryl sulphate (MLS), conversely quite popular as a primary detergent in Europe, but little used in the US, offers greater ease of thickening and generally does not have the same propensity to darken with age as TLS. Substitution of one of these amine-neutralised lauryl sulphates for the SLES in the dry-hair shampoo formulation shown in Table 2.2 will give a product with much denser foam, although there will be less of it.

It is fashionable (and usually desirable) to adjust the pH of shampoos to be mildly acid (e.g. pH 6–6.5) This is essential when using ALS, in order to avoid hydrolysis with the concomitant liberation of ammonia. Many alkanolamides contain appreciable amounts of free amine, giving rise to quite high

Table 2.3 Effect of changing the cation on the physical properties of sodium lauryl sulphate (SLS)

	SLS	ALS	MLS	TLS
Solubility	——— increasing ———→			
Viscosity	←——— increasing ———			
Ease of thickening	←——— increasing ———			

SLS = sodium lauryl sulphate
ALS = ammonium lauryl sulphate
MLS = monoethanolamine lauryl sulphate
TLS = triethanolamine lauryl sulphate

pHs in the final product. Consequently when using these with ALS, a slight excess of acid should be present before adding the alkanolamide in order to prevent the pH from rising above 7 at any time. Other lauryl and laureth sulphates are occasionally encountered (e.g. magnesium, said to possess a particularly low irritation potential and to be an excellent foamer).

Practical alternatives to the lauryl ether sulphates are few and include α-olefin sulphonates (mainly in the USA) and sulphosuccinates [28, 29] (generally very mild, but significantly more expensive). Without doubt, one of the success stories in recent years has been betaines, mainly cocoamidopropyl betaine:

$$R-\overset{\overset{\displaystyle O}{\|}}{C}-NH-(CH_2)_3-\underset{\underset{\displaystyle CH_3}{|}}{\overset{\overset{\displaystyle CH_3}{|}}{N^+}}-CH_2-COO^-$$

These amphoteric surfactants (see chapter 1), once considered to be somewhat esoteric specialities, have become commodity items and prices have fallen, making them readily available to the formulator of all but the cheapest products.The addition of a small amount of betaine to the dry-hair shampoo formulation in Table 2.2 would much improve it by making it less stripping and offer a practical marketing proposition as a basic or 'family' shampoo (Table 2.4), performing quite well over a wide pH range.

Cost considerations usually dictate that speciality surfactants assume a secondary role. Table 2.5 gives some formulations illustrating typical use levels of some of these materials. Along with the betaines, most speciality surfactants will improve the quality of the lather, the mildness (as measured by skin and eye irritation), and, most important, the afterfeel of the hair [30–33].

A particularly interesting amphoteric is sodium carboxymethyl tallow polypropylamine-4, for which a considerable amount of data relating to its very low eye irritation has been amassed [34, 35]. This molecule is rather unusual in having a much larger and more mobile hydrophilic 'head' than more conventional amphoterics and it is to this configuration that Lomax [34, 35] attributes many of its properties.

Table 2.4 Typical formulation for a family shampoo

Sodium laureth sulphate, 27%	27
Cocoamidopropyl betaine, 30%	5
Coconut diethanolamide	3
Preservative	*q.s.*
Perfume	*q.s.*
Colour	*q.s.*
Sodium chloride	*q.s.*
Deionised water	to 100

Table 2.5 Shampoo formulations illustrating typical use levels of selected speciality surfactants

Sodium lauryl sulphate 30%	7	—	—	—	—
Sodium laureth sulphate, 27%	21	—	—	24	20
Ammonium lauryl sulphate, 30%	—	30	—	—	—
Monoethanolamine lauryl sulphate, 33%	—	—	35	—	—
Cocoamidopropyl betaine, 30%	4	3	—	—	—
Cocodimethylamine oxide, 35%	2	—	—	—	—
Sodium lauroyl sarcosinate, 30%	—	—	—	—	6
Cocoamphocarboxy glycinate, 30%	—	—	5	6	—
Sodium carboxymethyl tallow polypropylamine, 27%	—	—	—	3	—
TEA coco-hydrolysed animal protein 30%	—	3	—	—	—
Alkanolamide (usually coconut dialkanolamide)	2.5	2.5	3	2.5	4
Preservative	*q.s.*	*q.s.*	*q.s.*	*q.s.*	*q.s.*
Perfume	*q.s.*	*q.s.*	*q.s.*	*q.s.*	*q.s.*
Colour	*q.s.*	*q.s.*	*q.s.*	*q.s.*	*q.s.*
Thickener	*q.s.*	*q.s.*	*q.s.*	*q.s.*	*q.s.*
Deionised water	to 100% →				

There is a tendency for 'frequent use' shampoos to have higher levels of milder secondary surfactants, but an overall lower level of active matter. The examples shown in Table 2.5 could be described as suitable for 'frequent use', either at full strength, or diluted to about 80% strength. All of these formulations are very much 'middle of the road' in terms of overall quality. Experimenting with different combinations of these secondary surfactants can sometimes give excellent results. Many secondary surfactants exhibit lower irritation to skin and eyes than the commonly used primary surfactants and in some cases an apparently synergistic effect exists whereby a comparatively small addition of the secondary surfactant can lead to a significant decrease in the level of irritation of the overall blend. A mechanism explaining this effect is proposed by Lomax [36].

The surfactants used in a shampoo need to be selected on the basis of a whole range of properties, including:

- Cost
- Foam height [37]
- Foam texture
- Detergency
- Irritancy (lack of)
- Ease of handling and mixing
- Compatibility with other ingredients
- Colour
- Odour
- Purity
- Biodegradability (not usually a problem) [38–40]

The relative importance of these parameters will depend on the end use, e.g.

irritancy will be of paramount importance in a baby shampoo, while ease of handling might be critical in a plant equipped with only simple mixing facilities.

2.3.2 *Thickeners and foam stabilisers*

There are several methods by which shampoos may be thickened:

- alkanolamides or their alternatives
- polymeric materials
- electrolytes

Many of these ingredients are multifunctional, and most products use more than one.

By far the most commonly used alkanolamide in the UK is coconut diethanolamide; in the US there has always been a preference for a narrower cut, generally lauryl or lauryl/myristyl blends. Both monoethanolamides and diethanolamides are widely used. Monoethanolamides are more effective, both as foam stabilisers and as thickeners, but have the disadvantage of being waxy solids requiring a hot process to incorporate them into the mix.

$$R-C(=O)-NH-CH_2-CH_2OH$$

Monoethanolamide

$$R-C(=O)-N(CH_2-CH_2OH)_2$$

Diethanolamide

Diethanolamides are very dependent on the method of manufacture and can vary considerably in composition from supplier to supplier. The so-called Kritchevsky amides, based on a 1:2 mole ratio of fatty acid: diethanolamine, usually contain quite significant amounts of a whole range of other materials, some undesirable, most of little benefit. Superamides, prepared from a 1:1 mole ratio, are purer. Quite large amounts of glycerine can be present in coconut diethanolamide prepared from coconut oil and, while this has the effect of making the material easier to handle and incorporate into a batch, the ultimate thickening effect is reduced, although a benefit is sometimes more effective solubilisation of perfumes.

While lauric, lauric/myristic and coconut mono- and diethanolamides dominate, other alkanolamides find more limited use, e.g. oleic and linoleic diethanolamide in gel shampoos based on SLES (where their thickening effect is remarkable), lauric monoisopropanolamide (probably the most effective thickener and foam stabiliser of the group), stearamides (as part of a pearlescing system) and ricinoleic diethanolamide, said to possess some conditioning properties.

Some surfactant systems are notoriously difficult to thicken, not responding well to electrolyte additions. Various raw materials manufacturers have developed alternatives to alkanolamides which overcome this difficulty. These include PEG-6000 distearate, PEG-55 propylene glycol oleate and PEG-120 methyl glucose dioleate (PEG stands for polyethylene glycol). These materials can be particularly effective in systems based on some sulphosuccinates, ethoxylated sorbitan esters, alkane sulphonates, lauryl sulphates with particularly low levels of free fatty alcohol, or conventional systems where a much higher than usual viscosity is required, e.g. the currently fashionable childrens' 'wobbly bath gels'.

Most surfactant systems are non-Newtonian in behaviour and can exhibit phenomena such as shear-thinning, shear-thickening, time-dependence and yield points, to name but a few. The ultimate test is whether the viscosity of the product is suited to the packaging in which it is sold, and meets the consumer's expectation. Thickness is invariably equated to concentration, value for money, and 'richness' in the layman's perception of the finished product.

Polymeric materials used for thickening shampoos include a variety of natural gums, such as guar, karaya, locust bean, carragheenan, and tragacanth. In practice these materials, in unmodified form, find little use in today's shampoos, and this class of thickeners is dominated by cellulosic derivatives, such as hydroxyethyl cellulose, carboxymethyl cellulose, and hydroxypropyl methylcellulose. These materials are usually dispersed in all or part of the water in the formulation before adding the other ingredients. A disadvantage with most of these materials when used on the production scale is the difficulty of obtaining complete solution rapidly; there is an unfortunate tendency for the gum to clump together and form 'golf balls' consisting of dry powder surrounded by partially hydrated gum which can seem to take an eternity to dissolve. Non-ionic cellulose derivatives, such as hydroxyethyl cellulose, exhibit an inverse cloud point (cf. nonyl phenol ethoxylates and other similar series) and this can be used to good effect by adding the gum to water heated to a temperature above the cloud point. The gum, insoluble at this temperature, will disperse to form a milky liquid which, on cooling, will give a clear and lump-free solution. Some gums are available in a surface-treated form, whereby each particle is coated with a solution-retardant chemical, thus allowing the particles to disperse before substantial hydration takes place and avoiding the formation of 'golf balls'. High-speed mixing is usually recommended to initially disperse these materials, which should be sprinkled into the vortex. Once initial dispersion is complete, stirring should be slowed down to avoid unnecessary aeration. Very high-shear mixing is best avoided since this can cause cleavage of the polymer molecules with consequent adverse effect on performance.

Gum-thickened products have a different rheology to otherwise similar electrolyte-thickened products. The temperature effect is less for gum systems than for an electrolyte-thickened system. This feature can be useful when formulating for climates where the product might be subjected to large temperature variations in storage, distribution and use. In addition, gums will often stabilise foam by strengthening the film at the air/liquid interfaces in the matrix of bubbles. This also has the effect of making the foam feel denser ('creamy') in use. This difference in tactile effect is not limited to the foam—it is sometimes apparent as soon as the shampoo is poured into the hand. There are, however, some disadvantages, namely high cost, lengthened processing time (with higher energy consumption if a hot process is used), and the difficulty of making viscosity adjustments afterwards. This last problem can often be overcome by using a small amount of electrolyte for final viscosity adjustment.

The effect of electrolytes on the viscosity of surfactant systems is a result of the increase in ionic density of the solution with its consequent effect on micelle size and shape. It is conventional to add an electrolyte with the same cation as that of the primary anionic surfactant in the system, e.g. ammonium chloride in ammonium lauryl sulphate based systems, and sodium chloride in sodium lauryl ether sulphate based systems. This is logical since most of the anionic surfactant is present in ionised form and the addition of, say, sodium chloride to ammonium lauryl sulphate solution could partly negate the reasons for choosing ALS instead of SLS.

Addition of electrolyte beyond a certain point will cause thinning, often quite rapid, and an increase in cloud point. The point at which the maximum in the viscosity/electrolyte curve occurs will depend on many factors, including concentration, presence of alkanolamides, level of free fatty alcohol or ether in the surfactants, and the effect of the perfume. Some typical curves are shown in Figure 2.1. Note that the peak viscosity occurs at a lower salt level when materials such as free fatty ether or alkanolamide are present. This is probably due to the formation of mixed micelles. A well-formulated shampoo will have a curve which is not too steep and which has a target viscosity range in the centre (i.e. the most linear) part of the ascending curve (see Figure 2.2). Shampoos which have gone 'over the top' of their curve can often be rescued by the addition of water, or a blend of water and the primary surfactant. Electrolytes should always be added as solutions (typically 25% w/w for NaCl) to avoid high local concentrations that might lead to gelation and, consequently, unnecessarily protracted mixing.

Occasionally one encounters shampoos which need thinning rather than thickening. This usually occurs when the active matter is high, when the raw materials already contain high levels of electrolytes or when large amounts of some amphoterics are present and is simply solved by the addition of a small amount of a short-chain alcohol or glycol. Ethanol or isopropanol may be used; the glycols (less odorous and not so flammable) are preferred. Propylene, butylene and hexylene glycols are effective and commonly available.

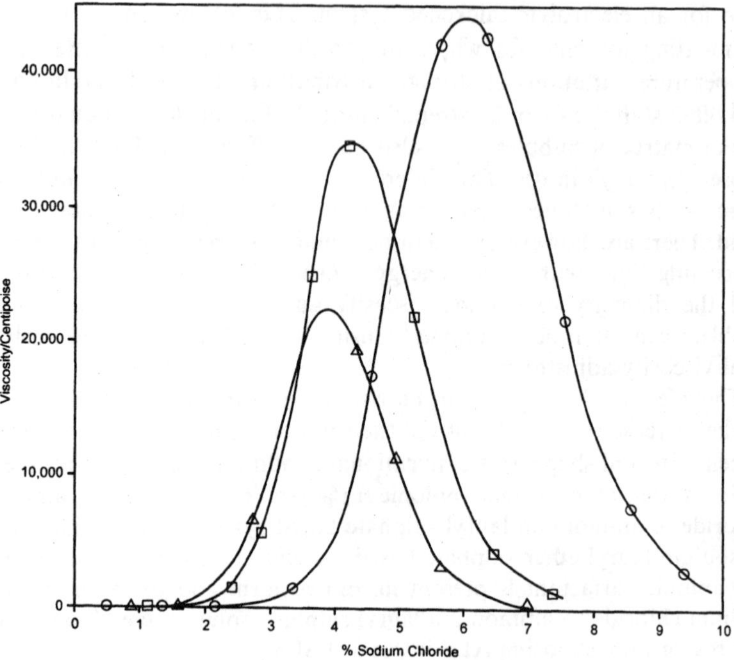

Figure 2.1 Typical viscosity/electrolyte curves for a surfactant with high and low levels of free fatty alcohol. △, sodium lauryl ether sulphate, 11% active; □, sodium lauryl ether sulphate, 11% active + 1% laureth-3; ○, sodium lauryl ether sulphate, 11% active + 3% cocamide-DEA (DEA stands for coconut diethanolamide).

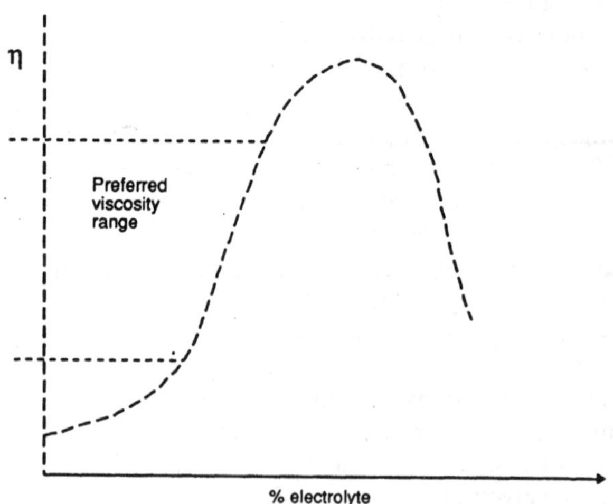

Figure 2.2 Viscosity/electrolyte curve for a well-formulated shampoo.

2.3.3 *Perfumes*

Perfumery is a complex subject in its own right (see chapter 8), but most of the detailed knowledge is required in the creation of the fragrances; the more mundane process of using these fragrance creations in the product formulation requires less ability, although selection may not be so easy [41–44].

Occasionally, one is required to develop a shampoo as a line extension to a fragrance-led range. Where the range already contains toiletry products it may only be necessary to include an existing perfume at the appropriate level. If, however, the range consists entirely of alcoholic-based perfumery products, it may be necessary for the fragrance house concerned to modify the perfume compound to achieve an acceptable performance in a detergent base. At the other extreme, in very cheap shampoos, the perfume is sometimes regarded as quite incidental. There have been attempts to market shampoos with high levels of fragrance (although not part of a fragrance-led range), even fragrances designed to persist on the hair after washing [45], but these have met with limited success.

Fragrance can often be simply added to the shampoo after the completion of any hot part of the process. If solubility is a problem, then the perfume can be premixed with a suitable solubiliser before addition, the ratio within the premix being determined by trial and error. There are many solubilisers to choose from but some, for example polysorbates may cause thinning of the product. In the author's experience, PEG-40 hydrogenated castor oil often works well. Where the manufacturing method permits, premixing the perfume with the alkanolamide, provided that this can be carried out at a moderately low temperature (say < 40°C), will usually suffice.

2.3.4 *Preservatives*

As with perfumes, this subject merits a book in its own right. The main concerns, as with all products destined for use on the human body, are resistance to spoilage and protection of the consumer. In the case of shampoos, it is particularly important to ensure that the product contains no pathogenic organisms, especially those capable of damaging the eyes. A variety of preservatives are available to the formulator and all of those in the following list have been commercially used in shampoos, either alone or in various combinations.

- Parabens (the shorter chain esters are the most soluble)
- Imidazolidinyl ureas
- 2-Bromo-2-nitropropane-1,3-diol
- 5-Bromo-5-nitro-1,3-dioxane
- Dimethyl dimethylol (DMDM)/hydantoin
- Methylchloroisothiazolinone and methylisothiazolinone

- Phenoxyethanol
- Diazolidinyl urea
- Methyldibromoglutaronitrile
- Quaternium-15
- Sodium iodate
- Glutaraldehyde
- Formaldehyde

Many of the surfactants used as raw materials for shampoo manufacture are already preserved, and the contribution of this portion of the preservative system must be taken into account. Provided that sufficient quantity is being ordered, most manufacturers of surfactants are happy to consider alternative preservative systems. Other ingredients, such as proteins and herbal extracts, may also be preserved and it is therefore quite conceivable that a newly developed shampoo may already contain a number of different preservatives. Apart from the various performance-based choices inherent in selecting a preservative system, which are well documented elsewhere [46, 47], there is also the problem of legislation; it is vital to know in which countries the product is to be sold in order to ensure that the product meets local requirements. This is true of all raw materials, but preservatives, particularly formaldehyde, (along with colours and ultraviolet absorbers) are a special case and are often heavily regulated. The chosen preservative system should pass a recognised (e.g. BP, USP) challenge test in the final formulation, preferably when freshly prepared, and when aged. If such testing can be accompanied by meaningful preservative assays, so much the better.

2.3.5 *Opacifiers and pearlisers*

Opacification of shampoos is usually used for aesthetic reasons, although it is occasionally a useful technique to use when the product cannot be made completely clear. If a pastel colour is required, then opacification is a prerequisite. Opacification may be achieved simply by adding a small amount of fine, intensely white polymer dispersion. Styrene/acrylate copolymers predominate, but many quite complex blends are available. Most contain some surfactant, usually anionic, which should cause no problem in most shampoo bases. Trials with different opacifiers at different levels will determine how to achieve the desired effect. For ease of dispersion and subsequent product stability, dilution of the opacifier to a 10% solution with water before addition to the main mix is usually recommended.

The vast majority of opaque pearlescent products depend for their effect on the suspension of various stearate crystals in the liquid base (the term 'stearate' as used here may cover a broader fatty acid cut than just C_{18}). Different stearates give different effects; diethylene glycol monostearate, for instance, gives a flatter, more opaque pearl than monoethylene glycol monostearate, which has less opacity but more 'sparkle'. There are three commonly used ways of achieving pearlescence using stearates.

1. To buy a ready-made pearlised base that simply needs dilution and the addition of any other desired materials to make a finished shampoo. Such products are, of necessity, very thick.
2. To add the chosen pearlising agent (most commonly ethylene glycol mono-distearate) to the hot mix, or a substantial part of it, at a temperature of about 75°C and form the crystals *in situ* during the cooling cycle. This method can be quite cheap on raw material costs, but expensive on energy and tank time. The process is also sensitive to changes in, for example, mixing equipment, stirrer speed and cooling rate, leading to variable results.
3. To buy a highly concentrated pearlising agent in a liquid or semi-liquid form that may be added to the batch as a cold mix. This is usually the most cost-effective route and as little as 2% of such a concentrate can give a good effect.

The main factors that affect the appearance of the 'pearl' are:

(i) Composition of the stearate ester
(ii) Presence of alkanolamides and other materials
(iii) Rate of cooling
(iv) Shear rate of stirring
(v) Composition of the base

An effective concentrated pearlising agent can be made in the laboratory to the following formula:

	% *w/w*
Ethylene glycol mono/distearate	20
Coconut monoethanolamide	20
Sodium lauryl ether sulphate, 27%	60
Hexylene glycol (to reduce viscosity)	*q.s.*

This formula, prepared with high-shear mixing, will give high opacity and low sparkle. The use of low-shear mixing will have the reverse effect. Rapid force-cooling will also favour opacity, while slow cooling will favour sparkle, i.e. any conditions which favour the growth of larger crystals will increase sparkle at the expense of opacity. Changing the alkanolamide also has an effect; coconut diethanolamide will give less sparkle and lauric mono-isopropanolamide more. When left undisturbed for some time, a large visible difference between the above variants is not apparent but, on pouring from one container to another, the difference becomes more marked as the thin plate-like stearate crystals align themselves with the direction of flow. Although most of the pearl concentrates on the market are based on sodium lauryl ether sulphate, some that are based on amphoterics or on non-ionics are also available, thus increasing the scope for the use of these materials.

2.3.6 *Conditioning agents*

The concept of a conditioning shampoo is a paradox. The majority of active conditioning agents are cationic surfactants and are therefore incompatible with the anionic surfactants that are the basis of nearly all shampoos. To take an extreme example, the addition of cetyl trimethyl ammonium chloride to a shampoo based on sodium lauryl sulphate will result in the formation of cetyl trimethyl ammonium lauryl sulphate, a very large and fairly useless molecule which will do its very best to precipitate out of solution. The properties of both the anionic and cationic molecules present can be modified to make them more compatible; the base itself can also be made more amenable to the inclusion of both of these seemingly antagonistic species.

1. The base can be made less anionic in character by including a reasonably high percentage of amphoteric surfactant. (The amphoteric can be thought of as analogous to a co-solvent, capable of uniting otherwise incompatible materials, e.g. ethanol as a co-solvent for water and diisopropyl adipate.)
2. The charge density on the cationic and anionic can be reduced. Here,

Table 2.6 Formulations for conditioning shampoos using polymeric cationic materials in predominantly anionic bases

Ammonium lauryl sulphate, 30%	—	—	—	30	15
Sodium lauryl ether sulphate, 27%	30	27	35	—	15
Triethanolamine lauryl sulphate, 42%	5	—	—	—	—
Cocoamidopropyl betaine, 30%	6	—	6	8	5
Pearl concentrate (SLES based)	2.5	—	2	—	—
Cationic guar gum	0.2	—	0.5	—	—
Quaternised hydrolysed protein	—	—	—	2	—
PEG-15 tallow polyamine	1	—	—	—	2.5
Cocoamphocarboxyglycinate, 30%	—	8	—	—	—
Sodium carboxymethyltallow polypropylamine-4, 27%	—	4	—	—	—
Cocoamidopropylamine oxide, 30%	—	—	2	—	—
Coconut diethanolamide	3	2	—	2.5	1.5
Lauric diethanolamide	—	—	3	—	—
Perfume	*q.s.*	*q.s.*	*q.s.*	*q.s.*	*q.s.*
5-Bromo-5-nitro-1,3-dioxane, 10%	0.2	0.1	—	—	—
Methylchloroisothiazolinone and methylisothiazolinone, 1.5%	—	0.05	0.1	0.1	0.05
Sodium chloride	*q.s.*	*q.s.*	*q.s.*	—	*q.s.*
Ammonium chloride	—	—	—	*q.s.*	—
Hydroxypropyl methyl cellulose	—	—	—	—	0.5
Colour	*q.s.*	*q.s.*	*q.s.*	*q.s.*	*q.s.*
Deionised water	to 100 →				
Citric acid	*q.s.* to pH 5.5	*q.s.* 6.0	*q.s.* 6.0	*q.s.* 5.5	*q.s.* 6.5

ethoxylation is the universal panacea; sodium lauryl ether sulphate has better compatibility with cationics than does sodium lauryl sulphate, while ethoxylated cationics are generally more compatible with anionics.

3. A variation on theme (2) is the use of a polymeric cationic, with the overall effect one of reduction of charge density. Some of the polymeric cationics available are surprisingly substantive and can also provide good afterfeel when used in products such as hair and body shampoos [48, 49].
4. The level of active matter can be made high enough for any anionic/cationic complex to be solubilised by the large excess of anionic present [50].
5. A sufficiently high level of non-ionic (really another variation on the theme of coupling agents) will assist compatibility.

Some formulations where polymeric cationic materials are used in predominantly anionic bases are shown in Table 2.6. The examples shown in Table 2.7 illustrate the use of non-polymeric cationics in conditioning shampoos.

Other means of conditioning, not based on cationic materials, are available.

Table 2.7 Formulations for conditioning shampoos using non-polymeric cationics

Sodium lauryl ether sulphate, 27%	33	—	—	—
Ammonium lauryl ether sulphate, 27%	—	40	15	36
Ammonium lauryl suphate, 30%	—	—	30	—
Sodium lauryl ether (10 EO) carboxylate, 30%	15	—		
Cocoamidopropyl betaine, 30%	7	5	—	—
Cocoamphocarboxyglycinate, 30%	—	—	6	4
PEG-3 cocoamide	3	—		
Coconut diethanolamide	—	2.5	—	3
Lauric monoisopropanolamide	—	—	2	—
Pearl concentrate	—	2.5		
Ethylene glycol mono-distearate	—	—	2	—
Stearamidopropyl dimethylamine lactate	—	2.5	—	—
Cetyl trimethylammonium chloride, 30%	1	—	—	—
N-(2-hydroxyethyl)-*N*,*N*-dimethyl-*N*-2-hydroxyethyl ammonium chloride, 30%	—	—	2	
γ-Gluconamidopropyl dimethyl-2-hydroxyethyl ammonium chloride, 60%	—	—	—	2
D-Panthenol	0.3			—
Dimethicone copolyol	—	—	0.6	—
Perfume	*q.s.*	*q.s.*	*q.s.*	*q.s.*
5-Bromo-5-nitro-1,3-dioxane, 10%	0.15	0.2	—	0.2
Methylchloroisothiazolinone and methylisothiazolinone, 1.5%	—	—	0.05	
Hydroxypropyl methyl cellulose	—	—	0.8	—
Ammonium chloride	—	*q.s.*	—	*q.s.*
Sodium chloride	*q.s.*	—	—	—
Colour	*q.s.*	*q.s.*	*q.s.*	*q.s.*
Citric acid	*q.s.*	*q.s.*	*q.s.*	*q.s.*
	to pH 5.5	6.0	5.5	6.0
Deionised water	to 100	⟶		

The substantivity of cationics is an electrostatic effect restricted to the outside of the hair shaft where the charge lies. Cationics, therefore, will not penetrate the hair shaft, the surface electrostatic attraction working against this possibility. Consideration of molecular size leads to the conclusion that deep penetration of the hair shaft by large species is improbable. Where there is significant damage to the hair shaft, however, the story is different. Partially hydrolysed proteins can be shown to be effective in 'repairing' split ends. In this case, the large protein molecules are effectively bridging the gaps and filling in the holes caused by physical damage. The implication of the above argument is that a sufficiently small molecule of low polarity should be able to penetrate the hair shaft, and this proves to be the case. Both panthenol (a pantothenic acid precursor) and amino acids have been shown to penetrate the undamaged hair shaft and, once there, to impart worthwhile conditioning effects. Panthenol is well-documented and, properly formulated into a

Table 2.8 Formulations for a conditioning shampoos using non-quaternary conditioners

	A	B	C	D	E
Sodium lauryl sulphate, 30%	5	—	—	—	—
Sodium lauryl ether sulphate, 27%	30	36	—	—	—
Monoethanolamine lauryl sulphate, 33%	—	—	—	40	—
Ammonium lauryl sulphate, 30%	—	—	10	—	6
Ammonium lauryl ether sulphate, 30%	—	—	35	—	36
Cocoamidopropyl betaine, 30%	—	9	7	—	—
Cocoamidopropyl hydroxysultaine, 30%	5	—	—	—	4
Sodium carboxymethyltallow polypropylamine-4, 27%	—	—	—	6	—
Lauric diethanolamide	2	2.5	—	—	—
Ricinoleic diethanolamide	—	—	—	3	—
Pearl concentrate	—	—	—	—	3
PEG-55 propylene glycol oleate and propylene glycol	—	—	3	—	2.5
Quaternium-80	—	—	0.2	—	—
Dimethicone copolyol	—	—	0.3	0.4	0.3
Cetyl dimethicone copolyol	—	—	0.2	—	—
PPG-5-ceteth-10-phosphate	—	5	—	—	—
Hydrolysed collagen protein	1	—	—	2	—
Keratin amino acids, 50%	—	—	—	1	—
D-Panthenol	2	0.5	—	—	0.3
Dimethiconol and cyclomethicone	—	—	—	—	1
Perfume	*q.s.*	*q.s.*	*q.s.*	*q.s.*	*q.s.*
Colour	*q.s.*	*q.s.*	*q.s.*	*q.s.*	*q.s.*
Ammonium chloride	—	—	*q.s.*	—	*q.s.*
Sodium chloride	*q.s.*	*q.s.*	—	*q.s.*	—
DMDM-hydantoin	0.2	—	—	—	—
2-Bromo-2-nitro-1,3-propandiol	—	—	0.05	—	0.05
Methyldibromoglutaronitrile and phenoxyethanol	—	0.1	—	0.1	—
Citric acid	*q.s.* to pH 7.0	*q.s.* 7.0	*q.s.* 6.5	*q.s.* 6.0	*q.s.* 6.5
Deionised water	to 100 ⟶				

product, can justify claims for moisturising and hair thickening, as well as conditioning. Sufficient active ingredient must be used to justify any claims made.

A variety of other materials have been used as conditioning agents in shampoos, including various vegetable oils, vitamins, lanolin and its derivatives, herbal extracts and some speciality silicones. The speciality silicones merit examination since they are currently popular, offer an alternative to cationic materials and should be less prone to 'build-up' or 'over-conditioning'. Silicone glycol copolymers can assist gloss and conditioning and seem to be substantive, to a degree, from shampoo bases, although they do reduce viscosity. More recently, very high viscosity dimethiconol dissolved in cyclomethicone has become more popular, although 'build-up' problems with shampoos using this ingredient have been reported. Build-up, together with the relative merits of various silicones and silicon derivatives in shampoos, is discussed in more detail by Sejpka [51]. An even more recent system advocates the use of several silicones together, at quite low levels, to give an optimised conditioning effect (see formulation C in Table 2.8).

2.3.7 *Colours and colour fading*

A large number of colours are available, but the list is restricted by the regulations governing their use wherever the product is sold. Whilst higher purity colours are more expensive than technical grades, it is preferable to use colours which meet both the EC and US specifications whenever possible; such colours will be acceptable virtually anywhere in the world. The cost premium is minimal.

Some aspects of colour stability, e.g. stability to pH variation and light, can be predicted but, unfortunately, the number of possible interactions involving the colour in a base as complex as a modern shampoo means that colour stability (especially to light) must be individually evaluated for each formulation. This can be carried out in natural sunlight or under accelerated conditions in a purpose-built apparatus using a xenon arc lamp or similar, with suitable filters to give simulated high-intensity sunlight. The instrument may be calibrated by using a product of known stability as a standard or by using blue wool to BS 1006 [52].

Colour fading can usually be minimised by incorporation of a suitable UV absorber. Water-soluble absorbers usually work best in shampoos, the most popular being benzophenone-4 and benzophenone-2. However, it is always worth trying others, both UV-A and UV-B absorbers. Usage levels are low, generally 0.05–0.1%.

Colours should be added in solution and not as solid material. Aqueous solutions of colours require preservation—an aqueous/alcoholic base is safer.

2.3.8 *Other additives*

Sequestrants, such as EDTA (ethylene diamine tetra-acetic acid) salts, are sometimes added at low levels to aid rinsing in hard water and boost the efficacy of preservatives. The majority of other shampoo additives are not truly active ingredients, being included for marketing reasons rather than performance.

Anti-dandruff shampoos have been a growth area in recent years, with various active ingredients having been used. Selenium sulphide is sometimes encountered, but is mainly restricted to prescription products. Coal-tar extracts and phenolic derivatives have been used but are not very effective and traditionally find their place in 'medicated' shampoos. Their use is declining due to doubts about the safety of coal-tar derivatives. Derivatives of undecylenic acid, in particular the sodium salt of undecylenic acid monoethanolamide sulphosuccinate, have been found to have some degree of effectiveness. The widespread adoption of zinc pyridinethione as an active ingredient has substantially increased the use of anti-dandruff shampoos. This ingredient is usually supplied as a 48% dispersion and is a very dense solid which needs to be suspended in the product. To obtain stability this requires a suitable clay- or gum-type suspending agent, with fairly high yield-point, and restricts the formulator to producing an opaque product. A more recently available anti-dandruff agent is piroctone olamine, which is water-soluble and therefore much easier to use.

The nature of dandruff, and its causal relationship with scalp flora, particularly pityrosporum ovale, has given rise to much speculation. The exact nature and cause of dandruff, and its definition as a clinical condition, has not been fully established [53–58]. Consequently, the classification of anti-dandruff products as cosmetics or pharmaceuticals remains unresolved, except in the USA where they are clearly OTC drugs. The formulations shown in Table 2.9 illustrate various types of anti-dandruff shampoo.

Baby shampoos form another important category. Babies' hair does not require a high level of detergency to clean it and, above all, the products must be non-irritant to eyes and skin, or as close to this ideal as possible. In this sense, baby shampoos have a close affinity with frequent shampoos, the latter often lying between baby shampoos and conventional shampoos in terms of formulation. Baby shampoos are discussed in more detail in chapter 5.

Other specialist shampoos are shown in Table 2.10 and include swimmers' shampoos, containing a reducing agent (usually sulphite or thiosulphate) to neutralise the chlorine from the water in swimming pools, and UV-protection shampoos. Unfortunately, most UV absorbers are not very substantive to the hair when applied from a shampoo base. Dimethyl *p*-amino benzoic acid (PABA) ethyl cetearyldimonium tosylate, however, is a modified PABA derivative that is cationic in nature and can give better results, although its poor solubility characteristics are somewhat restrictive [59].

The final class of shampoos that will be mentioned here—anti-build-up

Table 2.9 Typical formulations for anti-dandruff shampoos

Sodium lauryl ether sulphate, 27%	40	—	25	—	—
Ammonium lauryl ether sulphate, 30%	—	—	—	20	—
Ammonium lauryl sulphate, 30%	—	—	—	15	30
Triethanolamine lauryl sulphate, 40%	—	42	15	—	—
Cocoamidopropyl betaine, 30%	9	—	8	—	—
Cocoamphocarboxy glycinate, 30%	—	—	—	5	6
Lauric/myristic diethanolamide and propylene glycol (80/20)	—	4	—	—	—
Coconut diethanolamide	2.5	—	1.5	2	3
Diethylene glycol mono-/distearate	—	—	3.5	—	—
Pearl concentrate	2.5	—	—	—	—
Sodium undecylenic acid monoethanolamide sulphosuccinate, 50%	—	—	—	5	—
Zinc pyridine thione, 48%	2	4	2	—	—
Piroctone olamine	—	—	—	—	0.7
Hydroxyethyl cellulose	—	0.5	—	—	—
Magnesium aluminium silicate	—	1	—	—	—
Bentonite	1	—	—	—	—
Carbomer 934	—	—	0.25	—	—
Sodium hydroxide	—	—	*q.s.*	—	—
to pH			5.5		
Citric acid	—	—	—	*q.s.*	*q.s*
to pH				6.5	6.0
Perfume	*q.s.*	*q.s.*	*q.s.*	*q.s.*	*q.s.*
Colour	*q.s.*	*q.s.*	*q.s.*	*q.s.*	*q.s.*
Ammonium chloride	—	—	—	*q.s.*	*q.s.*
Preservative	*q.s.*	*q.s.*	*q.s.*	*q.s.*	*q.s.*
Deionised water	to 100	→	→	→	→

shampoos—is perhaps the simplest. As the name suggests, these shampoos are formulated to help remove excess build-up of conditioning agents and styling aids. Consequently, they are usually quite simple systems (Table 2.10), with good detergency and containing no conditioners. Since most conditioners (whether from shampoos, conditioners or styling aids) are applied to the hair under acid conditions, anti-build-up shampoos are left alkaline. (This can also help remove acid resins.) For the same reason, 'lacquer-removing' shampoos, the ancestors of this class of product, usually contained an excess of mild alkali (e.g. TEA). A means of measuring the efficacy of these products in removing deposited polymers from the hair shaft is discussed by Sendelbach *et al.* [60].

Before leaving shampoos, the question of impurities should be considered. Most surfactants contain small amounts of numerous by-products. Of particular concern are the nitrosamines and 1,4-dioxane.

Nitrosamines are potent carcinogens derived from the reaction of nitrite and secondary amines. The most important in cosmetics and toiletries is nitrosodiethanolamine (NDELA), which is formed, under some circumstances, from diethanolamine. To avoid this, lower grades of triethanolamine (which can contain substantial amounts of diethanolamine) should be replaced with a higher grade (99% + purity). As mentioned earlier, some coconut

Table 2.10 Formulations for swimmers', UV-protection, and anti-build-up shampoos

	Swimmers shampoos	UV-protection shampoos		Anti-build-up shampoos	
Sodium lauryl ether sulphate, 27%	36	40	40	24	20
Sodium lauryl sulphate, 30%	—	—	—	10	—
Sodium lauryl sarcosinate, 30%	—	—	—	6	—
Triethanolamine lauryl sulphate, 42%	—	—	—	—	25
Cocoamphocarboxy glycinate, 30%	6	—	—	—	—
Cocoamidopropyl betaine, 30%	—	6	10	—	—
Cocamine oxide, 35%	—	—	—	—	3
Pearl concentrate	2.5	—	2	—	—
Coconut diethanolamide	2.5	3	4	—	2
Lauric diethanolamide	—	—	—	2	—
Panthenol	—	—	—	—	0.4
Polyquaternium-7, 8%	—	2	—	—	—
Octyl dimethyl PABA	—	0.3	—	—	—
Benzophenone-4	—	0.2	—	—	—
Dimethyl PABA ethyl cetearyl dimonium tosylate	—	—	1	—	—
Urea	2	—	—	—	—
Sodium thiosulphate	0.3	—	—	—	—
Citric acid	*q.s.*	*q.s.*	*q.s.*	—	—
	to pH 7.0	6.5	6.0		
Hexylene glycol	—	—	*q.s.*	—	—
Triethanolamine	to pH—	—	—	8.5	8.5
Sodium chloride	*q.s.*	*q.s.*	—	*q.s.*	*q.s.*
Methylchloroisothiazolinone (and) methylisothiazolinone, 1.5%	—	0.03	—	—	0.05
Phenoxyethanol (and) methyldibromoglutaronitrile	—	—	0.05	0.05	—
Propylene glycol (and) 5-bromo-5-nitro-1,3-dioxane	0.15	0.1	—	—	—
Deionised water	to 100	→	→	→	→

diethanolamides, especially the Kritchevsky type, can contain quite large amounts of diethanolamine and it is advisable to replace these with other alkanolamides or alternative thickeners/foam boosters.

1,4-Dioxane is a cyclic dimer of ethylene oxide and can arise whenever ethoxylation is used in the manufacture of a surfactant. Like NDELA, it has been identified as a carcinogen and some European countries are now tending to use non-ethoxylated surfactants to minimise dioxane levels. Control can be exercised at each stage of surfactant manufacture and it is now possible to buy 27% active SLES with dioxane < 25 ppm [61–65]. A brief review of some of the more modern shampoo ingredients is given by Woodruff [66].

2.4 Conditioners

Harry [67] attempts to distinguish between conditioners, rinses and tonics, but such classifications have been overtaken by events, with terms such as rinse, conditioner, deep conditioner, balsam, hair mask, cream rinse, being used interchangeably by the public and industry alike. Classification requires a definition for 'conditioning' and, since this term means many things to many people, e.g. reduction of fly-away, gloss, sheen, manageability, ease of handling, or simply general overall appearance, the issue is best avoided by grouping all such products together.

The basic mode of action of hair conditioners has been touched upon in section 2.3.6. In a conditioner, the properties of cationic surfactants can be fully exploited without having to consider the question of compatibility with anionics. Consequently, cationics with high charge density, and properties undiluted by ethoxylation or polymerisation, feature prominently among the preferred active ingredients. Before examining the formulation of hair conditioners in detail, the properties that hair conditioner should impart will be considered.

(i) Improved wet combing
(ii) Improved dry combing
(iii) Reduced fly-away (i.e. antistatic)
(iv) Increased gloss
(v) Increased volume

In addition, more specific claims, e.g. improved curl retention, repair of damage of the hair shaft, or moisturising, may be required.

The most common physical form for a conditioner is the thick opaque liquid, although clear liquids have recently become popular. Gels sprays and non-pourable creams are available, while 'leave-on' conditioners (usually in spray form), as opposed to the more common 'rinse-off' products, are also in demand.

Classification of the raw materials into neatly defined categories is not as straightforward as with shampoos, but a suggestion is as follows:

- Primary surfactants (nearly always cationic)
- Polymers
- Bodying agents
- Auxiliary emulsifiers
- Oily components
- Other 'active' ingredients
- Thickeners
- Perfumes
- Preservatives
- Diluents (usually water)
- Colours
- Other non-functional ingredients

Many of the ingredients are multifunctional within the formulation, and not all need be present.

2.4.1 *Cationic surfactants*

The most common principle active ingredients in hair conditioners are quaternary ammonium compounds (or 'quats') corresponding to the formula

$$\left[R_1-\overset{R_2}{\underset{R_4}{N}}-R_3\right]^+ X^-$$

where X, the non-surface active anion, is commonly chloride and R_1, R_2, R_3 and R_4 are alkyl or other groups. Normally two or three of these groups are methyl, and one or two are derived from long chain fatty acids, e.g. tallow or coconut oil. Typical examples are

$$\left[C_{16}H_{33}-\overset{CH_3}{\underset{CH_3}{N}}-CH_3\right]^+ Cl^-$$

Cetyl trimethylammonium chloride

$$\left[C_{16}H_{33}-\overset{CH_{16}H_{33}}{\underset{CH_3}{N}}-CH_3\right]^+ Cl^-$$

Dicetyl dimethylammonium chloride

$$\left[C_{18}H_{37}-\overset{CH_3}{\underset{CH_3}{N}}-C_6H_5\right]^+ Cl^-$$

Stearyl dimethyl benzylammonium chloride

It is notable that, while the optimum length for most anionic surfactants for shampoo use is around C_{12}/C_{14}, in the case of cationics for conditioners, this rises to C_{16}/C_{18}. More recently, there has been considerable interest in quats with longer chains, e.g. C_{22}, behenyl. Fuller details of the many available quats are given by Hunting [68], the CTFA Cosmetic Ingredient Dictionary [69], and elsewhere [9, 70–73]. Some quats are relatively poorly biodegradable. A more recently introduced type of quat, the 'ester quat' is claimed to be better in this respect. As the name suggests, the side chains contain ester linkages:

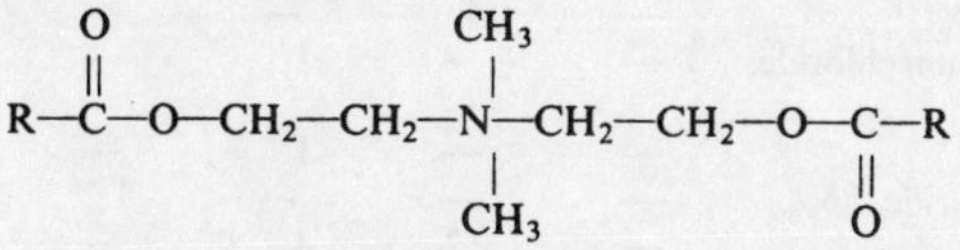

$$R{-}\overset{\overset{\displaystyle O}{\|}}{C}{-}O{-}CH_2{-}CH_2{-}\underset{\underset{\displaystyle CH_3}{|}}{\overset{\overset{\displaystyle CH_3}{|}}{N}}{-}CH_2{-}CH_2{-}O{-}\underset{\underset{\displaystyle O}{\|}}{C}{-}R$$

, Where R = alkyl, usually C_{16-18}

Clearly, the properties of the various quats differ and this is governed by some rules of thumb. The properties can often be related to charge density; a highly charged cationic will be more strongly attracted to the negatively charged hair surface but, once on the hair surface, the number, shape and size of the fatty chains in the molecule are the determinants. A 'quat' with two fatty chains will provide more lubricity than a quat with one, while a long chain may be a more effective lubricant than a shorter one. As a result, it is quite common to encounter single fatty-chain quats in products designed to be used on greasy hair or for frequent use, and twin fatty-chain quats in products intended for dry or damaged hair. The fatty chains may be derived from more exotic sources such as mink oil, lanolin acids, glucamic acid (contains many hydroxyl groups and is very water-soluble), oleic acid, isostearic acid, etc., as well as from coconut oil or tallow. At the other end of the molecule, there may be a number of nitrogen atoms, among which the positive charge centre may be distributed. The amidoamines are a form of 'do-it-yourself' quat that can be neutralised *in situ* by an acid of the formulator's choice. The conditioning effect is quite light, comparable with a conventional single fatty-chain quat. Examples are stearamidopropyl dimethylamine and stearamidoethyl diethylamine. An interesting comparison between the properties of various types of quat is made by Allandic and Gummer [74].

A hair conditioner might be made by simply diluting down a suitable quat with water. The stability and effectiveness may, however, be limited, therefore it is common to use quats in conjunction with fatty alcohols, usually C_{16}–C_{18}. Very thick emulsions can be obtained with quite low solids content, and an effective product made at low cost. Formulae A and C in both Tables 2.11 and 2.12 illustrate this type of product. There are some drawbacks, including:

(i) The appearance of the emulsion, which may be coarse and grainy in texture.
(ii) The low opacity. While this need not be considered disadvantageous, the translucent nature of such emulsions can be unattractive.
(iii) The rheology of the system, normally highly non-Newtonian, with a high yield-point making controlled use of the product difficult. A further problem is the frequently encountered increase of viscosity with time.

Table 2.11 Formulations for conventional cationic conditioners

	A	B	C	D
Behenyl trimethylammonium chloride	1	—	—	—
Cetyl trimethylammonium chloride, 30%	—	4	—	—
Dimethyl dihydrogenated tallow ammonium chloride, 75%	—	—	1	—
Stearamidoethyl diethylamine	—	—	1	—
Stearyl dimethyl benzylammonium chloride, 75%	—	—	—	1.2
Cetyl alcohol	—	2.2	4	—
Cetostearyl alcohol	3.5	—	—	2
Mineral oil	—	1.2	—	0.4
Cetomacrogol	—	0.3	0.3	—
Glyceryl stearate	—	—	—	0.5
Methyl paraben	—	0.1	0.1	—
2-Bromo-2-nitro-1,3-propanediol	—	—	0.03	0.05
Methylchloroisothiazolinone and methylisothiazolinone, 1.5%	0.05	0.03	—	—
Citric Acid	to pH 4.5	5.0	4.5	0.25
Colour	*q.s.*	*q.s.*	*q.s.*	*q.s.*
Perfume	*q.s.*	*q.s.*	*q.s.*	*q.s.*
Deionised water	to 100 ⟶			

Table 2.12 Formulations for conditioners based on less common cationics

	A	B	C	D	E
Steapyrium chloride	1.8	—	0.8	—	—
Quaternium-70, 50%	—	1.4	—	—	—
Stearamidopropyl dimethylamine	—	0.8	—	—	—
Amodimethicone, 35%	—	—	2	—	—
Quaternium-52, 50%	—	—	—	—	2.5
PEG-5 stearyl ammonium chloride, 30%	—	—	—	4	—
Quaternium-33, 50%	—	—	—	—	1
Cetyl alcohol	1.8	3.2	3.5	—	—
Cetostearyl alcohol	—	—	—	4	3
Stearic acid	—	—	—	—	1
Ethylene glycol stearate	—	—	—	—	1.5
Cetomacrogol	—	0.25	—	—	0.4
PEG-5-ceteth-10-phosphate	—	—	—	0.8	—
Mineral oil	—	—	—	1.5	0.8
Panthenol	—	0.5	—	—	—
Hydrolysed protein	—	1.5	—	—	0.5
Quaternium-15	—	—	0.1	—	—
DMDM hydantoin	0.1	—	—	—	—
2-Bromo-2-nitro-1,3-propanediol	—	0.04	—	0.05	—
Methylchloroisothiazolinone and methylisothiazolinone, 1.5%	—	—	—	—	0.05
Citric acid	to pH 4	0.18	3.5	4	3.5
Sodium chloride	0.4	—	—	—	—
Colour	*q.s.*	*q.s.*	*q.s.*	*q.s.*	*q.s.*
Perfume	*q.s.*	*q.s.*	*q.s.*	*q.s.*	*q.s.*
Deionised water	to 100 ⟶				

(iv) The sensitivity of the product to changes in manufacturing method, in particular variations in high-shear stirring input, small changes in which can produce very large viscosity variations.

These problems can usually be minimised in a number of ways.

(i) Introduction of other waxy components, such as esters. The structural viscosity arising from hydrogen bonding with the hydroxyl groups of the fatty alcohols is thereby reduced.

(ii) Addition of a small amount of oil, which smooths out the emulsion and increases the opacity, while reducing the overall melting point of the oil phase. This facilitates processing and makes it easier to obtain repeatable viscosities from batch to batch.

(iii) Addition of a non-ionic emulsifier. In most hair conditioners the cationic has a dual role: that of an active ingredient and that of an emulsifier. A small amount of non-ionic emulsifier can assist greatly in the emulsification, freeing the quat for optimum use in its primary role. The increase in stability resulting from the cationic/non-ionic blend may be due to the formation of mixed micelles and to the presence of mixed surfactants at the oil/water interface of the emulsion. Formulations B and D in Table 2.11, and B, D and E in Table 2.12 illustrate the use of such forms of stabilisation.

2.4.2 *Cationic polymers and other active ingredients*

This class of compounds, used extensively in conditioning shampoos, has also become widely used in hair conditioners, alone or together with ordinary quats. The 'backbone' structure of these polymers, on to which are grafted

Table 2.13 Formulations for polymer-based conditioners

Polyquaternium-22, 40%		2.5	—	—	—
Polyquaternium-6		—	2	—	—
Methylvinylimidazolium chloride/vinyl pyrrolidone copolymer, 40%		—	—	1.5	—
PEG-15 tallow polyamine, 50%		—	—	—	3
Cetostearyl alcohol		5	4.2	3.5	2
Glyceryl stearate		—	—	1.2	—
Glyceryl stearate and PEG-100 stearate		—	—	—	5
Cetomacrogol		0.5	0.4	1.6	0.4
Mineral oil		—	1.4	4	—
Methyl paraben		—	—	0.1	0.1
Quaternium-15		0.1	—	—	0.07
2-Bromo-2-nitro-1,3-propanediol		—	0.08	0.04	—
Behenyl trimethylammonium chloride		1	—	—	—
Citric acid	to pH	4	4.5	0.15	4
Perfume		*q.s.*	*q.s.*	*q.s.*	*q.s.*
Colour		*q.s.*	*q.s.*	*q.s.*	*q.s.*
Deionised water		to 100% ⟶			

quaternary ammonium groups and, where necessary, fatty chains, may have various derivations, including partially hydrolysed protein, cellulose, starch, guar gum, silicones, chitosan and various synthetics. The synthetics may be homopolymers such as polydimethylaminomethyl methacrylate or dimethyl diallyl ammonium chloride, or copolymers such as dimethylaminoethyl methacrylate/vinyl pyrrolidone or hydroxyethylcellulose/dimethyl diallylammonium chloride. The positively charged sites occurring at intervals along the polymer molecules permit an electrostatic attraction to the hair shaft in the same way as monomeric quats. However, because the polymer may be modified by the inclusion of more or less quaternary groups and a variable number and length of side chains, its properties may be tailored to emphasise any particular aspect such as charge density, substantivity and lubricity.

Polymeric quats can possess one quality that is absent in their monomeric

Table 2.14 Formulations for composite conditioners

Dimethyl distearylammonium chloride, 75%	0.8	—	1.2	—	—	—
Stearamidopropyl dimethylamine	1	—	—	—	—	—
Cetyl trimethylammonium chloride, 30%	—	4	—	2	1.2	—
Hydroxypropyltrimonium guar	—	—	—	0.5	—	—
Polyquaternium-22, 40%	—	—	—	—	—	1
Quaternium-80	0.8	0.5	—	—	—	—
Quaternium-22, 60%	—	—	—	—	—	2.5
Dimethicone propyl PG-betaine, 50%	—	—	—	—	2.5	—
Steartrimonium hydrolysed protein	—	1	—	—	—	—
Mineral oil	—	1.5	1.5	—	0.5	—
Cetomacrogol	0.4	0.3	0.5	—	1.2	0.3
Cetostearyl alcohol	4.5	4	5	3.5	1.8	4
Glyceryl stearate	—	—	—	—	2.7	—
Vitamin E acetate	—	—	0.4	—	—	—
Vitamin A palmitate, 1 mIU/g	—	—	0.1	—	—	—
Stearic acid	—	—	1	—	—	—
Hydrolysed protein	—	—	1	—	—	—
Keratin amino acids, 50%	—	1	—	—	—	—
Panthenol	0.5	—	—	—	0.2	—
DMDM hydantoin	—	—	0.1	—	—	0.1
Methyl paraben	0.1	0.15	0.15	—	—	0.1
2-Bromo-2-nitro-1,3-propanediol	—	0.03	—	—	0.05	—
Methylchloroisothiazolinone and methylisothiazolinone, 1.5%	0.05	—	—	0.08	—	—
EDTA	—	0.1	—	—	—	—
Perfume	*q.s.*	*q.s.*	*q.s.*	*q.s.*	*q.s.*	*q.s.*
Colour	*q.s.*	*q.s.*	*q.s.*	*q.s.*	*q.s.*	*q.s.*
Citric acid	0.25	—	—	0.2	*q.s.*	*q.s.*
to pH					4.5	4.0
Deionised water	to 100 →					

siblings—that of film forming; indeed, for this reason, many also find a use in hair-fixative products. Conditioners, particularly when used by thóse with naturally greasy hair, can leave the hair rather lank and with a tendency to re-soil rapidly. To some extent, this can be overcome by formulating with cationic polymers with few, if any, fatty side chains. The polymeric quats rarely have the emulsification power of the monomers and will therefore often require the help of an auxiliary emulsifier. In most other respects they can be dealt with as ordinary quats. The formulations in Table 2.13 illustrate their use. Much more information about these materials is available, predominantly from manufacturers' published literature [75].

The formulator may well wish to utilise the properties of both polymers and conventional quats, which can be freely mixed to get the best of both worlds. Conventional quats will often provide the necessary additional emulsifying power. The formulations in Table 2.14 are examples of conditioners formulated with more than one active ingredient.

Although the vast majority of conditioners are based on the established properties of cationics, a few products are not. Anionic systems are occasionally encountered (Table 2.15, formulation B), as are systems based on high levels of protein (Table 2.15, formulation C). A variety of exotic oils,

Table 2.15 Formulations for non-quaternary conditioners

	A	B	C
Carbomer 934	0.5	—	—
Hydroxypropyl methylcellulose	—	—	0.7
Sorbitan monopalmitate	0.4	0.2	—
Polysorbate 40	0.6	0.25	—
Polysorbate 20	—	—	3
Mineral oil	3	3.5	—
Lanolin oil	1	—	—
Avocado oil	—	3	—
Wheatgerm oil	—	3	—
Paraffin wax	—	2.5	—
Sodium cetyl sulphate	—	0.6	—
Stearic acid	—	2.5	—
Triethanolamine	0.5	—	—
Acetamide *MEA and LAMEA†	2	1	—
Panthenol	1	0.4	—
Hydrolysed protein	2	—	8
Keratin amino acids	—	0.5	2
Glycerine	—	3	—
Methyl paraben	0.1	0.2	0.15
Propyl paraben	0.1	0.1	—
Quaternium-15	—	0.1	0.2
Methylchloroisothiazolinone and methylisothiazolinone, 1.5%	0.04	—	—
Deionised water	to 100	⟶	⟶

* MEA, mono ethanolamide;
† LAMEA, lactamide mono ethanolamide.

vitamins and other materials have been proposed, but their substantivity from non-ionic or anionic emulsion bases must be considered questionable. Amino acids and panthenol are widely used, as in shampoos. A variety of humectants, for example acetamide MEA, lactamide MEA, polyols and sodium pyrrolidone carboxylate, are often included but 'moisturising' as applied to hair is not as well defined or documented as when applied to skin. The argument that, since hair consists of dead tissue, hair and skin must be treated quite differently, is frequently used. Nevertheless, many ingredients with non-proven action continue to be widely used in hair preparations that apparently satisfy consumer demands. A concise up-to-date review of active ingredients used in hair conditioners is given by Woodruff [76], who also illustrates their use with example formulations.

2.4.3 *Bodying agents*

As with shampoos, conditioners are perceived to be more effective when thick and creamy. This is usually achieved by using quite high levels of fatty alcohols along with other waxy esters, and hair conditioners containing only lipids that are solid at room temperature are common. The function of the oils and waxes is not merely one of giving body. A comparison between a simple formula, such as formula A in Table 2.11, and an aqueous solution of a quat will clearly demonstrate significant improvement in handle of the hair, especially wet combing, indicating that the fatty alcohols have substantivity to the hair when applied from a cationic emulsion.

2.4.4 *Auxiliary emulsifiers*

Depending on the hydrophilic lipophilic balance (HLB) and the steric properties of the molecule, some quats are good emulsifiers, while others are not. Their use with anionic emulsifiers is clearly prevented by incompatibility; where sufficient emulsion stability cannot be obtained with the cationics alone, non-ionics are the preferred addition. There is a huge range to choose from, but ethoxylated fatty alcohols seem to be particularly effective. Structurally akin to the fatty alcohols present, they can form a harmonious blend, very much in line with the work of Griffin [77], who suggested blends of two emulsifiers (one with high and one with low HLB) chosen from a series whose members had related chemical structures.

2.4.5 *Oil components*

The effect of oils on ease of manufacture and product stability has already been referred to. In most formulations, such non-polar ingredients, used at typically low levels, can be interchanged without greatly affecting stability. Small quantities of exotic oils with attractive names may be included, although any sound technical reason for doing so is not always obvious. There has been a recent fashion for 'oil-free' conditioners; in the case of emulsion-based products this need have little, if any, impact on product performance.

2.4.6 *Thickeners*

Cationic emulsions, which are inherently thick for any given solids content, do not usually require thickeners. They may be readily manipulated by modifying manufacturing conditions. Although they usually respond to salt in a similar way to shampoos, this method of thickening must be treated with extreme caution if irreversible emulsion breakdown is to be avoided. The following rules should be observed.

(i) Add only very small amounts of salt solution at one time, since the viscosity will peak at much lower salt levels than that of typical shampoos.
(ii) Add salt solution only when the emulsion is cold (maximum temperature 30°C).
(iii) Use dilute salt solution (10% maximum) and add very slowly with constant stirring.
(iv) Investigate the effect of the rate of stirrer shear when the salt solution is added. A combination of a small amount of salt solution and high-shear mixing will often achieve a viscosity that cannot be obtained by one method alone.
(v) If working on a full-scale batch, always try a sample in the laboratory first and be careful to take into account any effects of scaling-up.

Gum-type thickeners, the norm in clear conditioners (see section 2.4.11) may also be used. Cellulose derivatives dominate and generally work very well. Carbomers are rarely encountered because of their established incompatibility with most cationics.

2.4.7 *Perfumes*

Conditioners usually require only low levels of perfume. Typically, a perfume which is present in shampoo at 0.5% will perform equally well in conditioner at only 0.2%.

A number of quats on the market use isopropanol as a solvent and this can sometimes leave a persistent odour which may be hard to mask. The best remedy is to select a grade of quat which does not suffer from this problem. Conditioners are fairly easy to perfume, but the stability of the perfume in the often quite strongly acidic environment of hair conditioners must be checked.

2.4.8 *Preservatives*

Most of the remarks made in section 2.3.4 on shampoos also apply here, except that few ingredients used in conditioners will normally contain preservatives (proteins and herbal extracts are exceptions). Many, indeed most, cationic surfactants are active to some degree against micro-organisms. Consequently, conditioners are fairly low-risk from a microbiological viewpoint and serious problems are mercifully rare.

2.4.9 *Colours*

The addition of colours to shampoos is perfectly straightforward using stock solutions (which must be adequately preserved) added to the batch. Most of the widely used colours are, however, anionic in character and, if added in the same way to the conditioner, will give rise to intensely coloured spots. There are several ways, alone or in combination, of avoiding this.

(i) Add the colours when the bulk is still hot. Unfortunately the product will normally whiten on cooling, especially if high-shear stirring is used, making the colour appear lighter.
(ii) Use warmed colour solutions.
(iii) Use very dilute colour solutions ($<0.1\%$ strength) and add slowly, avoiding the creation of high local concentrations.

A few colours, such as rhodamine, because of their cationic nature, are fully compatible with conditioners.

2.4.10 *Manufacture*

Traditionally, emulsions are made by heating oil and water phases separately, combining them when hot, and then cooling them. Many conditioners can, however, be made by adding all of the remaining ingredients to the heated water and stirring fairly vigorously until an emulsion is formed, before cooling as usual. Furthermore, because of their high water phase/oil phase ratio, these products lend themselves well to low-energy emulsification methods as advocated by Lin and co-workers [78–82]. In this method, only part of the water is heated and a concentrated emulsion is first formed. The remaining water is added cold under controlled conditions. A high-shear mixer, whether bottom or top entry, or in-line, is always useful when making these products.

2.4.11 *Clear conditioners*

These are usually thickened, aqueous solutions of quats, polymers, or both, often containing other active ingredients such as panthenol and water-soluble silicone derivatives. Many of these products develop quite a strong lather when worked into wet hair. If desired, this can be enhanced by the addition

Table 2.16 Formulations for clear hair conditioners

	A	B	C
Cetyl trimethyl ammonium chloride, 30%	—	4.5	3
Cetyl pyridinium chloride	1.5	—	—
Cocoamidopropylamine oxide, 35%	—	—	3
Polyquaternium-22, 40%	—	—	2.5
Panthenol	0.5	—	—
Hydrolysed protein	0.5	1	—
Dimethicone copolyol	—	0.5	1
Glycerine	3	—	—
Hydroxyethyl cellulose	—	1	—
Hydroxypropyl methylcellulose	1.2	—	1.2
EDTA	—	0.1	0.1
2-Bromo-2-nitro-1,3-propane diol	0.02	—	—
Phenoxyethanol (and) methyldibromoglutaronitrile	—	0.04	0.04
PEG-40 hydrogenated castor oil	—	0.6	—
Perfume	0.1	0.15	—
Colour	*q.s.*	—	—
Citric Acid	to pH 4.5	4	4
Deionised water	to 100	⟶	

of amine oxides or betaines as foaming agents. Both will be predominantly cationic in character at low pH and may well also contribute to the conditioning effect. The formulations in Table 2.16 are examples of this type of product. Formulation C in Table 2.15 is also clear, but is a non-quaternary based product.

2.4.12 *Hair thickeners*

These products vary in popularity according to the dictates of fashion. Hair can be made thicker by causing it to swell, or by coating it to increase its diameter. Hair will swell in water, more so under alkaline conditions and although direct application may be impractical, powerful humectants may be used to increase the moisture content of the hair. Swelling of the hair shaft as a result of the application of panthenol solution is documented [83]. Almost any polymer exhibiting a degree of substantivity will cause the hair-shaft diameter to increase, particularly in a non-rinse-off product. Examples of thickener formulations are given in Table 2.17.

2.4.13 *Leave-on-conditioners*

Some of the early popular conditioners were 'leave-on' products presented in the crude pump-spray packs available at that time. This type of product is now returning to the shelves, greatly assisted by considerable advances in pump-spray technology, which have been promoted by concern over ozone depletion. For ease of dispensing, a leave-on conditioner should be water-thin and, to avoid formulating a product which will be too 'heavy', i.e. leave too

Table 2.17 Formulations for hair thickeners and leave-on conditioners

	Thickeners		Leave-on conditioners		
Vinylpyrrolidone/vinyl acetate copolymer, 60/40	2.5	2	—	—	—
Polyquaternium-11, 50%	1.5	—	—	—	—
Steartrimonium hydrolysed protein	2.5	2.5	—	—	—
Polyquaternium-22, 40%	—	2	—	—	1
Quaternium-80	—	—	—	—	0.5
Hydroxyethyl cellulose	0.9	1.1	—	—	—
Dimethicone copolyol	—	0.5	—	—	—
Oleth-20	0.5	—	—	—	0.4
Cetyltrimethylammonium chloride, 30%	—	—	0.25	0.2	—
Keratin amino acids, 50%	—	—	0.4	—	—
Lactamide-MEA	—	—	—	1.2	—
Panthenol	0.25	—	—	1.2	0.1
Propylene glycol	—	—	—	2	—
Colour	*q.s.*	*q.s.*	*q.s.*	*q.s.*	*q.s.*
Perfume	0.15	0.1	*q.s.*	-	*q.s.*
Methylchloroisothiazolinone (and) methylisothiazolinone, 1.5%	0.05	0.05	—	—	0.05
2-Bromo-2-nitro-1,3-propandiol	—	—	—	0.05	—
Methyldibromoglutaronitrile and phenoxyethanol	—	—	0.1	—	—
Deionised water	to 100 →				

much residue on the hair, a low solid content must be used. Water contents therefore tend to be extremely high and raw material costs low. The greater emphasis on moisturisers in this product category is quite logical, since omission of rinsing enables sufficient contact time for such materials to perform and ensures that 100% of material applied remains on the hair. There are some leave-on products in the form of thick liquids and creams but these are usually restricted to intensive treatments for severely damaged or extremely dry hair. Some examples of formulations for pump-spray, leave-on conditioners are given in Table 2.17.

2.4.14 *'Hot oils', tonics and other conditioners*

2.4.14.1 *Hot oils* Most 'hot oils' are neither hot nor oily. While the misnomer might be construed as misleading, most of the products actually perform quite well. A simple blend of oils, including 'exotics' such as avocado, wheatgerm, rosehip and jojoba, can be effective when applied hot to extremely dry hair. However, most of the products on the market are water solutions based on conventional quats. The formulations already given to illustrate clear conditioners could equally well be used with less thickener.

A few self-heating products have begun to appear on the market. These are based on the exothermic dissolution of hydroxy compounds such as alcohols and glycols in water. The products are, of necessity, anhydrous and are usually based on polyethylene glycols or PEG/polypropylene glycol (PPG) copolymers. Quats, panthenol and other suitably soluble materials can be incorporated. The products are applied directly to wet hair and give an immediate and readily perceivable temperature rise.

2.4.14.2 *Tonics* Many conditioners can work quite well as pre-shampoo rather than post-shampoo treatments, provided that the subsequent shampooing is not too fierce. Some products of this type have been marketed with limited success as hair and scalp conditioners, or as tonics; formulations A and B in Table 2.18 are examples.

In the days before strict controls there were numerous 'hair tonics' on the market, many making quite outrageous claims, often promising cures for baldness, or at least implying this. The widespread recognition that treatment for hair loss by topically applied products has a long history of abject failure means that these products now represent only a tiny fraction of the total hair-care market (except in Japan [84]). In any case, such treatments are outside the remit of the cosmetic chemist.

Not all hair tonics are so blatantly misleading and, in some cases, they form convenient vehicles for the application to the scalp of a wide range of beneficial additives. These include vitamins such as Vitamin E acetate and nicotinate inositol, Vitamin A palmitate, Vitamin B6, panthenol, pantothenates and panthenyl esters, D-biotin, vasodilators such as nicotinic acid esters, coal-tar extracts, herbal extracts, camphor, menthol, allantoin and sulphur. The positioning of many hair tonics places them firmly in the growing

Table 2.18 Formulations for hair tonics

	A	B	C
Panthenol	1	1.5	0.25
D-Biotin	0.02	—	—
Vitamin E nicotinate	0.2	—	—
Methyl nicotinate	—	—	0.05
Allantoin	—	—	0.05
Vitamin E acetate	—	0.4	—
Camphor	—	—	0.1
Menthol	—	—	0.05
α-Bisabolol	0.1	—	—
Carbomer 940	0.4	—	—
Carbomer 934	—	0.3	—
Polyquaternium-11, 20%	—	0.3	—
Triethanolamine	0.45	0.25	—
Oleth-20	—	1	—
Nonoxynol-9	3	—	—
Capric/caprylic triglyceride	—	3	—
Propylene glycol	5	4	—
Ethanol B96	25	—	50
Colour	*q.s.*	*q.s.*	*q.s.*
Perfume	*q.s.*	*q.s.*	*q.s.*
Methyl paraben	—	0.1	—
2-Bromo-2-nitro-1,3-propanediol	0.03	0.04	—
Deionised water	to 100	⟶	

grey area of 'cosmoceuticals', and the rationale for including such products in a purely cosmetic range is questionable. The use of hair tonics as dandruff treatments is more established. Non-rinse-off products are well suited for this application and recent work describes use of piroctone olamine as the active ingredient [85]. Hair tonics in general are dealt with at greater length in Harry [86].

2.4.14.3 *Other conditioners* An interesting treatment of the conditioning concept has been proposed [87], in which it is claimed that a hydrolysed keratin protein containing cystine/cysteine residues, applied part-way through the permanent-waving process, covalently bonds with the hair via broken disulphide bonds in the hair keratin.

UV absorbers may be added to conditioners. The few studies carried out indicates that efficacy varies with the nature of the sunscreen [88, 89]. In order to be effective, a significant amount of sunscreen would need to be deposited on the hair and this would seem to be practical only from a leave-on product such as a styling aid. The effect of UV radiation on hair is discussed more fully in section 2.5.1.6.

2.5 Styling aids

2.5.1 *Hairsprays*

The simplest of styling aids, these products are usually applied to dry hair and hold it in place by a combination of coating each hair with a thin deposit of stiff polymer, and 'gluing' hairs together at points where they cross. Such products evaporate rapidly and the effect is 'instant', i.e. the product is applied to the finished hairstyle to hold it in place. Hairsprays were revolutionised by the introduction of the aerosol with its concomitant huge increase in product performance. The pump-spray revolution has been altogether quieter, though no less effective, with concern over ozone depletion adding to already increasing sales. It is now possible to achieve a very fine spray, a choice of spray angles and different delivery quantities per stroke by careful selection of the most suitable pump. Further legislation to control the emission of volatile organic compounds (VOCs) may ultimately favour pump sprays at the expense of aerosols, although water-based aerosol sprays using lower levels of VOCs are being developed [90].

The degree of hold given by a hairspray may be varied quite easily and is dictated by fashion; more-natural styles require only a light degree of hold, while complex styles may require a very stiff finish to hold them in place. Various descriptive terms, such as light hold, medium hold, strong hold, ultrahold, megahold, have been used, the only apparent limitation being the degree of imagination of the marketing department. There is, of course, a

Table 2.19 Formulations for a medium hold hairspray

	A	B	C	D	E	F
Ethanol B96	to 100% →					
Octylacrylamide/acrylates/butylaminoethyl methacrylate	—	—	—	—	—	1.7
Ethyl ester of vinylmethyl ether/maleic acid, 50%	—	—	—	—	7	—
Butyl ester of vinylmethyl ether/maleic acid, 50%	—	6	3.5	—	—	—
Vinyl pyrrolidone/vinyl acetate copolymer, 30/70, 50%	—	—	2.5	8	—	—
Vinyl acetate/crotonic acid copolymer	4.5	—	—	—	—	—
Quaternium-43	—	—	—	—	0.1	—
2-Amino-2-methyl-propane-1-ol	0.4	—	—	—	0.1	0.3
Triethanolamine	—	0.25	0.15	—	—	—
Phenyl dimethicone	—	—	0.3	0.2	—	0.25
Panthenol	0.1	0.25	—	—	—	—
Dimethicone copolyol	0.1	—	—	—	0.4	—
Oleth-20	—	0.4	—	—	—	0.3
Silk protein hydrolysate	—	—	0.15	—	—	—
Perfume	0.2	—	—	0.15	0.2	0.15
Deionised water	—	2	6	7	5	—

limit to the degree of hold which may be obtained but, in practice, this limit is high enough to permit the formulation of a range of products covering normal usage requirements. Typical formulations for medium hold and strong hold hairsprays are given in Tables 2.19 and 2.20.

Table 2.20 Typical formulations for a strong hold hairspray

	A	B	C	D	E	F
Ethanol B96	to 100% →					
Octylacrylamide/acrylates/butylaminoethyl methacrylate	—	—	—	—	—	3.2
Ethyl ester of vinylmethyl ether/maleic acid, 50%	—	—	—	9	—	—
Butyl ester of vinylmethyl ether/maleic acid, 50%	10	—	—	—	—	—
Vinyl acetate/crotonic acid copolymer	—	—	—	—	6	—
Vinyl pyrrolidone/vinyl acetate copolymer, 30/70, 50%	—	—	14	—	—	—
Vinylcaprolactam/vinyl pyrrolidone/dimethylaminoethyl methacrylate, 37%	—	6	—	—	—	—
Lauryl pyrrolidone	—	0.1	—	—	—	—
Phenyl dimethicone	—	—	0.3	—	—	—
Diisopropyl adipate	0.3	—	—	—	—	—
Dimethicone copolyol	—	—	—	0.4	0.1	0.2
Panthenol	—	—	—	—	0.1	—
Quaternium-33	—	—	—	0.1	—	—
2-Amino-2-methyl-propane-1-ol	0.4	—	—	0.12	0.45	0.5
Perfume	0.2	—	0.25	0.2	0.2	—
Deionised water	—	6	—	5	—	—

The main ingredients in a hairspray are:

- Polymer
- Solvent
- Plasticiser
- Neutraliser
- Perfume
- Other additives

The available polymers will be considered first.

2.5.1.1 *Polymers* Early hairsprays used shellac in either its natural or its dewaxed form to provide the hold (hence the name lacquer). Shellac gives a brittle film that flakes off easily yet is difficult to remove completely. It has now been displaced by modern synthetics, most of which are derived from acetylene [91, 92]. The first of these to make a big impact on the hair-care market was polyvinyl pyrrolidone (PVP). This is widely available in different degrees of polymerisation, which are usually referred to by the *K* number, derived from the viscosity of a dilute solution of the PVP relative to that of water. *K* numbers between 30 and 90 are common.

PVP is very hygroscopic and absorbs sufficient water from humid atmospheres to appreciably soften the deposited film and render it unacceptably tacky. Copolymerising the vinyl pyrrolidone with a harder, more hydrophobic material, such as vinyl acetate overcomes this, and copolymers in which the ratio of VP to VA can be varied at will, e.g. from 80:20 to 20:80, are freely available. At levels above 40% VA, the copolymer loses water solubility. However, higher VA levels are commonly used in hairsprays.

More recently, resins have continued to improve on the balance of properties obtainable, but such optimisation always gives rise to trade-offs of properties. Ideally, a hairspray polymer should be soluble in ethanol, and be hydrophilic enough to be easily removed from the hair by shampooing, but hydrophobic enough to be compatible with hydrocarbon propellants for aerosol use. It should give powerful hold without brittleness, yet be easily removed by combing or brushing, and must have good adhesion, yet not be sticky. All these properties preferably should be combined with antistatic and conditioning properties, and a competitive price. The following polymers are commercially available and widely used.

- Vinyl pyrrolidone/*t*-butyl acrylate copolymer
- Vinyl pyrrolidone/*t*-butyl acrylate/methacrylic acid copolymer
- *N*-*t*-butylacrylamide/ethyl acrylate/acrylic acid copolymer
- Vinyl acetate/crotonic acid copolymer
- Vinyl acetate/crotonic acid/vinyl neodecanoate copolymer
- Octylacrylamide/acrylates/butylaminoethyl methacrylate copolymer
- Ethyl ester of vinyl methyl ether/maleic acid copolymer

Table 2.21 Effect on most polymers of varying the degree of neutralisation

	Low percent neutralisation		High percent neutralisation
Water solubility		increasing →	
Hydrocarbon compatibility		← increasing	
Firmness of hold/brittleness		← increasing	
Stickiness in high humidity		increasing →	
Ease of removal by shampooing		increasing →	
Ease of removal by brushing		← increasing	

- Butyl ester of vinyl methyl ether/maleic acid copolymer
- Vinyl pyrrolidone/dimethylaminoethyl methacrylate copolymer
- Vinylcaprolactam/vinyl pyrrolidone/dimethylaminoethyl methacrylate copolymer

The last of these polymers is particularly good where a virtually colourless product is required. Although most of the above resins initially give a very light colour in solution, the development of colour with age is very common and difficult to overcome. Formulation B in Table 2.19 illustrates a water-white, fragrance-free product based on this material.

Some polymers have free acidic groupings in their molecules and can be fully or partly neutralised with a suitable base, modifying their properties considerably. The bases most commonly used are 2-amino-2-methyl-propane-1-ol (AMP), 2-amino-2-methyl-1,3-propanediol (AMPD), triethanolamine (TEA), triisopropanolamine (TIPA) and tris (hydroxymethyl) amino methane (Tris). The effects on most polymers of varying the degree of neutralisation are shown in Table 2.21. In most cases, a fairly low percent neutralisation gives the best combination of properties for a hairspray (but not necessarily for other styling aids based on the same resin).

For any given polymer system, an increase in solids will give a firmer hold. However, the relationship is not always linear and, where very firm holds are required, better results can sometimes be obtained from a mixture of polymers. The viscosity of the formulation should be kept as low as possible and a suitable pump spray chosen (the lower the viscosity, the easier it is to obtain a fine spray).

Hairspray polymers can be quite difficult to handle. Many are sold as concentrated, highly viscous solutions (usually in ethanol), which are hard to empty out of drums and have a remarkable propensity to stick to anything and everything. Those supplied in solid form are more amenable, although more efficient stirring may be required and batch processing times may be longer.

2.5.1.2 *Solvents* For non-aerosol products, there is really only one solvent—ethanol either alone or diluted slightly with water. It has a suitably rapid evaporation rate, inoffensive odour that is not too difficult to mask, excellent solvent properties for most resins, and is reasonably priced. Isopropanol has been used in some countries with a high excise duty on ethanol, but has a slightly slower evaporation rate and an objectionable smell. Ethanol B, denatured with Bitrex (CTFA name denatonium benzoate) and marked with *t*-butyl alcohol, is now standardised throughout the EC. In some countries (e.g. Norway and Sweden) additional denatonium benzoate must be added to comply with requirements for alcohol denaturing.

The addition of water to the solvent system may be necessary:

(i) to improve the solubility of an ingredient not fully soluble in ethanol;
(ii) to retard evaporation rate;
(iii) to reduce cost; and
(iv) to prevent precipitation from an acid-resin based system that has been neutralised to a high level. A high percent neutralisation will often increase water solubility and reduce alcohol solubility and, in some cases, a colloidal solution results. In the long term, precipitation can occur when the colloidal particles coalesce.

2.5.1.3 *Plasticisers* Plasticisers are used to modify the properties of the polymer film, making it more flexible. Only small quantities are normally required; 5% of the dry weight of polymer is a good rule of thumb. A huge number of different materials have been used as plasticisers, including a variety of esters (usually liquid), various silicones (phenyl methyl silicones have good alcohol solubility and are especially useful), silicone glycol co-polymers, proteins, polyols, lanolin derivatives, etc. It should be remembered that other materials in the formulation, for example perfume, may act as plasticisers, even if that is not the purpose of their inclusion.

A poor choice of plasticiser may have adverse effects such as weakening and dulling of the film. This can occur when the plasticiser is not fully miscible with the resin, leading to a discontinuous film with high local concentrations of plasticiser. A useful check is to examine some evaporated product on a glass microscope slide to check whether the deposited film is dull or discontinuous. Although plasticisers are not always necessary, the stiffer the polymer film, the more plasticiser is likely to be needed.

2.5.1.4 *Neutralisers* The amount of neutraliser required for a given percent neutralisation can either be obtained from polymer manufacturers' literature, or simply be calculated from the acid value of the resin. The polymer properties may be modified by the neutraliser used, and it is wise to follow the polymer manufacturers' recommendations.

2.5.1.5 *Perfume* Hairsprays are not difficult to perfume, in that there is not usually a solubility problem. Perfumes should be strong in odour and, to avoid interference with the plasticising of the resin, should be used at low levels. The raw odour of the ethanol may be quite difficult to mask and the residual odour on the hair may need careful control.

2.5.1.6 *Other additives* Preservatives are unnecessary in ethanol-based products such as hairsprays. Many other additives, such as vitamins, proteins, amino acids and herbal extracts, are used. Most of these are present in very small amounts and cannot seriously be expected to dramatically affect product performance. Ultraviolet filters have become a fashionable addition and some work has demonstrated that they may be of value in leave-on styling products [88, 93, 94], although some of the studies have used unrealistically high levels of UV filters.

There seems to be little doubt that hair can be damaged appreciably by ultraviolet light (and other environmental influences). The nature and extent of this damage, which includes photo-oxidation, loss of mechanical strength, increased alkaline solubility and colour changes are examined by Dubrief [94].

2.6 Setting lotions

The fundamental difference between hairsprays and setting lotions is that the latter are intended for application to wet hair. Water-soluble polymers are therefore used and the alcohol content, tends to be lower. Completely water-based products are possible since drying time is not so critical. Use

Table 2.22 Formulations for a setting/blow-dry lotion

Ethanol B96	40	40	18	30	—
Polyquaternium-11, 50%	2	—	—	—	—
Hydrogenated tallow dimethyl benzyl ammonium chloride, 75%	0.2	—	—	—	—
Sodium polystyrene sulphonate, 50%	—	—	2.2	—	—
Vinyl pyrrolidone/vinyl acetate copolymer, 60/40	—	—	—	1.2	2
Vinyl acetate/crotonic acid copolymer	—	1.2	—	—	—
Quaternium-22, 60%	—	0.4	0.2	—	—
Methylvinylimidazolium chloride/vinyl pyrrolidone copolymer 95/5, 40%	—	—	—	0.5	0.25
Nonoxynol-9	0.25	—	—	—	0.3
Perfume	0.1	0.05	0.15	0.1	0.1
Panthenol	—	0.2	—	0.1	0.2
Diisopropyl adipate	—	0.2	—	—	—
Oleth-20	—	—	0.5	0.25	—
Denatonium benzoate, 2.5% solution	—	—	0.005	—	—
2-Bromo-2-nitropropane-1,3-diol	—	—	0.03	0.01	—
Methylchloroisothiazolinone and methylisothiazolinone, 1.5%	—	—	—	—	0.05
Deionised water	to 100 →				

of acidic polymers with a higher degree of neutralisation is another widely practised option. Some products are marketed specifically as blow-dry lotions, but are essentially similar. Since the action of these products is not 'instant' in the manner of a hairspray, and since levels of hold are generally lower, there is more scope to incorporate conditioning ingredients; a blend based on a cationic polymer plus a hairspray-type resin can give good results. A strongly anionic resin, sodium polystyrene sulphonate, has been successfully used since its highly conductive film is effective in reducing static. This can be important during blow-drying, when, towards the end of the process, the hair is virtually dry and can exhibit 'fly-away', making styling difficult. This problem is not as obvious with other forms of heating devices such as tongs and heated curlers. Typical formulations for a setting/blow-dry lotion are shown in Table 2.22.

2.7 Other styling aids in spray form

Names such as 'spritz', 'spray gel', 'sculpting spray', etc. are usually hairspray or setting lotion variants of one sort or another. Often a 'wet-look' or glazed effect is desired, and this usually demands a high resin content in a water or water/ethanol base. Since styling of the wet hair is carried out, the product must not dry too quickly and must be compatible with water. Especial care must be taken with the high-gloss products not to reduce gloss by injudicious choice of plasticisers. Likewise, the high solids contents of some formulations may restrict pump-spray choice. Typical formulations for wet-look styling sprays are given in Table 2.23. The products are popular in other physical forms such as gels and creams (see sections 2.8 and 2.9).

A few other spray products, claiming specific properties such as volume enhancement and gloss, are available (see Table 2.24). Almost any hair-styling aid enhances volume by virtue of the deposited polymer, which increases the diameter of the hair shaft, This is much more obvious in the case of products applied to dry hair; 'wet-look' and sculpted styles tend to reduce volume by sticking adjacent hairs together in parallel lines, whereas a more random orientation of the individual hairs gives a more bulky appearance. Over a

Table 2.23 Formulations for wet-look styling sprays (applicable to wet hair)

Deionised water	to 100	⟶
Vinyl caprolactam/vinyl pyrrolidone/dimethylaminoethyl methacrylate, 37%	6.5	6
Vinyl pyrrolidone/dimethylaminoethyl methacrylate, 20%	—	4
Lauryl pyrrolidone	0.3	—
Cetyl trimethyl ammonium chloride, 30%	—	0.3
2-Bromo-2-nitropropane-1,3-diol	0.04	—
Methyldibromoglutaronitrile (and) phenoxyethanol	—	0.05

Table 2.24 Formulations for miscellaneous spray products

	Volume spray	Gloss sprays			Non-aerosol mousse
Vinyl pyrrolidone/vinyl acetate copolymer, 60/40	2.5	—	—	—	—
Steartrimonium hydrolysed protein	0.75	—	—	—	—
Panthenol	1	—	—	—	—
Vinyl caprolactam/vinyl pyrrolidone/ dimethylaminoethyl methacrylate, 37%	—	—	—	—	6
Lauric diethanolamide	—	—	—	—	0.3
Ammonium nonoxynol-4 sulphate, 30%	—	—	—	—	4
Phenyl dimethicone	—	20	—	—	—
Dimethicone copolyol	—	—	—	10	—
Ethanol B96	40	80	—	77.7	8
Perfume	0.1	—	0.4	0.3	—
Deionised water	to 100	—	—	10	to 100
2-Bromo-2-nitro-1,3-propanediol	—	—	—	—	0.05
Light mineral oil	—	—	40	—	—
C_{7-8} isoparaffin	—	—	59.6	—	—
Quaternium-80, 50%	—	—	—	2	—

period of time, panthenol can cause swelling of the hair shaft, and this contributes to 'volume'.

Gloss, a fashion-related characteristic, may be achieved either by careful selection of polymers, as in wet-look products, or by the application of a light, oil-based spray after styling. Aerosol products containing straight mineral oil are sometimes used but silicone-based sprays are probably more popular. Dimethicones with volatile silicones as diluents, or a phenyl methyl fluid or a dimethicone copolyol in an alcohol base, may be used.

A final category of spray product is the non-aerosol mousse (see Table 2.24). These products depend on the product being mixed with air when the bottle is squeezed. The addition of a suitable foaming agent to the formulation gives a foam of rather coarse and wet texture. Such foams are aesthetically quite inferior to those generated by aerosols, which continue to dominate this sector of the market.

2.8 Hair gels

Traditionally male-oriented, gels are nevertheless also widely used by women. The earlier types of gel, aimed specifically at men, were oils gelled by the use of metallic stearates or silica. Such gels behave like hair cream, giving high gloss and low levels of hold. The chosen base is mineral oil, which has low odour and colour, good oxidative stability, and is cheap. Transparent colloidal gels, invariably relying on high levels of non-ionic surfactant, are an alternative, but high levels of non-ionics, can give irritation, especially

Table 2.25 Formulations for oil-based hair gels

	A	B	C	D
Mineral oil	60	to 100	15	50
C_8/C_{12} triglyceride	33.4	—	—	—
Fumed silica	6	—	—	—
Aluminium stearate	—	5	—	—
Microcrystalline wax	—	4	—	—
Oleth-5	—	—	10	5
DEA-oleth-3 phosphate	—	—	7	—
2-Octyl dodecanol	—	—	—	25.2
Carbomer 940	—	—	—	2
Propylene glycol	—	—	6	—
Perfume	0.5	*q.s.*	*q.s.*	*q.s.*
Butyl paraben	0.1	0.05	—	—
Methyl paraben	—	—	0.2	—
2-Bromo-2-nitro-1,3-propanediol	—	—	0.05	—
Deionised water	—	—	to 100	—
Cocamine	—	—	—	5.8
Ethanol B96	—	—	—	12

eye irritation. The irritation potential of individual ethoxylates may vary considerably (e.g. oleth-10 is much more irritant than oleth-5 or oleth-20) and careful checking of final formulations is necessary. Carbomers, neutralised with long-chain fatty amines instead of the more usual materials [e.g. triethanolamine (TEA) and aminomethyl propanol (AMP)] can thicken oils to gel consistency, provided that sufficient polar solvent, such as ethanol, is present. Note that oils gelled with metallic stearates (see formulation A in Table 2.25) require very high processing temperatures (up to 130°C), and appearance and stability can vary with cooling rate and stirring.

The majority of gels on the market are aqueous, or, occasionally, aqueous/alcoholic. Carboxyvinyl polymers are the most important and, while their prime function is to create the clear gel base, they also have some fixative powers and contribute to the overall hold of the formulation. A wide range of polymers are normally included as the primary film-formers. The basic requirements are for good water solubility, clarity in solution and compatibility with the carbomer resins. The final property, although precluding the use of highly cationic polymers, allows some polymers of relatively low charge density to be used, thus enabling conditioning properties to be added. Typical formulations for styling gels are shown in Table 2.26. Formulations A and B are for 'soft hold' gels, while formulations C to F give a firmer hold.

Carbomer 940, which gives the clearest gels, is compatible with many other polymers only when fully or partly neutralised. This must considered when manufacturing such products. Polymers used should be diluted with water before addition and added slowly after about two thirds of the neutraliser has been added to the carbomer, (see earlier general remarks on gum dispersion).

Table 2.26 Formulations for styling gels

	A	B	C	D	E	F
Carbomer 940	0.8	0.7	0.8	0.9	0.6	0.65
Vinyl pyrrolidone/vinyl acetate copolymer, 60/40	—	—	—	—	—	3.3
Polyquaternium-11, 20%	—	1.5	—	—	5.5	1.5
Polyvinyl pyrrolidone, K30	1.8	—	—	3.5	—	—
Vinyl pyrrolidone/vinyl caprolactam/dimethyl aminoethyl methacrylate, 37%	—	—	6	—	—	—
Benzophenone-4	0.05	0.05	0.05	0.05	0.05	0.05
EDTA	0.1	0.1	0.1	0.1	0.1	0.05
Phenoxyethanol and methyldibromoglutaronitrile	—	0.03	0.05	—	—	0.05
Methyl paraben	0.1	0.1	—	0.1	0.1	—
Quaternium-15	0.07	—	—	0.07	0.07	—
Triethanolamine	0.96	0.8	0.9	1.1	0.7	0.7
Perfume	0.1	0.1	—	0.15	0.1	0.25
Polysorbate 20	—	—	—	0.5	—	0.75
Nonoxynol-9	0.4	0.5	—	—	0.5	—
Deionised water	to 100 ⟶					

Full exploitation of clear gel aesthetics requires transparent packaging, which poses problems since carbomers are degraded by UV light. Loss of viscosity and clarity results. This problem may be overcome by including a UV absorber (benzophenone-4 usually works well). Carbomer gels are sensitive to transition metal ions, which can catalyse gel degradation. EDTA or a similar sequestrant is an effective remedy. The vigorous agitation required to dissolve the unneutralised carbomer can cause aeration problems since, even prior to neutralisation, the solution has an appreciable viscosity and a relatively high yield-point. The viscosity is very pH dependent, therefore the inclusion of a UV absorber and a sequestrant can be turned to advantage by adding (strongly acid) benzophenone-4 and EDTA (in its free acid form) to water before the carbomer. A thin solution results, and entrained air is easily expelled by standing the mix for a while before neutralisation. Acrylic polymers, suitable for the formulation of clear gels, are available in liquid form (acrylates/steareth-20 methacrylate copolymer) and in pre-neutralised form.

Gels prepared from carbomers are influenced by the choice of neutralising agent. Sodium hydroxide gives a very stiff gel, while amines give a softer gel. The hydroxy amines, predominantly TEA, are the most widely used. Problems associated with carbomer gels, and suggested remedies are summarised in reference [95]. One such problem is uncontrolled aeration. A large number of very small air bubbles will make the product look opaque, but a small number of large bubbles can be attractive. Controlled aeration is difficult to achieve during mixing but easier to achieve during transfer from mixing vessel to intermediate bulk container, or from intermediate bulk container to filling machine. An air bleed can be introduced at a low pressure point such as the inlet side of a pump, where varying air flow rate and orifice size can give the desired effect.

2.9 Styling creams and glazes

Styling creams and glazes (see Table 2.27) are variations on the formulations shown in Table 2.26. Opacification to produce creams may be achieved either by the inclusion of a small amount of opacifier of the styrene/acrylate copolymer type, or by using a conventional oil phase plus emulsifier. The presence of a large amount of carbomer aids emulsion stability, and the choice of emulsifier type and level is not as critical as in a non-gum thickened emulsion. Oil phases inevitably have a plasticising effect and may result in reduced hold. Styling creams are often packed in polystyrene jars, therefore the inclusion of esters such as isopropyl myristate, which might react with the plastic, should be avoided. The leaching-out of plasticisers, or the sorption of ingredients due to product/packaging interaction can occasionally lead to interesting effects such as a loss or increase of product opacity around the edge of the jar. Glazes or hair-sculpting gels tend to contain high amounts of polymer (analogous to their spray counterparts) and aim at a high gloss or wet-look finish.

Table 2.27 Formulations for styling creams and glazes

	Styling creams			Glazes	
Vinyl pyrrolidone/dimethylaminoethyl methacrylate copolymer, 20%	—	—	—	—	6
Vinyl caprolactam/vinyl pyrrolidone/ dimethylaminoethyl methacrylate copolymer, 37%	—	3	—	8	—
Hydroxyethyl cellulose	—	—	—	0.7	—
Cetyl trimethylammonium chloride, 30%	—	—	—	0.6	—
Polyquaternium-11, 20%	4	—	—	—	—
Carbomer 940	0.7	0.8	0.7	—	1
Styrene/acrylate copolymer	—	—	0.2	—	—
PVP K30	—	—	2.2	—	—
Mineral oil	1.5	0.5	—	—	—
Sorbitan monopalmitate	0.1	—	—	—	—
Polysorbate-40	0.18	—	—	—	—
Glyceryl stearate (and) PEG-100 stearate	—	2	—	—	—
Phenoxyethanol and methyldibromoglutaronitrile	—	0.05	0.03	—	0.05
Ethanol B96	25	—	—	—	—
Perfume	*q.s.*	*q.s.*	0.15	0.15	0.1
Nonoxynol 9	—	—	0.45	0.45	0.4
2-Bromo-2-nitro-1,3-propanediol	—	—	—	0.05	—
Methyl paraben	0.1	—	0.1	—	—
EDTA	0.1	0.1	0.1	—	0.1
TEA	0.85	0.9	0.85	—	1.15
Deionised water	to 100 →	→	→	→	→

2.10 Hair oils/brilliantines/pomades/styling waxes

Hair oils and brilliantines are alternative non-spray gloss enhancers consisting principally of mineral oil or petroleum jelly, depending on the final physical form of the product. Other oils, such as castor oil, recognised as imparting good gloss, are also used [96]. For a liquid product, the main ingredient is mineral oil of the appropriate viscosity, coloured and perfumed. For a pomade, petroleum jelly with similar additives would be used. In the latter case the product is filled into jars in its liquid state, at elevated temperature. After solidification it is customary to 'flash' the surface by passing the product under a heater. This gives a smooth concave meniscus. These products, however, are considered old-fashioned and represent a dwindling sector of the market. A more modern approach is to produce a styling wax by adding ingredients that will give some degree of hold and make the product harder. In the examples shown in Table 2.28, methyl hydrogenated rosinate and lanolin wax are used for this purpose. In these formulations, perfume should be added as late as possible (just before filling) to minimise perfume degradation, which occurs at the temperatures encountered (typically > 70°C).

Table 2.28 Formulations for styling waxes

Methyl hydrogenated rosinate	10	12
Lanolin	4	—
Petroleum jelly	30	—
Castor oil	25.1	20
Microcrystalline wax	30	20
Lanolin wax	—	5
Perfume	0.8	1.5
Mineral oil	—	21.4
Butyl paraben	0.1	0.1
Paraffin wax	—	20

2.11 Hair creams

Hair creams are emulsion products providing modest hold and high gloss, rather old-fashioned, but made in surprisingly large quantities—mostly for export to countries that are not subject to the rapidly changing fashions of western Europe. An effective hair cream is an emulsion that breaks down quite readily on application. The emulsion must, however, be stable for the anticipated shelf-life of the product under variable storage conditions, often at extremes of temperature. Achieving such a delicate balance is not always easy and, perhaps surprisingly, the ancient beeswax/borax and beeswax/lime water systems perform well in this context. Hair creams, which may be O/W or W/O systems, rarely contain 'exotic ingredients' and generally sell at low price points. Consequently, formulations need to be produced cheaply.

Table 2.29 Formulations for hair creams

	A	B	C	D	E
Mineral oil	35	40	30	10	16
Lanolin	3	—	—	2	1
Beeswax	2.2	3	4.5	—	—
Lanolin alcohols	—	—	2	—	—
Petroleum jelly	—	—	30	—	—
Paraffin wax	—	5	—	—	2
Stearic acid	1.2	—	—	5	—
Myristyl myristate	—	—	—	3	—
Glyceryl stearate	—	—	—	3	3.6
Cetostearyl alcohol	—	—	—	—	1.6
Dimethicone	—	—	—	2	—
Sorbitan sesquioleate	—	2.5	—	—	—
Ceteareth 20	—	—	0.4	—	0.5
Borax	—	0.7	0.5	—	—
Calcium hydroxide	0.12	—	—	—	—
Carbomer 934	—	—	—	0.5	—
Triethanolamine	—	—	—	1.5	—
Glycerine	—	—	—	5	—
PEG-400	—	2	—	—	—
Magnesium sulphate heptahydrate	0.7	—	—	—	—
Deionised water	to 100	→	→	→	→
Methyl paraben	0.1	0.15	0.1	0.1	0.1
Propyl paraben	0.1	—	0.05	0.05	—
DMDM hydantoin	—	0.1	—	0.1	—
Quaternium.15	—	—	0.07	—	0.1
Perfume	*q.s.*	*q.s.*	*q.s.*	*q.s.*	*q.s.*

Occasionally, polymers are used to increase hold but, more frequently, the natural tackiness of the oil phase is relied on to provide the required effect.

Typical formulations for hair creams are shown in Table 2.29. Formulations A, B and C are W/O emulsions and are therefore made by adding water phase to oil phase, very slowly at first. A limited amount of high-shear mixing will improve the appearance of the emulsion but over-application of high shear may cause emulsion breakdown. Formulation D is an O/W system in which the carbomer should first be dissolved in the water phase (see earlier), with a small portion of the water being kept back to dilute the triethanolamine, this solution being added immediately after the oil phase. Formulation E is a conventional O/W emulsion. A further discussion of hair styling products, with example formulations is given by Alexander [97].

2.12 Permanent waving

Permanent waving is not truly permanent due to the fact that the hair is constantly growing. There is, however, a fundamental difference between permanent-waving preparations and the styling products that have already been discussed. This difference is the chemical reaction that occurs between the product and the protein of the hair shaft; covalent bonds cross-linking the protein chains are broken. Clearly, quite severe conditions are required to rupture such high-energy bonds and this can only be achieved with an element of risk. For this reason, permanent-waving products are primarily sold through professional outlets and are applied by qualified hairdressers. The home-perm market is a relatively small proportion of the total and the products sold for home-use tend to be weaker, thereby reducing the risk of severe hair damage if the process should go wrong.

Historically, various processes have been used. These include heat treatment, with or without various 'alkalis' [98–100], and, later, a whole series of reducing agents, [101]. However, the major reductant is thioglycollic acid, usually as the ammonium salt.

In simplistic terms, the mode of action is as follows:

(i) Reduction of –S–S– bonds between cystine amino acid groupings on adjacent polypeptide chains.
(ii) Rearrangement of the hair into its 'new' position by wrapping around rollers, or other mechanical means. The stresses imposed shift the polypeptide chains relative to each other.
(iii) Reforming of the –S–S–bonds by oxidation. At this point the various –SH residues will be linked to 'new partners' on the adjacent chains, and the strength of these new bonds will make the hair retain its new shape.

The chemistry involved is in fact much more complex, with a multitude of side reactions occurring at both the reduction and oxidation stages. These are discussed in more detail elsewhere [100–104]. Clearly, only a percentage of the bonds originally broken will be reformed; any residues from the reduction process that do not form new cross-links will be converted to a variety of small side chains on the polypeptide matrix. Quite apart from the chemistry, it is obvious that, from purely geometric considerations, not all of the bonds will be reformed. Typically, between 17% and 43% of the hair's cystine is reduced during processing, with a permanent degradation of up to 30% [102]. This damage includes lipid reduction [105] as well as protein damage [106], and can result in a reduction in strength of the hair fibres by 20% [107]. Cystine appears to be the only amino acid in the hair's polypeptide chains that is changed during the perming process [102]. The plotting of strain/stress curves for hairs before and after treatment can be useful. The elongation of hair under tension initially obeys Hooke's law, i.e. the response is linear. It then passes into a zone of high yield, and finally enters the post-yield zone in which the hair exhibits more resistance to further applied stress before untimately breaking. It has been suggested that analysis of the post-yield slope can help to optimise formulations [107].

The factors affecting the efficacy of the product are the same as those determining the potential hair damage.

(i) Processing time
(ii) Processing temperature
(iii) Concentration of reducing agent
(iv) Ratio of lotion to hair quantities
(v) Penetration of the lotion
(vi) pH
(vii) The nature and condition of the untreated hair

Other variables, such as the number and diameter of the rollers, the tension of the hair on the rollers and the speed at which the operation is carried out, are highly pertinent but are all related to hairdressing technique and are beyond the scope of this chapter. Each of the factors listed above will be considered in turn.

Processing time should be no longer than is necessary to achieve the required result. Home perms will generally be designed to allow for a longer processing time, since the home-user will not be as quick as a professional in manipulating the hair.

Processing temperature will be ambient for most products. Those designed for hot use are invariably restricted to the professional market. For these products, heat may be supplied either by external means (electrical or hot air), or by mixing the ingredients just before use to achieve an exothermic chemical reaction.

Concentration of active ingredients will be determined by the manufacturer and will not be changed during use except by dilution, either intentional or accidental (e.g. arising from excessive wetness after pre-shampooing). A wide range of concentrations is available, depending on the hair type for which the product is intended.

The *ratio of lotion to hair* will only become problematic when very long hair is involved, and when there is insufficient lotion to treat the whole head. This would appear to be more of a problem in Europe than in the US [108].

The *penetration of the lotion* will often be enhanced by the inclusion of a surfactant, of which a wide variety, mostly non-ionic and anionic, have been used. Penetration may also be enhanced by the inclusion of a hydrogen-bond breaker, such as urea, in the formulation.

pH can be critical—if it is too low, the product will not work well; if it is too high, severe hair damage might result. Although maximum bond cleavage is thought to occur at pH 9 [106], above pH 8.5 bond reformation is less complete. At pH < 7.5 the amount of bond cleavage is low but reformation is more complete. Apart from the protein damage and lipid depletion already referred to, there is an overall loss of sulphur from the hair, again more pronounced at higher pH. In most permanent-waving lotions, the thioglycollic acid is neutralised either by ammonia, or by a mixture of ammonia and a hydroxyamine (e.g. TEA). Some ammonia will evaporate during use of the product, leading to a gradual pH drop, which is a useful safeguard against overprocessing. In practice, initial pH is usually approximately 9.2 ± 0.2 for a conventional ammonium thioglycollate lotion.

pH is currently restricted by law to a maximum of 9.5 and total thioglycollate (expressed as free acid) to 8% (11% for professional products).

Hair type and condition must be considered by the formulator, and products can be designed for different hair types. Coarse hair will require a more severe treatment, and fine or damaged hair a gentler one. There will be considerable variation from person to person as well as between ethnic types. Hair that has been damaged should not be treated with waving products until it has recovered its strength and condition. Variation also occur across the scalp of an individual and from root to tip of individual hairs. Bleached hair tends to be very porous and should be treated with care.

Various mechanical means are used to protect parts of the hair and scalp during perming. These include the application to desired areas of an absorbent wrap coated in lotion, and pretreatment of susceptible areas with a conditioner or barrier-type product. Conditioners may be incorporated into the waving lotion; good results have been reported for amine functional silicones in this application.

The use of a high molecular weight keratin hydrolysate (with or without added fatty quaternary groups), used after rinsing out the waving lotion but before neutralisation, has been suggested [87]. Covalent bonding of the protein to the hair shaft is claimed to give a 'permanent' conditioning effect, i.e. lasting through a number of shampoos. This is an attractive concept and seems to work well in practice, although a criticism might be that any sites on the hair shaft used to bond with the protein are no longer available to bond with other sites on the hair.

Permanent-waving lotions are available in various physical forms, e.g. clear or cloudy liquid, cream or lotion, gel or thickened liquid. In each case it is important to use only low-conductivity, demineralised water and to use equipment made of 316 (or equivalent) grade stainless steel or polypropylene/polyethylene for all contact parts. Although other plastics may also be suitable, it is imperative to avoid heavy metal contamination, which may decompose the product and produce discoloration, both in the product and, in extreme cases, on the hair. For this reason, sequestrants are added, and trace heavy metals in other raw materials used should be minimised. In the case of emulsion or gel-type products requiring hot manufacture

Table 2.30 Formulations for permanent-waving lotions

	Clear lotion	Clear lotion	Conditioning	Cream	Opaque gel/cream	Gel
Ammonium thioglycollate, 60%	—	—	12	14	—	—
Thioglycollic acid	6.5	7.5	—	—	5	5
Dithiodiglycollic acid, 60%	—	—	1.5	—	—	—
Ammonia, 0.880	10.8	11.6	1.8	2.2	8.0	8.4
	to pH 9.2	9.3	9.2	9.1	9.1	9.2
Ammonium bicarbonate	—	4	—	—	—	—
Laureth-12	—	0.5	0.75	—	—	—
Stearamidopropyl dimethylamine lactate	—	—	0.5	—	—	—
Oleth-20	0.6	—	—	—	0.4	—
Laneth-75	0.6	—	—	—	—	1
Cetyl alcohol	—	—	—	0.8	0.4	—
PEG-15 cocamine	—	—	—	—	0.5	—
Mineral oil	—	—	—	5	2	—
Lanolin oil	—	—	—	1	—	—
Glyceryl stearate (and) PEG-100 stearate	—	—	—	2.5	—	—
Carbomer 941	—	—	—	—	1.6	1.6
Magnesium aluminium silicate	—	—	—	1.2	—	—
Disodium EDTA	0.1	—	—	—	0.1	0.1
Sodium heptonate	—	0.15	0.1	0.1	—	—
Perfume	0.3	—	0.25	0.5	0.25	—
Deionised water	to 100	→	→	→	→	→

processing, thioglycollate should be added last, after cooling the mix. These products can be extremely difficult to perfume since (i) the product is at high pH and strongly reducing—a formidable combination; and (ii) the sulphide odours emitted during processing of the hair are very hard to mask (see chapter 8).

The examples shown in Table 2.30 are based on thioglycollates, but various other active materials, such as glyceryl thioglycollate, thioacetic acid, 2-amino thioethanol, sulphites and bisulphites, have been added. Cysteine derivatives, particularly *N*-acetyl-1-cysteine are used as primary actives, but are almost exclusively found in products designed for the Japanese market. Their history and mode of action are reviewed by Iwasaki [109].

2.12.1 *Neutralisation*

The reforming of the –S–S–bonds by oxidation is quite straightforward, although any bonds that are unable to realign may be oxidised to form various residual side groups on the hair's polypeptide chains. Oxidation is commonly carried out by hydrogen peroxide, although sodium perborate, sodium or potassium bromate, urea peroxide, and sodium or ammonium persulphates have been used. Hydrogen peroxide may be used either as a straightforward solution (sometimes buffered and frequently stabilised by the addition of materials such as phenacetin or sodium stannate), or with a

Table 2.31 Formulations for neutralisers

Hydrogen peroxide, 35%	4.5	6	—	6	5
Sodium bromate	—	—	10	—	—
Phenacetin	0.02	0.02	—	0.04	0.02
Sodium stannate	—	0.005	—	0.005	0.005
Phosphoric acid	0.06 (to pH 3)	—	—	to pH 3	to pH 3
Citric acid	—	to pH 3.2	—	—	—
Nonoxynol-9	—	1	1	—	0.5
Ethanol B96	—	3	—	—	—
Hydrolysed protein	—	—	2	—	—
Hydroxyethyl cellulose	—	—	0.5	—	—
Glyceryl stearate (and) PEG-100 stearate	—	—	—	6	—
Sodium styrene/acrylates/divinylbenzene copolymer (and) ammonium nonoxynol-4 sulphate	—	—	—	—	0.2
Deionised water	to 100 →				

surfactant added to enhance penetration. The solution may also be clouded or produced in the form of a pourable emulsion. Bromates (the second most popular type of 'neutraliser') may also be supplied in solution but are more frequently presented in solid form in sachets, the contents of which are then dissolved in water just prior to use. This is a popular approach with home perms. A large excess of neutraliser is normally used, and this ensures that the equilibrium point of the reaction is shifted far enough to enable as complete an oxidation of the broken bonds as possible. Other important factors are the contact time, temperature, concentration of oxidising agent, and penetration. Air oxidation has been proposed, and products based on this principle have been marketed. However, the process is very slow and the results leave much to be desired. Examples of some typical types of neutraliser are shown in Table 2.31.

Since both waving lotions and neutralisers usually have quite extreme pHs and exhibit considerable redox potential, they are not particularly susceptible to microbiological contamination. Consequently, products may sometimes be unpreserved if good manufacturing practices are adhered to. If preservatives are used, however, their stability under the conditions just described should be checked carefully.

The effectiveness of perms may be evaluated by various techniques, of which salon-based tests are the most important. Alternatives and supplementary methods have been suggested [110, 111], but professional hairdressers' opinions must predominate. A general review of permanent waving is given by Alexander [112, 113].

2.13 Bleaches

Bleaches are virtually always based on hydrogen peroxide, rarely at concentrations above 12%, and generally lower. To maintain stability,

Table 2.32 Formulation for a bleach base (to be mixed 1:1 with hydrogen peroxide)

Oleic acid	22
Ammonia solution, 0.880	6.6
Isopropanol	7
Coconut diethanolamide	8
Deionised water	to 100

hydrogen peroxide solutions must be acidic, and additional stabilisers are usually added. Most of the formulations already given for peroxide-based, permanent-wave neutralisers (Table 2.31) will work equally well as hair bleaches. The extra stability conferred on the peroxide by the acid medium reduces its effectiveness as a bleach. (To bleach by oxidation it must of necessity decompose.) As with most chemical reactions, the process is more rapid in the presence of heat. Acidic solutions of this kind are used as spray-in products (non-aerosol), the gentle bleaching effect of which is accentuated by the use of a hair drier or by the sun's heat.

A stronger effect is achieved by mixing the peroxide solution with ammonia solution just prior to application (see Table 2.32). This increases the alkalinity. If the final blend is in the form of a lotion, cream or gel, it is much less likely to drip or run and is easier to use. A solution of an ammonium soap plus free ammonia, sometimes containing coconut or oleic/linoleic diethanolamides in a suitable solvent, and formulated to thicken to a soft gel when diluted with the hydrogen peroxide solution, may be used. Hydroxyamines (e.g. TEA) may be used instead of ammonia, but have the disadvantage of not evaporating off (and therefore lessening the risk of hair damage) in the event of overlong processing.

Even stronger bleaching action may be achieved by mixing per-salts (e.g. ammonium or potassium persulphate) with the peroxide/ammonia bleach. The per-salts are normally supplied in sachets, blended with diluents which may also act as alkaline buffers; sodium silicate is often used (see Table 2.33). These blends are extremely powerful oxidising agents and care must be taken to exclude organic material (e.g. some colours) from the mix and to select packaging with adequate moisture barrier properties.

Strong bleaching of hair is a harsh process and should not be carried out too frequently. As with permanent waving, some degradation of the hair occurs at each treatment, and hair damage, especially near the tips, is the

Table 2.33 Formulation for a bleach 'booster' sachet

Ammonium persulphate	30
Potassium persulphate	30
Sodium silicate	40

inevitable result of over-bleaching. As well as lightening the hair, bleaching often causes some reddening. This is because the hair pigmentation consists mainly of eumelanin (black) and pheomelanin (red), and the latter is less easily oxidised.

Even if there is not obvious damage, some additional porosity will result, and the use of a conditioner, preferably protein based, is strongly recommended. As with permanent waving, various mechanical means are used to protect the hair when only part of it is to be treated (e.g. for tipping or streaking). These include plastic hoods through which portions of the hair can be pulled, and the use of bleach-soaked porous pads which may be applied selectively.

Before bleach and permanent-wave products are marketed, careful trials must be carried out to check their safety. In addition, the consumer package must carry suitable instructions for use, and warnings against misuse.

2.14 Hair dyes

Hair dyes are conveniently grouped into temporary, semi-permanent and permanent, the distinction usually arising more as a result of the dye system used, rather than the longevity of the colour.

2.14.1 *Temporary dyes*

Almost any soluble colour can be used here [111]. Penetration of the hair shaft by the colour does not occur to any significant extent, and the overall effect is to add only a slight additional coloration to the hair. A means of keeping the colour on the hair is required, and a setting lotion base is a popular choice—all of the examples quoted earlier might be used for the purpose. Complete removal by a single shampooing is usually possible.

Another means of temporarily colouring the hair is to apply finely ground metals by means of a 'puffer' spray. Such metals, which include brasses, bronzes and aluminium, both untreated and anodised in various colours, are widely used by the paper and board converting industry, and by printing-ink and paint manufacturers. They provide a metallic effect when applied to hair, and are used for highlighting rather than whole-head treatment. Hairspray is used to prevent the powder brushing off easily. Organic and inorganic pigments can be suspended in gum-thickened or emulsion bases and streaked on to the hair. Bright reds, blues, yellows and greens are possible and application may be by means of mascara-type brushes, nail-varnish type packs, or even in the form of sticks.

Temporary hair colouring (or, on highly bleached hair, semi-permanent colouring) may also be achieved by using the leuco derivative of a basic dye such as crystal violet. This is formed by reducing an aqueous/alcoholic solution of the dye with sulphur dioxide (an objectionable process, which should be carried out with due regard to operator safety). The product is either applied directly to the hair, or is diluted to form a rinse. This is followed by rapid air oxidation to give a 'blue rinse' effect. Complexes of acid dyes with cationic surfactants have also been used [114].

Temporary dyes represent only a small portion of the hair colourants market, which is dominated by permanent and semi-permanent colourants.

2.14.2 *Semi-permanent colourants*

A wide variety of dyestuffs have been tried over the years [114], including azo dyes, either preformed or formed *in situ* by coupling on the hair, and reactive dyes [115–117]. This latter class is based on groups such as mono- or dichlorotriazine, which can react directly with the hair and to which an appropriate chromophore is attached. A good effect is difficult to obtain under practical application conditions. Other systems involving the use of strong reducing agents (e.g. permanent-waving lotions) to break bonds within the hair shaft and provide a reaction site for hair dyes have been proposed, but have not proved to be commercially realistic.

In practice, most semi-permanent hair dyes are based on basic dyestuffs, whose cationic character gives them a natural affinity for the hair, or on metallised dyestuffs, often in combination with nitro derivatives of aromatic diamines or aminophenols. Shampoo is the most commonly used base and presents no problems with metallised dyes provided that the active level of anionic surfactant, the presence of which retards the dyeing process, is low. Metallised dyes are not very compatible with salt, therefore a blend of amphoteric/cationic/non-ionic surfactants with low salt levels is preferred. Performance of all semi-permanent dyestuffs may be enhanced by the inclusion of solvents. Water causes considerable swelling of the hair shaft, but aqueous solutions of some solvents cause even more, leading to greatly increased dye uptake. Widely used solvents include benzyl alcohol, butyl alcohol, and methyl, ethyl and butyl ethers of monopropylene or dipropylene glycols. Levels are rarely above 5%. Various patents, some quite recent, cover such solvent-assisted systems, which have been in use for at least 25 years. Increased penetration may also be achieved by using the hydrogen-bond breaking properties of urea. Since hair tends to be more porous near the ends than at the roots, and bleached or permed hair tends to be more porous than untreated hair, quite large variations in dye uptake may occur. This can be overcome, as in textile dyeing, by using levelling agents; the inclusion of a small amount of non-ionic, e.g. oleth-20, often helps. Pohl [118] recommends the use of a mixture of high and low molecular weight dyestuff. Typical formulations for semi-permanent hair colourants are shown in Table 2.34.

In theory, virtually any colour could be achieved by using a combination of, say, suitable red, yellow and blue dyes. In practice, dyes absorb on the

Table 2.34 Formulations for semi-permanent hair colourants

	Liquid	Cream	Shampoo	Shampoo
Metallised dyes	—	—	—	
o-Nitro-*p*-phenylene diamine	—	—	—	total < 2.5	total < 2
p-Nitro-*o*-phenylene diamine	—	—	—	
Basic dyes	< 1.0	< 1.0	< 1.0	—	—
Ammonium lauryl sulphate, 30%	—	—	—	10	12
Cetyl trimethyl ammonium chloride, 30%	3	4	2	—	—
Cocoamidopropyl betaine, 30%	—	—	15		
Oleth-20	0.5	0.2	—	0.5	—
Coconut diethanolamide	—	—	2	—	2.5
Cetostearyl alcohol	—	3	—	—	—
Glyceryl stearate (and) PEG-100 stearate	—	3	—	—	—
Hydroxypropyl methylcellulose	0.8	—	0.5	0.5	0.4
Triethanolamine	to pH 8	to pH 8	to pH 8.5	—	—
Citric acid	—	—	—	0.3	0.25
Preservative	*q.s.*	*q.s.*	*q.s.*	*q.s.*	*q.s.*
Perfume	*q.s.*	*q.s.*	*q.s.*	*q.s.*	*q.s.*
Benzyl alcohol	3	—	—	2	—
PPG-2 butyl ether	—	—	—	2	—
Deionised water	to 100	to 100	—	to 100	to 100

hair shaft at different rates, and are removed at different rates by shampooing. Consequently, products containing a mixture of dyestuffs can produce different colours when applied under different conditions (time, temperature, pH, etc.). In addition, the colour achieved may vary considerably from one type of hair to another, and after application and subsequent shampooing. Minimising these variations can be achieved by using colours that are close together; this approach is less likely to produce any unexpected (and usually unwanted) colours.

Basic dyes are more substantive from slightly alkaline bases, and metallised dyes from quite strongly acidic ones. Since salt can reduce substantivity, thickening, where required, is usually achieved by means of gums. Light-fastness is not usually a problem with semi-permanent colours (it can be with some of the temporaries) but scalp staining can occur, more so with solvent-assisted systems.

2.14.3 *Permanent hair dyes*

Permanent, or oxidation, dyes are fundamentally different to the other classes of dyestuffs that have already been considered. The dyes are formed during the dyeing process and are not present, as such, in the solutions before application. The products consist of two parts—a dye intermediate solution and an oxidising agent, the latter almost always being hydrogen peroxide. Dye intermediates are blends of primary intermediates and coupling agents, or modifiers, in a suitable base. During the permanent dyeing of hair the dye intermediate solution and the oxidising solution are mixed and applied to the hair. The primary intermediates are gradually oxidised and then undergo coupling reactions with the modifiers. The primary intermediates are all small molecules that can, to a degree, penetrate the hair shaft, particularly under the wet, alkaline conditions prevailing during the dye application. The subsequent oxidation and coupling reactions produce much larger molecules, many of which are then 'trapped' within the hair shaft, thereby making the effect 'permanent'. The primary intermediate should be an aromatic compound with at least two electron-donating groups in the 1,2 or 1,4 positions. The most effective combinations are either two amino groups or one amino and one hydroxyl group, attached to a benzene or toluene ring. Other substituents may be present, either on the ring or *N*-substituted, and the ring itself may be non-benzene derived (e.g. pyridine). The most commonly used are:

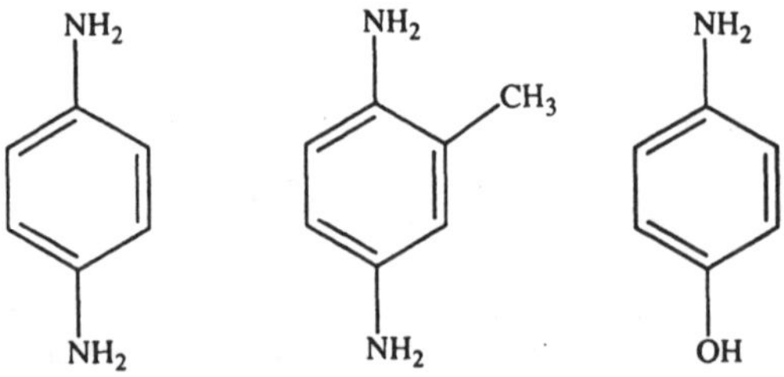

NH_2 NH_2 NH_2

CH_3

NH_2 NH_2 OH

p-phenylene diamine *p*-toluylene diamine *p*-aminophenol

Oxidation then takes place to form a quinonimine, e.g.

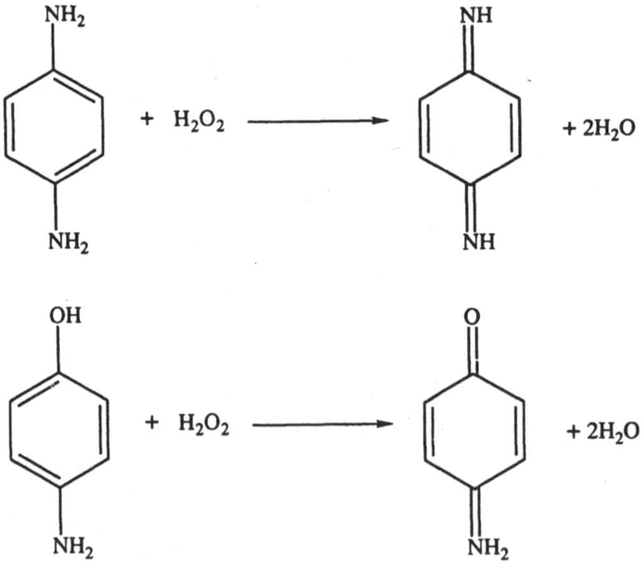

NH_2 NH

+ H_2O_2 ⟶ + $2H_2O$

NH_2 NH

OH O

+ H_2O_2 ⟶ + $2H_2O$

NH_2 NH_2

Coupling agents require the substituting groups to be in the 1, 3 positions. They are not easily directly oxidised, but will undergo oxidative coupling with the quinonimines formed from the primary intermediates.

Coupling agents may be based on benzene rings, (substituted or otherwise), multiple ring systems, (fused or unfused), or heterocyclic rings. More than one stage of oxidative coupling may occur, sometimes leading to some very large molecules [118]. The possible reactions, even starting from a blend of relatively few primary intermediates and coupling agents, are numerous. The hair itself may modify or even take part in some of the reactions. The original theory that Bandrowski's bases were intermediates in the dying of hair with oxidation dyes is now widely discounted. As well as reaction within the hair shaft, oxidation and coupling simultaneously take place in the solution outside the hair. The multitude of reactions and reaction rates may give rise to some strange intermediate colours in the solution (and sometimes on the hair!). For this reason, it is important to determine and stick to, the correct length of contact time. Unwanted shades are more likely to arise from underprocessing than from overprocessing.

The complex chemistry and reaction kinetics involved mean that the development of specific shades is to a large extent empirical, although general rules covering certain combinations of primary intermediate and modifier exist. Yellow, orange and red shades can be obtained with the nitro derivatives already referred to in section 2.14.2; otherwise it is quite unusual to mix oxidative dyes with other types. Permanent dye systems are able to colour hair to a lighter shade than the original. Peroxide and ammonia, present in excess to ensure complete and controlled oxidation of the dye intermediates, simultaneously bleach the hair. A further increase in the peroxide emphasises the bleaching effect. Permanent dyes are also capable of 'evening-out' the differences in colour between individual hairs. This is especially effective on grey hair, which consists of mixed coloured and white hairs, and is impossible to colour evenly by any other means.

A high pH is necessary and is usually achieved with ammonia, often buffered. If significant lightening is not required (i.e. for dark shades), hydroxyamines such as TEA can be used, and the pH kept a little lower. As with the semi-permanent dyes, solvents are sometimes used to increase hair swelling and dye penetration. Some cream products are available but shampoo-in bases with good wetting properties, obtained using a wide variety of surfactants (often in conjunction with ammonium oleate), ensure even dye distribution and are most popular.

Some of the dye intermediates and modifiers are very readily oxidised in alkaline solution, even by the air and, for this reason, water-soluble reducing agents such as sulphites, bisulphites, dithionites or ascorbic acid are added to the base. Batches of these products are often prepared under a blanket of nitrogen. Opaque or brown glass packaging helps to disguise the inevitable batch-to-batch colour variation of the dye solution.

Permanent colourants cause some hair damage; an increase of alkaline solubility from 6% (natural hair) to 10% (after five applications of oxidative dye) has been reported [119] but is really quite minor. The incorporation of conditioning agents into the base, and the use of post-dyeing shampoos and conditioners allows very good aesthetic effects; leaving the hair not only coloured (lighter or darker) but glossy and with good handle. Oxidative dyes are able to smooth out unevenness in natural hair colour, although caution should be exercised since extra porosity is conferred by bleaching or permanent waving. Hair should not be dyed until several weeks after perming or bleaching.

In developing a range of shades, testing should be carried out on various different hair types, both bleached and unbleached. A variety of colours of human hair tresses can be purchased for this purpose. Initial screening may be carried out on wool, which has a structure closely akin to hair, but final colour tests need to be carried out on real hair. Perceived colour will vary according to the hair type, style, gloss, etc. and also between different ethnic groups [120]. Evaluation is usually carried out by eye, preferably in a light cabinet fitted with 'north daylight' fluorescent tubes to provide standardised viewing conditions. Instrumental colour measurement may also be used [121].

The range of possible intermediates and modifiers is numerous and constantly changing. Some materials once used are now banned, while others are restricted. In the UK, the most up-to-date legislation covering this is The Cosmetic Products (Safety) Regulations 1989, S.I. No. 2233 (1989) and the Cosmetic Products (Safety) (Amendment) Regulations 1993. These are the latest enactments of the EC Cosmetics Directive and include Schedule 1 (materials not permitted) and Schedule 2 (restricted substances). Currently the total concentration of most of the dye intermediates is restricted both individually and in combination. In the USA, the use of colours in cosmetics is controlled by the FDA. Phenylene diamines, diamino phenols, α-naphthol, pyrogallol and resorcinol are all, for instance, restricted. Oxidation hair dyes are also required to carry a warning since many of the intermediates and modifiers are allergens and sensitisers. The warning must appear in the

Table 2.35 Formulations for oxidation hair-dye bases

	A	B
Isopropanol	7.5	—
PPG-2 methyl ether	—	1.8
Oleic acid	22	4
Coconut diethanolamide	10	11
Sodium lauryl ether sulphate, 27%	—	36
Salt	—	*q.s.*
Deionised water	to 100	to 100
Ammonium, 0.880	6.6	2

statutory form as prescribed by the Regulations. Hair dye intermediates have generated a huge number of patents and due regard must be paid to these when deciding a formula.

Table 2.35 gives typical formulations for oxidation hair-dye bases. In formulation B the salt level must to be adjusted to give the required viscosity on dilution with an equal volume of hydrogen peroxide solution. The dye intermediates themselves can have a considerable impact on the viscosity, and base may need to be varied from shade to shade to compensate for this.

2.14.4 *Other hair dyes*

Some hair colourants do not fall neatly into the temporary/semi-permanent/permanent classification. These include the vegetable dyes, mostly based on henna [122]. The dried leaves of the henna plant are powdered and applied as a paste to the hair, either alone or mixed with other plants such as indigo or camomile. (Diluents, fillers, thickeners, etc. can also be added.) A less satisfactory alternative is the preparation and use of an aqueous extract of henna leaves as a base. The use of henna as a hair dye has a number of disadvantages—the limited range of shades attainable (all reddish when using henna alone), build-up on repeated use (leading to rather unnatural colours), a long contact time, and the need to apply the product hot to obtain the best effect. Advantages are that the product is 'natural' and has a good record of non-irritancy. The main active ingredient is lawsone (2-hydroxy-1,4-naphthaquinone), also present in walnuts. The optimum pH is mildly acid.

Most women, when faced with greying hair, will simply dye it to the colour of their choice. Most men in the same situation will either persuade themselves that greying hair looks distinguished or assuage their vanity by using a lead–sulphur dye. These products consist of a solution of lead acetate and precipitated sulphur, both at 1–1.5% in an aqueous base, usually with a small amount of a wetting agent, humectant, and sometimes a little solvent. The products are applied regularly and left on the hair, where a slow reaction takes place, leading to a gradual build-up of a brownish colour. The effect is not always very natural looking and the hair may be left a little dull. The exact mechanism of the reactions is not clear. The products are toxic and should be handled with care.

2.14.5 *Dye removers*

No dye removers are completely successful. Semi-permanents, by their very nature, will eventually wash out, whereas permanent dyes will usually only fade slightly as they 'grow out' naturally. A limited degree of success may be obtained with sulphites and other related reducing agents, in particular sodium formaldehyde sulphoxylate. The use of activated charcoal can improve the effectiveness of such bases but is unsightly and messy [117].

2.15 Product evaluation and testing

The experienced formulator will normally have a fairly accurate idea of the performance that might be expected from a newly formulated product. The less-experienced will have difficulty. Even when working in known territory, there can be surprises—perhaps hitherto unreported synergism between raw materials, unexpected incompatibilities, or simply the fact that the formulator's assessment of product performance differs considerably from that of the end-user. Methods for testing and evaluating products are essential to confirm that products are meeting the requirements of the original brief. Such methods can be grouped into two main types.

(i) *In vitro* methods, which are laboratory-based, and can be quite objective. Although instruments can often be used to produce a quantifiable result, the substrate is not the same as hair growing on the head, and the conditions of the tests may be too far removed from real conditions of use.
(ii) *In vivo* methods, which overcome the problems associated with *in vitro* methods but at the expense of producing less-easily quantifiable results under less-reproducible conditions.

A combination of both approaches is often used, with laboratory-based instrumental techniques for initial screening, followed by salon or consumer testing for final selection. Many test methods and protocols, covering many aspects of product performance, have been used [37, 123–127].

2.15.1 *Stability testing*

The EC cosmetics directive specifies that products should remain safe and effective for two and a half years. Clearly, product testing for this duration is not commercially acceptable, and accelerated tests are widely used. The legal requirement is not concerned with changing attributes, for example loss or degradation of colour or perfume, although in practice, these are most likely to suffer. Many test methods, such as high- and low-temperature storage, freeze/thaw cycling, centrifugation and high-intensity light irradiation, are available to the development chemist. Such tests and their interpretation, although straightforward, require skill and experience. Monitoring of the first production batches should complement work carried out at the development stage. Specifications for production batches are mostly based on laboratory stability results and should be confirmed on large-scale production [128–130]. Compatibility testing should also be carried out. Two sets of tests run in parallel, one set in the actual containers to be used (or containers made from the same material), the other in inert containers, usually glass, give valuable comparative data.

2.15.2 *Claims justification*

If specific claims are to be made for the product, the manufacturer must be able to justify them. This can often be based on known properties of the raw materials used, provided that usage levels and conditions of use are consistent with those of the study on the raw material. Other cases may require laboratory testing, salon trials, or consumer panel tests to verify that the proposed claims are indeed true. In any panel test, the number and type of subjects used must be adequate, and the results must be subjected to an appropriate significance test. When selecting the best test method, each case must be treated individually. However, the previous comments made about the relative merits of different evaluation techniques are equally relevant. This whole area is one which will doubtless attract more attention when the forthcoming 6th Amendment to the EC Cosmetic Directive comes into force, with its requirement for creating a 'product information package' for each product.

2.15.3 *Product safety*

Last, but certainly not least, every development laboratory should have a review procedure to ensure the safety of new products. This procedure should include the examination of all formulations for safety. If a formulation is a 'standard' one, no further action may be necessary. If new raw materials or unusual combinations of materials are used, or if any ingredient is present at an unusually high level, the recommendation may be that the formulation should be externally vetted. The opinion of a qualified dermatologist, ophthalmologist or toxicologist may be an invaluable back-up to the cosmetic chemist's judgement. In cases where there is real doubt about the total safety of the product (including foreseeable conditions of misuse), further testing may be required, preferably in the form of a human volunteer study. Large companies may have departments dedicated to this, but it is essential that smaller companies can, and do, ensure the safety of people coming into contact with the product, not least the end-user.

2.16 Summary

This chapter has concentrated on the most important products in a concise manner. For those interested in acquiring more in-depth information, the references given will lead to more references, including a wealth of patents. Patents have not been quoted extensively in the chapter, since much of the information contained therein relates to formulations which have never been commercially produced. A huge amount of published material, including formulation guidelines is available from the majority of reputable raw material suppliers. The best of this material offers an excellent technical support service.

References

1. Harry, R.J. (1982) *Harry's Cosmeticology*. 7th edn. Chapter 23, p. 397. Longmans, London.
2. Alexander, P. (1990) The structure of the hair. *Manufacturing Chemist* **61**(2) 20.
3. Shaw, D.M. (1979) Hair lipid and surfactants. *Int. J. Cosmet. Sci.* **1**(6) 317.
4. Weigmann, H.D. and Kamath, Y.K. (1986) Modification of human hair through fibre surface treatments: characterisation by wettability. *Cosmet. Toilet.* **101** (6) 37.
5. Tolgyesi, E. (1983) Weathering of hair. *Cosmet. Toilet.* **98**(10) 29.
6. Swift, J.A. (1990) Fine details on the surface of human hair. Presented at the SCS symposium on *The Future of Hair Care Technology*, Harrogate, November 1990.
7. Bories, M.F., Martina, M.C., Bobin, M.F. and Cotte, J. (1984) Influences des variations thermiques sur la structure du cheveu. *Int. J. Cosmet. Sci.* **6**(5) 201.
8. Bories, M.F., Martina, M.C., Bobin, M.F. and Cotte, J. (1984) Influences des variations thermique sur les propriétés mecaniques du cheveu. *Int. J. Cosmet. Sci.* **6**(5) 213.
9. Stamm, R.F., Garcia, M.L. and Fuchs, J.J. (1977) The optical properties of human hair: I Fundamental considerations and goniophotometer curves. *J. Soc. Cosmet. Chem.* **28**, (9) 571.
10. Stamm, R.F., Garcia, M.L. and Fuchs, J.J. (1977) The optical properties of human hair: II: The lustre of hair fibres. *J. Soc. Cosmet. Chem.* **28**(9) 601.
11. Henderson, G.H., Karg, G.M. and O'Neill, J.J. (1978) Fractography of human hair. *J. Soc. Cosmet. Chem.* **29**(8) 449.
12. Report on 16th IFSCC Congress (1991) *Soap, Perfumery Cosmet.* **64**(1) 37.
13. Maes, D., Leduc, M., Nadvorkin, I.M., Reinstein, J.A., Turef, B.A. and Vieu, M. (1979) Some aspects of hair regreasing. *Int. J. Cosmet. Sci.* **1**(3) 169.
14. Knott, C.A., Daykin, K. and Ryan, J. (1983) *In vivo* procedures for assessment of hair greasiness. *Int. J. Cosmet. Sci.* **5**(3) 77.
15. Hunting, A.L.L. *The Encyclopaedia of Shampoo Ingredients*. London Micelle Press.
16. Rieger, M. (1988) Surfactants in Shampoos. *Cosmet. Toilet.* **103**(3) 59.
17. Milwidsky, B. (1989) Soapless shampoo actives. *Household and Personal Products Industry* **26**(9) 58; **26**(10) 52; **26**(11) 55.
18. Lomax, E. (1989) An introduction to surfactants. *CTMS* **XIV** 15; **XV** 21; **XVI** 15; **XVII** 31.
19. Harry, R.J. (1982) *Harry's Cosmeticology*. 7th edn. Chapter 24, p. 433. Longmans, London.
20. Donaldson, B.R. and Messenger, E.T. (1979) Performance characteristics and solution properties of surfactants in shampoos. *Int. J. Cosmet. Sci.* **1** 71.
21. Alexander, P. (1986) The Miranol amphoteric surfactants, carboxylates and sulphonates, their chemistry and application. *Soap, Perfumery Cosmet.* **59**(9) 493.
22. Lomax, E. (1988) New polyfunctional amphoterics. *CTMS* **1**(4) 18.
23. Mizayawa, K. and Tamura, U-hei. (1993) *N*-Acyl-*N*-methyl laurates. *Cosmet. Toilet.* **108**(3) 81.
24. Salka, B. (1993) Alkyl polyglycosides, properties and applications. *Cosmet. Toilet.* **108**(3) 89.
25. Kamegai, J. and Onitsuka, S. (1993) Alkylsaccharides, use in shampoo. *Manufacturing Chemist* **64**(9) 20.
26. *Shell Routes to Detergents*. Shell Chemicals leaflet, Shell.
27. *High Active Surfactants*. Albright Wilson booklet, Albright & Wilson, White Haven, UK.
28. Milwidsky, B. (1987) Sulphosuccinates and related compounds. *Household and Personal Products Industry* **24**(8) 52.
29. Schoenberg, (1989) Sulphosuccinate surfactants. *Cosmet. Toilet.* **104**(11) 105.
30. Petter, P.J. (1983) Fatty acid sulphoalkyl amides and esters as cosmetic surfactants. Presented at the symposium on *Surfactants in Cosmetics and Toiletries*, Coventry, April 1983.
31. Dominguez, J.G., Balaguer, F., Parra, J.L. and Pelejero, C.M. (1981) The inhibitory effect of some amphoteric surfactants on the irritation potential of alkylsulphates. *Int. J. Cosmet. Sci.* **3**(2) 57.

32. Surfadone surface active pyrrolidones (1988) *CTMS* **1**(4) 33.
33. Miyazawa, K., Ogawa, M. and Mitsu, T. (1984) The physico-chemical properties and protein denaturation potential of surfactant mixtures. *Int. J. Cosmet. Sci.* **6**(1) 33.
34. Lomax, E. (1989) Irritation and micellar properties of surfactants. *CTMS* **XI** 13.
35. Lomax, E. (1988) New amphoteric formulations. *Soap, Perfumery Cosmet.* **61**(11).
36. Lomax, E. (1992) The mechanism of surfactant detoxification. *Soap, Perfumery Cosmet.* **65**(11) 37.
37. Domingo Campos, F.J. and Druguet Tantina, R.M. (1983) Evaluation of the foaming capacity in shampoos; efficacy of various experimental methods. *Cosmet. Toilet.* **98**(9) 121.
38. Larson, R.J. and Bishop, W.E. (1988) New approaches for assessing surfactant biodegradation in the environment. *Household and Personal Products Industry* **25**(5) 84.
39. Carson, H.C. (1990) Surfactants—environmental issues. *Household and Personal Products Industry* **27**(6) 41.
40. Gilbert, P.A. and Pettigrew, R. (1984) Surfactants and the environment. *Int. J. Cosmet. Sci.* **6**(4) 149.
41. Grandval, G. (1986) Evaluating and choosing fragrances for functional products. *Cosmet. Toilet.* **101**(6) 69.
42. Harding, N. (1986) Evaluation of perfume submission from expert to consumer—A review of approaches. *Cosmet. Toilet* **101**(6) 73.
43. Aarts, P. (1986) Evaluation of fragrances from both captive and external sources. *Cosmet. Toilet.* **101**(6) 79.
44. Beerling, J. (1993) Formulating and fragrancing multi-functional shampoos. *Manufacturing Chemist* **64**(11) 18.
45. Blakeway, J.M. and Fabrizi, G. (1983) Substantivity of perfume materials to hair. *Int. J. Cosmet. Sci.* **5**(1) 15.
46. Orth, D.S. (1989) Microbiological considerations in cosmetic formula development and evaluation. *Cosmet. Toilet.* **104**(4) 49.
47. Cosmetics preservatives encyclopaedia—antimicrobials (1990) *Cosmet. Toilet.* **105** 49.
48. Busch, P., Hase, C., Schutze, R., Hensen, H. and Lorenz, P. (1987) Hair conditioning effect of guar hydroxypropyl trimethylammonium chloride. *Cosmet. Toilet.* **102**(5) 49.
49. Smith, L.R. and Gesslein, B.W. (1990) The functions and formulations of conditioning agents in shampoos—past, present and future. Presented at the SCS Symposium on *The Future of Hair Care Technology*, Harrogate, November 1990.
50. Patel, C.U. (1983) Antistatic properties of some cationic polymers used in hair care products. *Int. J. Cosmet. Sci.* **5**(5) 181.
51. Sejpka, J. (1993) The silicone story. *Soap, Perfumery Cosmet.* **66**(4) 29.
52. Needs, S.G. (1984) Methods of light fastness testing. Presenting at the SCS Symposium on *Product Stability*, Coventry, November 1984.
53. Van Abbe, N.J., Head, D., Reed, J.V., Murrell, E.A. and Baxter, P.M. (1986) Dandruff: infection or not? *Int. J. Cosmet. Sci.* **8**(1) 37.
54. Van Abbe, N.J., Baxter, P.M., Jackson, J., Bell, M.A. and Dixon, H. (1981) The effect of hair care products on dandruff. *Int. J. Cosmet. Sci.* **3**(5) 233.
55. Francois-Poncet, G.A. (1983) Trends in antidandruff preparations. *Cosmet. Toilet.* **98**(9) 101.
56. Kligman, Lowe and Maibach (1984) Scalp disorders. *Soap, Perfumery Cosmet.* **57**(11) 569.
57. Lansdown, A. (1986) Observations on zinc pyrithione as an effective treatment. *Soap, Perfumery Cosmet.* **59**(9) 499.
58. Shuster, S. (1988) Dandruff, seborrhoeic dermatitis and pityrosporum ovale. *Cosmet. Toilet.* **103**(3) 87.
59. *Hair Sunscreen Study VD-01-85*. Van Dyk & Co.
60. Sendelbach, G., Liefke, M., Schwan, A. and Lang, G. (1993) A new method for testing removability of polymers in hair sprays and setting lotions. *Int. J. Cosmet. Sci.* **15**(4) 175.
61. *Colipa: Document 88/085* (1988) Traces of NDELA in cosmetic products.
62. Dunnett, P.C. and Telling, G.M. (1984) Study of the fate of bronopol and the effects of antioxidants on *N*-nitrosamine formation in shampoos and skin creams. *Int. J. Cosmet. Sci.* **6**(5) 241.
63. Telling, G.M. and Dunnett, P.C. (1981) The determination of NDELA at trace levels in shampoos and skin creams by a simple rapid colorimetric method. *Int. J. Cosmet. Sci.* **3**(5) 241.

64. Taylor, P. (1989) Nitrosamines—occurrence, toxicology and analysis; avoiding their presence in cosmetics. Presented at the CTPA Autumn Conference, 1989.
65. Komp, B. and Reng, A.K. (1989) Developing ether sulphate-free surfactant formulations. *Cosmet. Toilet.* **104**(4) 41.
66. Woodruff, J. (1994) Shampoo takes a professional lead. *Manufacturing Chemist* **65**(2) 18.
67. Harry, R.J. (1982) Harry's Cosmeticology. 7th edn. Chapter 26, p. 498. Longmans, London.
68. Hunting, A.L.L. *Encyclopaedia of Conditioning Rinse Ingredients.* CTFA (1991) London Micelle Press.
69. CTFA (1991) *International Cosmetic Ingredient Dictionary*, 4th edn. CTFA, Inc., Washington, DC.
70. Milwidsky, B. (1990) After shampoo preparations. *Household and Personal Products Industry* **27**(5) 78.
71. Idson, B. and Lee, W. (1983) Update on hair conditioner ingredients. *Cosmet. Toilet.* **98**(10) 41.
72. Quaternary ammonium compounds and their application in hair conditioners (1988) *CTMS* **1**(4) 6.
73. Lunn, A.C. and Evans, R.E. (1977) The electrostatic properties of human hair. *J. Soc. Cosmet. Chem.* **28**(9) 549.
74. Allandic, A. and Gummer, G. (1993) Hair conditioning—quaternary ammonium compounds on various hair types. *Cosmet. Toilet.* **108**(3) 107.
75. Alexander, P. (1987) Cationic polymers for skin and hair conditioning. *Manufacturing Chemist* **88**(7) 24.
76. Woodruff, J. (1994) Staying in condition. *Manufacturing Chemist* **65**(3) 18.
77. Griffin, W.G. (1949) *J. Soc. Cosmet. Chem.* **1** 311.
78. Lin, T.J. (1978) Low energy emulsification I: principles and applications. *J. Soc. Cosmet. Chem.* **29** 117.
79. Lin, T.J. (1978) Low energy emulsification II: evaluation of emulsion quality. *J. Soc. Cosmet. Chem.* **29** 745.
80. Lin, T.J., Akabori, T., Tanaka, S. and Shimira, K. (1980) Low energy emulsification III: emulsification in the high alpha range. *Cosmet. Toilet.* **95**(12) 33.
81. Lin, T.J., Akabori, T., Tanaka, S. and Shimira, K. (1981) Low energy emulsification IV: effect of emulsion temperature. *Cosmet. Toilet.* **96**(6) 31.
82. Lin, T.J., Akabori, T., Tanaka, S. and Shimira, K. (1983) Low energy emulsification V: mechanism of enhanced emulsification. *Cosmet. Toilet.* **98**(10) 67.
83. Driscoll, W.R. (1975) *Drug Cosmet. Industry* **11** 642.
84. Oba, K. (1988) Cosmetic products affecting hair growth. *Cosmet. Toilet.* **103**(5) 69.
85. Futterer, E. (1983) Antidandruff tonic containing piroctone olamine. *Cosmet. Toilet.* **103**(2) 49.
86. Harry, R.J. (1982) Harry's Cosmeticology. 7th edn. Chapter 26, p. 498. Longmans, London.
87. Chester, J. and Mawby, G. (1987) Permanent hair conditioning. *Manufacturing Chemist* **58**(4) 53.
88. Giesen, Hollenberg, Hubbuch, Kalhöfer, Maier, Martin, Münzig, Schwan, Sperling and Tennigkeit. *UV Filters for Hair Protection.* Faschgruppe, Haarbehandlung Publication BASF.
89. Saettone, M.F., Giannaccini, B., Morganti, C., Persi, A. and Cipriani, C. (1986) Substantivity of sunscreens: an appraisal of some quaternary ammonium sunscreens. *Int. J. Cosmet. Sci.* **8**(1) 9.
90. Tazi, M. and Login, R.B. (1990) Progress in water-based hairspray. Presented at the SCS symposium on *The Future of Hair Care Technology*, Harrogate, November, 1990.
91. Petter, P.J. (1987) Trends in hair care products. *Soap, Perfumery Cosmet.* **60**(95) 29.
92. Petter, P.J. (1989) Acetylene derived polymers and their applications in hair and skin care. *Int. J. Cosmet. Sci.* **11**(1) 35.
93. Georgalas, A. (1993) Photoprotection of hair. *Cosmet. Toilet.* **108**(3) 107.
94. Dubrief, C. (1992) Experiments with hair photodegradation. *Cosmet. Toilet.* **107**(10) 95.
95. *Carbopol Trouble-Shooting Guide* (1988) B.F. Goodrich leaflet, B.F. Goodrich, Middlesex, UK.
96. Harry, R.J. (1982) Harry's Cosmeticology. 7th edn. Chapter 25, p. 483. Longmans, London.
97. Alexander, P. (1991) Dressing up hair. *Manufacturing Chemist* **62**(3) 26.

98. Alexander, P. (1988) Permanent waving. *Manufacturing Chemist* **59**(4) 42; **59**(5) 61.
99. Lee, A., Bozza, J.B., Huff, S. and de la Mettrie, R. (1988) Permanent waves: an overview. *Cosmet. Toilet.* **103**(5) 37.
100. Harry, R.J. (1982) Harry's Cosmeticology. 7th edn. Chapter 28, p. 555. Longmans, London.
101. Chesebrough-Ponds (1986) *European Patent 0 190 834.*
102. Gumprecht, J.G., Patel, K. and Bono, R.P. (1977) Effectiveness of reduction and oxidation in acid and alkaline permanent waving. *J. Soc. Cosmet. Chem.* **28**(12) 717.
103. Puri, A.K. (1979) Recent trends in the formulation of permanent waving products for the hair. *Int. J. Cosmet. Sci.* **1**(1) 59.
104. Chiodi, F. (1989) Modèle mathematique pour l'étude des equilibres physico-chimiques de permanentes. *Int. J. Cosmet. Sci.* **11**(6) 283.
105. Hilterhaus-Bong, S. and Zahn, H. (1989) Lipid chemical aspects of permanently waved hair. *Int. J. Cosmet. Sci.* **11**(4) 167.
106. Hilterhaus-Bong, S. and Zahn, H. (1989) Protein chemical aspects of permanent waving treatments. *Int. J. Cosmet. Sci.* **11**(5) 221.
107. Cannell, D.W. and Carothers, L.E. (1978) Permanent waving products: utilization of the post-yield slope as a formulation parameter. *J. Soc. Cosmet. Chem.* **29**(11) 685.
108. Marti, M.E. (1990) Comparison of permanent waving: USA vs Europe. *Cosmet. Toilet.* **105**(5) 113.
109. Iwasaki, A. (1994) Cysteine waving lotions. *Cosmet. Toilet.* **109**(3) 69.
110. Watanabe, K., Minel, M. and Horikushi, T. (1985) *US Patent 4,510,951.*
111. Harry, R.J. (1982) Harry's Cosmeticology. 7th edn. Chapter 27. Longmans, London.
112. Alexander, P. (1993) Waving hair with greater care (part 1). *Manufacturing Chemist* **64**(4) 33.
113. Alexander, P. (1993) Waving hair with greater care (part 2). *Manufacturing Chemist* **64**(5) 40.
114. Fishman, H.M. (1988) Non-permanent hair dyes, a review. *Household and Personal Products Industry* **25**(4) 62.
115. Broadbent, A.D. (1963) *American Perfumer and Cosmetics* **78**(3) 23.
116. Shansky, A. (1966) *American Perfumer and Cosmetics* **81**(121) 23.
117. Shipp, J.J. (1968) Unpublished work.
118. Pohl, S. (1988) The chemistry of hair dyes. *Cosmet. Toilet.* **103**(5) 57.
119. Bauer, D., Beck, J.P., Monnais, C. and Vayssie, C. (1983) Contribution to the quantification of the conditioning effects of hair dyes. *Int. J. Cosmet. Sci.* **5**(4) 113.
120. Bustard, H.K. and Smith, R.W. (1990) Studies of factors affecting light scattering by individual human hair fibres. *Int. J. Cosmet. Sci.* **12**(3) 121.
121. Gerrard, W.A. (1989) The measurement of hair colour. *Int. J. Cosmet. Sci.* **11**(2) 97.
122. Forrestier, J.P. (1981) Un séne, cassia obovata, utilise comme cosmétique: le 'henné neutré'. *Int. J. Cosmet. Sci.* **3**(5) 211.
123. Callingham, M. (1983) Shampoo evaluation, Parts I and II. *Soap, Perfumery Cosmet.* **56**(4) 177; **56**(5) 220.
124. Scandel, J., Reinstein, J.A. and Brudney, N. (1979) Studies on the cosmetic criteria of hair after shampoo. *Int. J. Cosmet. Sci.* **1**(2) 111.
125. Scandel, J., Reinstein, J.A. and Brudney, N. (1983) Shampoos and their aesthetic effects. *Int. J. Cosmet. Sci.* **5**(5) 157.
126. Ryan, J. (1990) Understanding product performance. Presented at the SCS symposium on *The Future of Hair Care Technology*, Harrogate, November 1990.
127. Joy, M. and Lewis, D.M. (1990) The use of FTIR in the study of surface chemistry of hair fibres. Presented at the SCS symposium on *The Future of Hair Care Technology*, Harrogate, November 1990.
128. Cadwallader, D.E. (1989) Stability Testing. *Cosmet. Toilet.* **104**(11) 87.
129. Toler, J.C. Preservation with safety. Presented at the SCS symposium on *Product Stability*, Coventry, November 1984.
130. Ford, D.M. (1983) The responsibilities and ethics of product testing. *Cosmet. Toilet.* **98**(2) 23.

3 Skin-care products

W.H. SCHMITT

3.1 Introduction

The technology of skin care is broad and differs from many other cosmetic categories because of the functional nature of many of the products. There are products that primarily have cosmetic effects and products that have very significant pharmacological effects. All products, however, are designed to interact and treat the largest organ of the body—the skin. To understand the need for the category, one must understand the business. The skin-care business is very large worldwide, with total sales of over 10 billion dollars divided between mass and prestige distribution. Mass is roughly 60% and prestige roughly 40%. The largest markets are Japan, at 3.5 billion dollars, the US at 2.5 billion dollars, and western Europe at 3 billion dollars.

Further divisions within the business are products based in sub-categories within the skin-care market. As an example, the US market is 23% hand and body moisturizers, 34% facial moisturizers, 13% suncare and 24% cleansers (excluding bar soaps). These ratios vary between mass and prestige, and from market to market, depending on national distribution patterns, usage habits and the number of products used by consumers on a regular basis. With the exception of hand lotions, sunscreens and lip protection products, and excluding bar soaps, the market usage and purchase is dominated by women, who account for over 80% of the usage and perhaps 90% of purchase.

3.2 Anatomy and physiology of the skin

The skin is the largest organ of the body. It is an organ that has diverse functions, including barrier function [1], protection from radiation, control of body temperature and transmission of stimuli from its abundant network of nerves.

An excellent reference text on the skin is provided by Montagna and Parakkal [2]. This text concentrates on the science of skin, which must be understood as a foundation before products affecting the skin can be appreciated or formulated. Our knowledge of the skin and its function expands every year, therefore it is imperative that one continues to study current literature, in order to remain informed. An excellent example of the progress of the science is the evolution of the knowledge that the outermost layer of skin, the stratum corneum is 'alive'. It had previously been accepted that the stratum corneum was a collection of dead cells, with no ongoing activity beyond barrier function. It is now known that many functions do take place and may originate in this layer.

In cosmetic chemistry and the business of skin care, we are primarily concerned with the outermost layers of skin (epidermis) and, in particular, those areas that are visible, such as face, hands and legs. To understand the overall anatomy and physiology of the skin, the cross-section shown in Figure 3.1 should be referred to throughout the following sections.

We are all familiar with our own skin. Closer examination reveals that it is varied in appearance, structure and function in different areas of the body. The skin is generally hairy, except for the palms and soles of the feet. Hair follicles vary in density and function, for example large terminal hair follicles on men's scalps and beards, to small vellus hair follicles on womens' faces and are all associated with sebaceous glands.

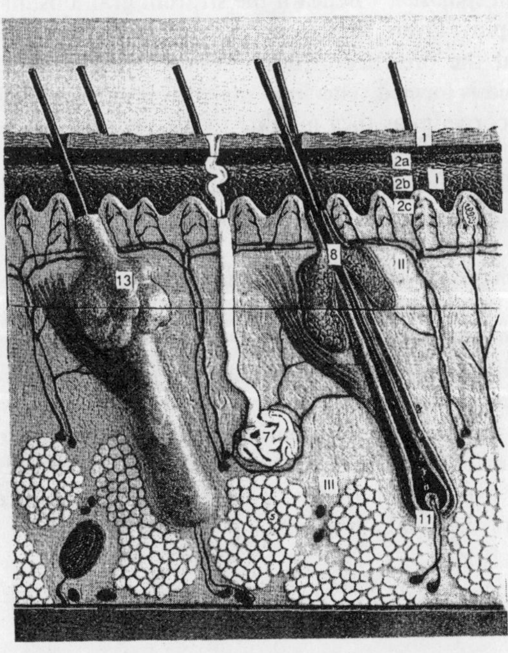

Figure 3.1 Cross-section of the skin showing epidermis (I); dermis (II); hypodermis (III); stratum corneum (1); stratum granulosum (2a); stratum spinosum (2b); stratum basale (2c); hair shaft (8); sebaceous gland (13); eccrine sweat gland (7); and hair bulb (11).

3.2.1 *Epidermis*

The epidermis consists of several layers and varies in surface topography and thickness. The epidermis on the soles of the feet and palms of the hands is thickened; on the scalp and face it is less thick, while on the trunk and the interior surface of the limbs it is thin.

3.2.1.1 *Stratum corneum* The outermost layer of the epidermis is the stratum corneum. This is the visible layer, and the one on which cosmetic chemists focus most attention. It consists of about 15 to 20 layers of flattened cells (keratinocytes), which become disassociated and ready for shedding or desquamation as they migrate to the surface. These cells are distinguished by not having nuclei or normal cell contents, and were thought to be dead, but they do have function and organization, and they can be controlled and managed.

3.2.1.2 *Stratum granulosum* The next layer within the epidermis is the stratum granulosum or granular layer. As cells progress to the surface, they form the characteristic granules of this layer.

3.2.1.3 *Stratum spinosum* Beneath the stratum granulosum is the stratum spinosum or spiny layer. This layer of cells is distinguished by a spiny appearance and the presence of desmosomes. This is the area where the intercellular lipid is formed. This lipid material is extruded between the cells of the stratum corneum as they migrate to the surface. The cells are looked upon as bricks, and the epidermal lipids as the mortar which helps hold the cells together. The lipids are similar to sebum, except for a lack of wax esters and squalene, and the presence of ceramides.

3.2.1.4 *Stratum basale* This is the innermost layer of the epidermis and is also known as the germinative layer. It is in contact with the dermis. It is in this layer that epidermal cells originate through the formation of keratinocytes, which differentiate as they migrate to the surface. This layer also produces Langerhans cells, which play a role in the immune system and the body's defense mechanisms. The entire cycle of formation, differentiation, migration and exfoliation from stratum basale to shedding by the stratum corneum takes from 40 to 75 days. The stratum corneum cycle is about one-third that of the total epidermis.

3.2.2 *Dermis*

The epidermis is dependent on the dermis for its source of nutrition. The dermis is a connective tissue that provides a support system for the epidermis. All of the skin's blood, nerves, lymph and external structures, such as hair follicles with sebaceous glands, and eccrine and apocrine sweat glands, are based in the dermis.

3.2.2.1 *Fibroblasts, collagen and elastin* The dermis has a fibrous component of various identifiable materials. The primary fibrous material is collagen, with a lesser content of elastin and smaller proportions of other fibrous structures. These fiber bundles and structures form an interwoven network that lends support, flexibility and suppleness to the skin. The source of these fibrous materials, the fibroblast, is the most numerous type of cell in the dermis.

3.2.2.2 *Mast cells* Mast cells are the second most numerous cells in the dermis. They are granular in nature and are most numerous near the capillaries. Mast cells participate in the release of enzymes and materials to control function or respond to injury.

3.2.2.3 *Blood supply* The blood supply for the dermis is abundant and consists of a complex of arterioles and capillaries that are drained by the venous system. The blood supply does not extend into the epidermis. The purpose of this complex network is not only for cellular nourishment, but also for temperature control, as demonstrated by flushing (dilation) and goose bumps (constriction) at opposite extremes of the body's temperature control.

3.2.2.4 *Hair follicles and sebaceous glands* The combination of hairs, and their follicles and attendant sebaceous glands is known as the pilosebaceous system. The hair follicle consists of epidermal tissue that extends into the dermis for its biological needs. There it obtains a source of nourishment from an abundant supply of blood, attaches to nerves and in the case of large terminal hairs, has an attachment to the arrectores pilorum muscle. This muscle can raise hairs in response to cold, or emotional signals.

In general, a sebaceous gland is associated with a hair follicle. For the purposes of this review, the activity of the sebaceous gland, and not the hair, will be focused on. The majority of sebaceous glands occur on the head and upper torso of the body, and are only absent from the palms and soles of the feet. Most sebaceous glands associated with hair follicles open inside the pilosebaceous canal. On the face, some large sebaceous glands either associated with fine velus hairs, or without hairs, can open directly on to the surface of the skin. The sebaceous glands form and secrete sebum, a unique blend of lipid materials not found anywhere else in the body. The size, maturity and amount of secretion is triggered by hormonal activity, which peaks in the teenage years [3]. Sebum is synthesized within the sebaceous gland and consists of triglycerides, fatty acids, wax esters, squalene, and cholesterol esters. The sebaceous glands, when active, produce a continuous flow of secretion without regard to the condition of the skin and hair, or the level of previously secreted sebum on the surface [4].

The sebaceous glands are a site for the development of acne. Even although the glands are controlled, and enlarge and develop by hormonal activity within the body, acne is dependent on many other factors beyond the development and activity of the glands [5]. This will be discussed later (section 3.9).

3.2.2.5 *Sweat glands* Two types of sweat glands are present in the human [6]. The most prevalent (more than two million) are the eccrine sweat glands, which have a primary function of controlling body temperature. The eccrine glands are also responsive to other triggers such as stress and sunlight. They extend from the epidermis into the dermis and, in the dermis, are surrounded by blood vessels. The sweat secreted by the eccrine system is relatively low in solids, consisting primarily of sodium chloride.

The apocrine glands are found primarily under the arms and associated with the sex organs. They are, like the sebaceous glands, primarily associated with hair follicles, usually opening into the pilosebaceous canal. These glands are much larger than eccrine glands but, like eccrine glands, originate in the dermis. The sweat produced, much less in quantity than that produced by eccrine glands, is higher in solids and of a more complex nature. It is viscous and milky, and contains high levels of protein, along with cholesterol, steroids and other lipids. It does not have a pronounced odor on secretion, but develops a distinctive odor due to enzymatic activity on the skin [7]. The purpose of aprocrine glands is unknown, but has been speculated to be sex-related, or a means of identification or communication through odor. This is the function of similar glands in other mammals.

3.2.3 *Skin color*

The color of the skin and hair is primarily due to the existence of pigment bodies known as melanin. Melanin is produced by the melanocyte cells present in the stratum basale. These melanocytes are present in this basal layer in great abundance and the number does not differ greatly in people with more or less melanin pigment or color in their skin. Melanin may be produced in either a yellow/red or black/brown color, and the color of the skin is determined by the amount and type of melanin produced. Melanin production is increased by exposure to the sun and plays a role in protecting the skin from sunlight.

3.3 Test methods

In addition to a basic understanding of the anatomy and function of the skin, a good foundation and understanding of various tests available for the function of the skin, as well as tests for product performance, is essential. The cosmetic chemist has been aided by many new test methods to help determine performance of skin-care products. These methods vary from *in vivo* observational or instrumental test methods to a wide variety of *in vitro* test methods.

3.3.1 *Efficacy testing*

The majority of skin-care products are sold to improve the appearance and feel of the skin, and are broadly classified as moisturizers. The condition and appearance of the skin is a function of its softness and flexibility, which is adversely affected by loss of water [8]. Tests for moisture content and moisture transmission, as well as the viscoelastic response, ultrasound techniques, and electrical properties of skin have been the subject of much study, with the development of increasingly sophisticated instrumentation to show the condition of the skin and its water content. Most of these new methods have evolved to help evaluate the potential performance of products, and are very useful in screening formulations, levels of ingredients and other stepwise formula variables. However, experience has shown that none of the available test methods totally replace *in vivo* application, professional observation and sophisticated consumer research utilizing both professional, and self- or consumer-evaluation of performance.

3.3.1.1 In vivo *observational tests for moisturizers* One of the primary and most valuable *in vivo* tests used, especially to test hand and body lotions, is the regression test method developed by Kligman [9]. This test is run in a location where the ambient relative humidity is conducive to development of dry skin. Panelists are 'dried down' by washing with a non-fatted bar soap for one week, without use of a moisturizer on the test site, which is either hands or lower lateral legs. The hands or legs are then observed and graded. The test protocol calls for use of product by some panelists, with others acting as non-treatment controls. At various times, during a two-week period, the condition of the skin is evaluated for dryness, flaking and redness. Treatment is then withheld for one week, and the rate of return to a dry condition is observed. This testing yields a good evaluation of the treatment potential of moisturizer products and their duration under severe conditions. The method has recently been improved as reported by Boisits *et al.* [10].

The condition of the face and its response to moisturizer products have also been evaluated by a variety of methods. A widely used method is the superficial facial line test, developed by Packman and Gans [11, 12]. This test has been developed for use by trained judges and also by untrained panelists. The standard method utilizes scoring, by trained judges, of superficial facial lines for depth of line in each of four areas in a half-face study. The other half of the face acts as a non-treatment control. The study is four days long, with no treatment on the first day and treatment twice a day on days 2, 3 and 4. The authors also discuss the use of self-evaluation by twelve untrained evaluators, who evaluate superficial facial lines in a similar test method.

3.3.1.2 *Cell turnover testing* A very useful test to determine cell turnover rate utilizes the dansyl chloride test [13], in which dansyl chloride, which fluoresces under a long-wave UV or Woods lamp, is applied to the skin. This material penetrates the stratum corneum to a uniform depth. The cell turnover rate, or transit time, is determined by disappearance of the stain compared to a non-treatment site.

3.3.1.3 In vivo *instrumental tests* An *in vivo* instrumental test that is proving to be of great value is profilometry, which is especially useful in the assessment of the changes to the skin's surface caused by the anti-ageing effect of retinoid therapy on sun damaged skin [14]. Two primary methods are used in profilometry. One method employs replicas and the other uses photographs. Fully hydrated skin has altered surface characteristics, as measured by both lower peaks and greater distance between peaks. This is also the case with skin that has fewer lines and wrinkles due to effects other than hydration. In evaluation with a replica, an instrument with a stylus produces a tracing of the topography of the skin, based on a replica produced by dental impression material. With two-dimensional photographs, computerized image analysis techniques are used to review the image [15].

The gas-bearing electrodynamometer (GBE), has been widely used to evaluate the viscoelastic properties of the stratum corneum, both *in vivo* and *in vitro* [16, 17]. More recently computer handling of data generated by this device has improved its utility [18]. The Twistometer® device measures resistance to torsion of the skin, as applied by a rotating disc inside an external ring, which acts as a stator. This device is used to evaluate the skin's softness and suppleness [19]. There has been a long history of the measurement of not only the water content of skin, but also the flow of moisture from the skin, which is known as the moisture vapor transmission rate (MVTR) or the moisture/water loss or transepidermal moisture loss (TEML). MVTR is a technique that has shown continual refinement. Early papers by Berube and co-workers outlined theory and practice [20–22]. Progress has continued to the more recent development of a well-accepted instrument, the Servo-Med® evaporimeter [23]. Most of the work on the direct measurement of water content in skin has utilized an infrared (IR) spectrophotometer. An excellent review is given by Potts [24].

3.3.1.4 In vitro *testing for moisturizers* Many *in vitro* tests are cited in the literature. An excellent overall review covering both *in vivo* and *in vitro* techniques has been written by Salter [25]. A technique for conducting the *in vitro* moisture vapor transmission rate was usefully outlined by Reiger and Deem [26]. Measurement of occlusivity by these techniques is useful and, in general, quite reproducible. The tests consist of the flow of moisture through a barrier, which is often a synthetic film with a transmission rate similar to that of skin. This technique (MVTR) combined with a water-holding technique to measure hydration, or the ability of the formula or test material to maintain humectancy, is very useful [27, 28] in evaluating or predicting potential *in vivo* results.

Test methods other than MVTR and water–holding capacity have been outlined by Quattrone and Laden [29]. They include scanning colorimetric techniques, scanning electron microscopy (SEM) and biomechanical properties of skin under various *in vitro* conditions. *In vitro* test methods can be instructive when their correlation to *in vivo* performance has been established. In the author's experience, a battery of tests and an excellent history of correlation are necessary for the tests to be of predictive clinical value. The problem is that, when *in vitro* tests are used with excised skin or a synthetic membrane, unlike the human skin, there is a lack of the biological underpinnings of the epidermis, which is anything but an inert, static or dead layer.

3.3.1.5 In vivo *test for sebum* An *in vivo* device for measuring the oiliness of the skin has been developed [30]. This device (Sebutape) is a film that visualizes and traps sebum. The tape can then be analyzed, and total relative sebum amount computed using image analysis. If quantification is not the only requirement, the sebum can be solvent-extracted, gravimetrically analyzed and broken down into component parts. This can be a very useful technique for assessing the instant removal of sebum by cleansers, or the effect on sebum production of various treatments.

3.3.1.6 In vivo *test for cleanser mildness* A wide variety of cleansers are available on the market and will be covered within this chapter. Tests for cleansers have focused primarily on testing soap or detergent formulations for mildness. In the past, mildness has been evaluated by the Soap Chamber Test [31]. In this test, dilute solutions of soaps or test solutions are maintained in a cup held against the skin of human volunteers. The skin is then evaluated for redness, scaling, and fissuring, often against high irritancy (non-fatted soap bar) and low irritancy (fatted sodium cocoisethionate bar) controls. A more recent modification [32] of this method calls for a shorter test, and measures the increase in MVTR by a Servo–Med® evaporimeter. This has shown close correlation with the longer duration visual assessment technique. The authors of the Soap Chamber Test cautioned as to the limitations of the test and recommended, in addition to this method, the use of exaggerated washing tests, commonly known as forearm wash tests [33]. These methods differ from the Soap Chamber Test in that the Forearm Test adjusts frequency for humidity, while the Flex Test [34] is reported to be unaffected by humidity. For evaluation of facial cleansers of all types for mildness and drying potential, the method outlined by Frosch [35] utilizes half-face testing with realistic use conditions. In order to obtain suitable dryness scores, this test is best run during periods of low relative humidity. Cleanser tests for efficacy, evaluating soil and make-up removal, as well as esthetic properties and after-cleansing skin-feel, must also be done to obtain a complete picture.

A combination of all these test methods, together with adequate clinical and consumer trial and observation, should enable one to have a great insight not only into the performance of products but, more importantly, as to how and why products work, and how variations in formula affect performance. This is necessary for the enlightened formulator.

3.4 Formulation

Throughout the remainder of this chapter, formulas will be presented as examples of particular formulation categories. These will be prototypical examples, taken from experience or from suppliers. They are only intended to give the formulator a general idea of typical formulas and constituents by category, and are neither tested for performance or stability, nor ready for the market. This would require considerable experimental work and testing. In addition, many new raw materials with functional and emotive benefits are available from raw material suppliers and should be reviewed on an ongoing basis during development.

Formulation is an art and a science, and requires close attention to detail and trial and error before proficiency becomes routine. The formulator must, first of all, make a product that has characteristics required by the ultimate consumer of the product. These requirements will vary depending on the specific usage, demographics of the consumer, expectations, claims, class of trade and packaging. The formula must be fit for use by the consumer at any point during its expected shelf-life. During formulation, this requires adequate stability trials and examination for microbiological integrity. It must also be possible to make the product using reasonable manufacturing equipment on a scale that is appropriate for anticipated demand. Further areas that are becoming increasingly important are safety under all conditions of use and potential misuse, and regulatory and environmental acceptability. These aspects of formulation will be discussed in the following sections.

3.4.1 *Formula information*

Much formula information, from a variety of sources, is available to the cosmetic chemist. Excellent background information and general formulation information may be obtained from the several reference books kept in most industrial libraries or large public libraries. The nature of these texts will generally give excellent historical data, but, the most up-to-date formulation data can be found from other sources. The first of these are the many periodicals that are published for the technologist in the cosmetic industry. These periodicals often have formulation articles and review articles covering a specific category, and also provide another valuable resource, i.e. identification of, and advertising from raw material suppliers. Raw material suppliers frequently publish information on starting formulas and give specific help with formulating. They are a rich source of ideas and guidance and some will even provide formulation assistance in their own laboratories. One must keep in mind, however, that the goal of the supplier is to sell the raw material.

An excellent starting point for the formulator is to review listed ingredients on the labels of finished formulas of market leaders' target products sold where ingredient disclosure is required, such as in the US where all cosmetic products must list all ingredients in descending order of content. It is helpful to review these ingredient lists, perhaps assay for a material or two, and speculate on the formula based on other sources of starting formulas. An assay for water, total lipid material and, if incorporated, humectant, will give much information about a formula that has already proven to be acceptable to consumers. The ingredient list in the US also uses accepted nomenclature from the US trade association, the Cosmetic Toiletry and Fragrance Association (CTFA) in its Cosmetic Ingredient Dictionary [36]. This is an excellent reference, as it lists not only the generic CTFA name, but provides technical information on the raw material, and lists suppliers of it. Another good source of information is to join and participate in local cosmetic chemists' organizations and attend their meetings. Seminars are held frequently and are attended by suppliers to the industry.

3.4.2 *Consumer testing*

Consumer testing and input is a must for the formulator. This should take place before and after product development. Once a prototype formulation has been developed, determined safe for use by consumers, and adequately preserved, it should undergo initial testing for consumer acceptance. Most commercial organizations have standards for consumer testing, which are usually organized by the marketing department. It is, however, very helpful to the formulator to seek consumer input at all stages during development of a product, for instance to develop an enlightened view of consumer requirements by talking to target consumers about their needs before development is initiated. This can be done informally or formally, either one-to-one, or with groups of consumers. Another useful technique is to give representative consumer users and non-users of the leading product in a category, an unidentified sample of the market leader to use for a week. Following this, the way in which the consumers used the product, the results of product use, and the likes and dislikes can be discussed on an open-ended basis. Attribute comments and areas of dissatisfaction with the target product should be noted. The information can then be used as a base line for later examination of a developed prototype. Another useful technique, known as a sequential monadic, is, at the next stage of development, to have target consumers use both the prototype formula and the market leader or competitive benchmark product. Half the panel first uses the prototype for a period of time, and then uses the competitive benchmark product or market leader; the rest of the panel uses the products in reverse order. With this technique, a good comparison of the prototype's performance versus the benchmark product on an attribute-by-attribute basis is obtained. At this stage, testing should be an on unidentified basis, since the requirement is for performance and attribute data as a guide to formulation.

In larger concerns, testing of finished products on an identified basis with advertising or concept, is usually handled by the marketing or market research department. It is possible, however, for the cosmetic chemist or formulator to have this work done for them by independent market research firms, or by advertising agencies.

3.4.3 *Stability testing*

Stability testing is done to ensure that a developed product will be fit for use during its expected life. Stability testing should be done early in the development cycle to remedy any problems before final testing. A well-run stability trial can provide much information about a product in a relatively short period of time. Prototype products in packaging, representing the ultimate trade package material, can be placed at ambient, and elevated (e.g. 37°C and 45°C) temperatures, refrigerated and cycled through freeze/thaw cycles, and placed in high humidity chambers. During these trials, testing should focus on attributes most important to the product's performance, or on the integrity of any active ingredients. In addition, there should be physical testing, for example pH and viscosity, chemical content of active ingredients, water content, etc. The formulator must not overlook attributes relating to consumer perception, such as fragrance, color, rub-in characteristics and appearance. Two or three months of successful elevated temperature testing and three or four freeze/thaw cycles will usually indicate that products will have an adequate shelf-life. It is important, however, to continue testing for longer periods at ambient temperature, to obtain an understanding of the product's ultimate shelf-life. Further testing whenever changes are made to the supply of raw materials or to the formulation is essential.

3.4.4 *Microbiological testing*

During the development phase, close attention should be paid to the microbiological integrity of the formulation. Professional advice should be sought from a well-trained industrial microbiologist. If this is not immediately available, input and guidance on preservatives from the suppliers' staff and consultants should be sought. Suppliers will often have more than one preservative and can make recommendations about particular types of formulation. Another method of finding appropriate preservatives is to review ingredient labeling on similar types of products that are available from reputable manufacturers and which have similar constituents, pH and package types. This, coupled with a microbiologist's or a preservative vendor's input can often give several trial systems. Finished products are usually tested using a challenge technique. If this essential testing is not available in-house, a qualified contract lab can assist. The test involves adding large numbers of a variety of organisms to the product and checking the ability of the product to reduce them to an acceptable level. To ensure that the product retains its ability to remain preserved, adequacy of preservation should also be checked during stability testing.

3.4.5 *Manufacturing trials*

A product cannot be a commercial success unless it can be made with large-scale equipment, therefore as soon as a prototype develops beyond its earliest stages, and at all subsequent stages of development, manufacturing should be considered. If an engineering group is available, it should be consulted, and should participate in a laboratory batch of the new product on a bench scale. This, if possible, should compare the compounding of the product, heating and cooling cycles, phase additions, etc., with a similar type of product that is already being manufactured. Any significant deviations from norm should be noted and investigated for possible change or for special attention during scale-up. Scale-up can represent large changes in batching times, time at temperature, etc. It is advisable to scale-up in increments, preferably not greater than 10 × at any stage. For instance, going from a 1 kg lab batch to a 10 kg lab batch, then to a 100 kg pilot batch and 1000 kg manufacturing batch would be a logical sequence. Any significant problem at each stage should be investigated and, at the very least, abbreviated stabilities should be run with product from each stage.

When final manufacturing is initiated, the development chemist or engineer should stay with the process until several batches have been made without problem. It is helpful, during the life of the product, to return to manufacturing and continuously improve the process. Time should be spent with veteran compounders, as they often have great insight into how a process can be improved. Product from first production trials should also be put through a complete stability test, comparing results to those obtained during development, and addressing any problems that may occur.

3.4.6 *Safety testing*

The product must be safe under all conditions of use and potential misuse. Safe products arise from a combination of careful formulating, good background data and adequate testing. A review of materials used in products for similar purposes is an excellent place to start. Most successful commercial products have been adequately tested and have a history of safe usage. Safety data for each raw material should be available from its manufacturer and should be reviewed. In the US, the CTFA has a Cosmetic Ingredient Review, which evaluates materials and also lists safety data for materials that have been reviewed. After a thorough background check, finished products should be tested through a competent third party. Many testing companies not only do safety testing, but also recommend appropriate testing that should be done for the class of product that has been developed. It is often useful to solicit testing protocols from two testing companies, asking them to explain and justify each test they recommend. It must be remembered, however, that they can only make enlightened recommendations based on input. Divulging as much information as possible at this stage, will give better, more appropriate testing input, and insight into possible untoward reaction.

After adequate clinical testing has been completed, and during all stages of consumer trial, it is important to check for unexpected reactions or patterns of misuse, while carefully extending exposure. Careful attention should be paid to reported reactions from usage or misusage situations. After launch, there should be a mechanism to check on reported reactions to determine if, on a broad exposure, these levels are higher than one would expect for the category. Category adverse response reporting data is available from the US Food and Drug Administration (FDA), and often from local trade associations.

3.4.7 *Regulatory and environmental requirements*

Regulatory requirements vary greatly from country to country, and can be ascertained by direct contact with the local regulatory authority. Excellent advice can frequently be obtained from local trade associations, who have insight into not only what is required by law or regulation, but also what is acceptable and expected. Membership of trade associations and participation in their activities is invaluable.

Environmental restrictions or concerns are often more elusive, or more difficult to determine. Trade associations, trade publications and local groups of cosmetic chemists can be of great help. Many companies also seek advice from consultants who are experts in this area, or, in some cases, have direct meetings and constructive dialogue with leading environmental groups. The person responsible for developing a product or formula should accept total responsibility for the product, and care and diligence should be exercised throughout every aspect of product development.

3.5 Skin cleansers

Skin cleansers constitute an important segment of the skin-care market. Cosmetic chemists are usually involved in the formulation of facial cleansers in forms other than bars. This, however, overlooks a large portion not only of the overall market, but also of consumers' usage patterns. Most consumers use bar products, either based on soap or synthetic detergent (syndet), for skin cleansing. All other cleanser products represent other types of usage and vary greatly from market to market depending on patterns of make-up usage, cleansing habits, etc.

In order to properly classify cleansers, it may be helpful to put them on a grid, roughly related to their physical nature, chemistry and functionality, and ranging from products that are anhydrous oils, at one end of the spectrum, to soap and water at the other (Table 3.1). Various product types have different functions and, while they have advantages for a specific condition or soil, none is universal. Most consumers use more than one type of cleanser, so an understanding not only of chemistry and formulation, but also of consumer usage patterns for their needs in a specific market or market segment, is key.

When classifying cleansers, the various soils on the skin must be considered. Soils come from a variety of sources, but can be classified broadly as oil-soluble, water-soluble and insoluble. Sources for oil-soluble soils can be sebum, residue from moisturizers, or waterproof make-ups. Water-soluble materials may also be residue from make-up or moisturizers, soluble skin

Table 3.1 Classification of skin cleansers according to physical nature, chemistry and functionality

More gentle to skin ←			More harsh to skin →			
Emollient and moisturizing ←			Foaming and astringent →			
Oleophilic ←			Hydrophilic →			
Mineral oil Vegetable oil Petroleum jelly	W/O cream or lotion	Non-foaming O/W lotion or cream	Fatted bars Mild foaming pastes Lotion	Fatted bar soap	Alcohol/ water	Bar soap Liquid detergent solution

soils, and soluble grime. Insoluble materials are represented by dead cellular matter, make-up pigments or, in hard-water areas, precipitated metallic soaps.

3.5.1 *Anhydrous oily cleansers*

The first category of cleansers is the most oleophilic group. Mineral oil, petroleum jelly, vegetable oils and esters are useful as non-drying skin cleansers for removal of waterproof make-up and oil-soluble soils. The disadvantage of these cleansers is that they are not readily rinsible and often have poor cosmetic and esthetic characteristics. More modern versions incorporate moderate to high levels of esters or gentle oil-soluble surfactant materials to make the products less greasy, more pleasant and, in some cases, rinsible. In addition, some formulations are thickened to form gels, which are easier to spread on to the skin and can be tissued off.

Typical formulation of an anhydrous oily cleanser

Ingredient	*Wt%*
Capric/caprylic triglyceride	12.0
PEG-400 dilaurate	6.0
Mineral oil, 70 viscosity	82.0

Procedure: Blend all ingredients at room temperature.

3.5.2 *Water-in-oil emulsions: cold creams*

The next category of products are water-in-oil (W/O) emulsion cream and lotion formulas, typified by cold cream [37]. Cold creams are referred to as W/O emulsions but, during processing and usage, are more complex water-in-oil-in-water (W/O/W) emulsions, or mixed emulsions. The partial or mixed external oil phase of these mineral oil/beeswax/borax systems dissolve oil-soluble materials and, because of the oily external phase and the beeswax/borax soap at the interface, they are able to solubilize, wet out and transport soil and waterproof make-up. Cold creams are generally not rinsible, are considered greasy and inelegant, and are tissued off the skin. They leave behind a film that has proven moisturizing characteristics. Classical cold cream formulas set the standard for mildness to skin and eyes.In recent years, some rinsible cold creams have emerged. These are cold cream and similar bases to which non-ionic emulsifiers have been added with an increased water phase. These rinsible systems then become oil-in-water (O/W) emulsions. They are not as satisfactory for waterproof make-up removal, but are rinsible and more cosmetically elegant.

Typical formulation of cold cream USP XXI

Ingredient	*Wt%*
Cetyl esters wax	12.5
White wax	12.0
Mineral oil	56.0
Sodium borate	0.5
Purified water	19.0

Procedure: Reduce the cetyl esters wax and the white wax to small pieces, melt them on a steam bath, add the mineral oil, and continue heating until the temperature of the mixture reaches 70°C. Dissolve the sodium borate in the purified water, warmed to 70°C, and gradually add the warm solution to the melted mixture, stirring rapidly and continuously until it has congealed. Pour into jars at 50°C and cool.

3.5.3 *Oil-in-water emulsions: cleansing milks*

Non-foaming O/W emulsions with greater than 50% water phase typify the next category of cleansers. These products are typically referred to as cosmetic milks. The primary cleanser and emulsifier is often a TEA*/fatty acid soap, or a detergent such as sodium cocoyl isethionate supplemented by anionic or non-ionic emulsifiers. Cosmetic milks are skin-friendly and have many similarities with O/W moisturizing lotions. However, they differ from moisturizing lotions in that they generally have less water in the water phase, excess TEA fatty acid, and higher levels of a secondary emulsifier such as self-emulsifying glycerol monosterate. Significant levels of mineral oils or esters may also be incorporated to ensure adequate levels of solubilization of oily materials without reducing rinsibility.

Typical formulation of an O/W cleansing lotion

Part	*Ingredient*	*Wt%*
A	Deionized water	56.8
	Xanthan gum	0.4
	Sodium cocoyl isethionate	25.0
	Preservative, EDTA*	0.4
B	Cocamide MEA*	3.0
	POE*-20 sorbitan monostearate	2.0
	Ceteareth-20	3.0
	Cetyl alcohol	2.0
	Ethylene glycol monostearate	2.0
	Isopropyl palmitate	2.0
	Poloxamer 338	2.0
	PEG*-7 glyceryl cocoate	1.0
C	Fragrance	0.4

* TEA, triethanolamine; EDTA, ethylene diamine tetra-acetate; MEA, mono ethanolamide; PEG, polyethylene glycol; POE, polyoxyethylene.

Procedure: Dissolve xanthan gum in water; heat to 65°C. Add sodium cocoyl isethionate, preservative, and EDTA. Premix part B and heat to 65°C. Add B to A with good mixing; then cool to 45°C. Add fragrance.

3.5.4 *Fatted mild syndet foaming bars and cleansers*

The next category is differentiated because of a foaming characteristic. This is the category of fatted syndet bars and gentle fatted foaming pastes and lotions. These products have good wetting properties but, when used in soap chamber or flex washing studies, are significantly more gentle than fatted soaps, detergent solutions or non-fatted soaps. They occur as bars (fatted sodium coco isethionate [38]), as pastes based on fatted built mono-alkylphosphates [39], and as pastes and lotions based on fatted and built sodium coco isethionate. Recent additions to this range of gentle foaming cleansers have been lotions based on sarcosanate surfactants, with high levels of glycerol, protein, and fatty materials to ensure a mild effect. Due to their gentle nature on the skin, mild bars, foaming pastes, and lotions are becoming more important on a worldwide basis.

Typical formulation of a mild syndet bar

Ingredient	*Wt%*
Sodium cocoyl isethionate	50.0
Stearic acid	34.0
Soap chip (80/20 tallow/coco)	10.0
Deionized water	5.0
Fragrance, pigments, etc.	1.0

Procedure: Blend all ingredients in amalgamator. Refine, extrude and stamp bars.

Typical formulation of a mild foaming syndet lotion

Part	*Ingredient*	*Wt%*
A	Deionized water	71.7
	Hydroxypropyl methylcellulose	0.5
	Triethanolamine	0.1
B	Sodium cocoyl isethionate	12.0
	Stearic acid	6.0
	Preservative, EDTA	0.4
C	Mineral oil, 200 viscosity	7.0
	Cetyl alcohol	2.0
D	Fragrance	0.3

Procedure: Disperse the hydroxypropyl methylcellulose in water. With mixing, add the triethanolamine to initiate hydration. Begin heating to 65°C. Add the part B ingredients. Premix part C; heat to 65°C. Add part C to the batch with good mixing. Cool to 45°C; add fragrance.

3.5.5 *Super fatted bar soaps*

Fatted bar soaps have greater cleansing characteristics, with less skin residue than fatted and built mild detergent pastes and lotions. The formulations contain varying amounts of fatty acids or other fatty or moisturizing materials to decrease their aggressive behavior on the skin. This results in a non-taut skin-feel upon rinsing, and less damage to the skin (see also chapter 9).

Typical formulation of a super fatted bar soap (cold cream)

Ingredient	*Wt%*
Soap chips 65:35 tallow:coco	97.65
Cold cream fat phase*	0.50
Lanolin	0.50
Antioxidant	0.10
Titanium dioxide	0.25
Fragrance	1.00

*See prior example cold cream USP XXI

Procedure: Blend all ingredients in an amalgamator. Refine, extrude and stamp bars.

3.5.6 *Astringents/toners*

Astringents and toners are a class of skin cleansers that have a special use and very specific formulations. They are hydroalcoholic solutions with an alcohol content from 20–70%. The products with lower levels of alcohol are developed for sensitive skins, while those with higher amounts are for oily skins. Astringents and toners often contain small amounts of emollients or humectants to decrease their defatting of the skin. They are generally used by oily-skinned consumers or teenagers as a supplement for cleansing and acne treatment. They are often the last cleansing step in a ritual to make the skin very clean in preparation for the use of a moisturizer.

Typical formulation of a hydroalcoholic astringent

Ingredient	*Wt%*
Deionized water	50.90
Ethyl alcohol	40.00
Glycerin	4.00
Sorbitol 70%	2.00
Menthol	0.10
Witch hazel extract	3.00
Fragrance	*q.s.*
Color	*q.s.*

Procedure: Dissolve menthol in alcohol. Add fragrance, water and other ingredients. Blend at room temperature.

3.5.7 *Bar soaps*

The final category in Table 3.1 consists of the minimally fatted or non-fatted bar soaps. These coco/tallow soaps are excellent cleansers, but can, if used excessively, irritate the skin [40, 41]. They enjoy considerable usage among consumers as part of an everyday cleansing ritual and, on a worldwide basis, are the most commonly used cleanser product (see also chapter 9).

Typical formulation of a bar soap

Ingredient	*Wt%*
Soap chips 80:20 tallow:coco	97.65
Water	1.0
Antioxidant	0.1
Perfume oil	0.75
Titanium dioxide	0.5

Procedure: Blend all ingredients in amalgamator. Refine, extrude and stamp bars.

3.5.8 *Particulate scrubs*

A specialty category not shown in Table 3.1 is the scrub creams, which contain particulate materials. These products are often O/W emulsions (if non-foaming) or gentle pastes (if foaming). They contain particles of polyethylene or other inert materials such as ground seed-husks. The purpose of these particulate materials is to remove loose flakes of stratum corneum and to polish the skin. Interesting research by Loden and Bengtsson [42] indicated that the effect of scrubbing with a particulate scrub was equal to that of 2.5 to 3 tape strippings. The use of particulate scrubs can be helpful if there is a need to remove a layer of stratum corneum. However, excessive use should be cautioned as, due to their mechanical action, they can be irritating.

Typical formulation of a particulate scrub

Part	*Ingredient*	*Wt%*
A	Polyethylene beads	5.0
	Mineral oil, 70 viscosity	20.0
	Octyl decanol	10.0
	Glyceryl stearate	4.0
	Ceteareth-12	1.5
	Ceteareth-20	1.5
	Glyceryl monooleate	1.0
	Propyl *p*-hydroxybenzoate	0.05
B	Deionized water	48.35
	Methyl *p*-hydroxybenzoate	0.1
C	Fragrance	8.5

Procedure: Heat A to 70°C. Heat part B to 75°C. Add part B to part A with agitation and cool to 40°C. Add part C and mix thoroughly.

The selection of cleansers and their constituents requires extensive testing to ensure suitability for use. The goal should be to remove soils and make-up with minimal damage to the skin. An interesting approach to the selection of surfactant materials and products is outlined by Komp [43], utilizing the Duhring Chamber Test, as well as the antecubital washing test [33]. As well as these exaggerated tests, it is very important to follow prototype formulation with controlled in-use testing for soil and make-up removal, and consumer evaluation use testing.

3.6 Moisturizers

Moisturizers are products that are usually emulsions, either O/W or W/O. There are two principal forms of these products: (i) semi-solid emulsions, known as creams; and (ii) flowable emulsions, known as lotions. Moisturizing products are differentiated not only by their emulsion type and/or physical form, but also by their functional use. For the purpose of this discussion, all-purpose creams, hand and body lotions and creams, facial moisturizing lotions, and facial 'night' creams, will be considered.

The purpose of moisturizing products is to restore and maintain the skin in a good-looking, fully moisturized condition. To maintain this condition, the stratum corneum must be in a fully hydrated condition that allows flexibility and elasticity. Early work by Middleton and Allen[44] and a review by Idson [45], show the relationship between water content in the skin, as affected by temperature and humidity. In most products, this moisturization is accomplished by a combination of hydrating by water followed by the actions of occlusives and humectants. Emulsion products of either O/W or W/O break down when rubbed out on the skin, and add water from their own composition to the surface layers of the stratum corneum. This hydrating effect by water accounts for the instant appearance benefits, including reduction of visible dry flakes and chapping (see Blank [8, 46]). Less-immediate effects occur through occlusivity and humectancy.

Occlusivity occurs when the transepidermal water loss is slowed through reduction of the moisture vapor transmission rate. Many fatty materials reduce the MVTR. Typical of these materials would be petroleum jelly, mineral oil, vegetable oil, silicone oils, waxes, fatty acids, alcohols, and esters of mineral, animal or plant origin, as well as many synthetic oily materials that are available to the cosmetic chemist.

Humectancy is a separate but related phenomenon, in which materials that have an affinity for water: (i) help bind water to the skin; (ii) resist evaporation from the skin; or (iii) under certain circumstances, attract water to the skin.

Typical of these materials are glycerin, sorbitol, sodium lactate and sodium pyrollidone carboxylate (the skin's naturally occurring humectant [47]). Such humectants have been shown to be valuable not only in hydrating the stratum corneum, but also in improving the elastic modulus and stress relaxation modulus, thus altering the viscoelastic behavior of the stratum corneum [48]. Extensive experimentation with glycerin [49] has shown that it moisturizes dry skin in a dose-related relationship dependent on the concentration of glycerin. It was postulated that the water–glycerin mixtures on the skin also assist in plasticizing the stratum corneum in a less than transient manner. Middleton [50] found that lactic acid was absorbed by the stratum corneum and that increased extensibility was retained by the stratum corneum.

3.6.1 *All-purpose creams*

All-purpose creams are typified by a W/O emulsion or by high oil content O/W emulsions. These products are for general face and body usage and generally have a heavy consistency and significant drag on rub-out. In addition, a small amount of product is able to cover a large area. Many all-purpose creams contain from 30–70% oil phase, resist wash-off, and are excellent protective as well as treatment products. They represent a very significant percentage of the European mass moisturizer market, and are often sold in flat packs (tins) or tubes. Performance characteristics and testing should focus on tests to maximize a reduction in the MVTR, resistance to wash-off and a prolonged improvement to the skin, as demonstrated by regression analysis. Lighter weight, more modern products are, however, becoming popular.

Typical formulation of a heavy all-purpose cream

Part	*Ingredient*	*Wt%*
A	Stearic acid	3.00
	Mineral oil, 70 viscosity	40.00
	Lanolin	7.00
	Petrolatum, USP	10.00
	Cetyl alcohol	2.00
	Microcrystalline wax	2.00
B	Magnesium aluminum silicate, 5% dispersion	5.00
	Triethanolamine	1.78
	Water	29.22
C	Fragrance and preservatives	*q.s.*

Procedure: Heat parts A and B to 70°C. Add part B to part A, mixing continuously. Mix and cool to 35–40°C and add part C. Continue mixing until dispersion is complete. Adjust evaporation losses with water.

Typical formulation of a modern all-purpose cream

Part	*Ingredient*	*Wt%*
A	Stearic acid	4.0
	Mineral oil, 70 viscosity	5.0
	Glyceryl monostearate	1.0
	Glyceryl monohydroxystearate	2.0
	Cetearyl alcohol	3.0
	Cetyl octanoate	10.0
B	Deionized water	60.0
	Glycerin	3.0
	Carbomer 941 (2% dispersion)	10.0
	Triethanolamine	1.5
C	Fragrance	*q.s.*
	Preservatives	*q.s.*

Procedure: Heat parts A and B separately to 70°C. Add Part A to part B with good agitation. Mix to 35°C and add part C. Continue mixing until dispersion is complete.

3.6.2 *Hand and body lotions*

Hand and body lotions are generally O/W emulsions, with typical lotion products containing 10–15% oil phase, 5–10% humectant, and 75–85% water phase. They are characterized by easy flow and pumpability, fast rub-in, and a lack of stickiness after rub-in. Most of these products are sold in bottles with or without pumps, or in tubes.Their primary performance characteristics are their ability to instantly hydrate skin and to relieve dry skin symptoms. In addition, they should have excellent profiles in MVTR reduction and prolonged improvement to the skin as shown by regression analysis.

Typical formulation of a hand and body lotion

Part	*Ingredient*	*Wt%*
A	Stearic acid	2.5
	Glyceryl monostearate	1.0
	Cetyl alcohol	1.0
	Petrolatum USP	1.0
	Mineral oil, 70 viscosity	2.0
	Isopropyl palmitate	2.0
	PEG-40 stearate	0.25
B	Deionized water	77.0
	Carbomer 934 (2% dispersion)	7.0
	Glycerin	5.0
	Triethanolamine, 99%	1.0
C	Preservatives	*q.s.*
	Fragrance	*q.s.*

Procedure: Heat parts A and B separately to 70°C. Add part A to part B with good agitation. Mix to 35°C and add part C. Continue mixing until dispersion is complete.

3.6.3 *Hand and body creams*

Most modern hand and body creams are O/W emulsions with greater levels of oil phase and possibly higher humectant levels than lotion products. They generally contain 15–40% oil phase and from 5–15% humectant phase. These products are easy to apply, have a reasonably fast rub-in, good esthetics, and often resist wash-off. They are typically sold in jars or tubes. Their performance characteristics are similar to hand and body lotions, often with the added benefit of more resistance to wash-off.

Typical formulation of a hand and body cream

Part	*Ingredient*	*Wt%*
A	Steareth-21	2.5
	Steareth-2	0.75
	Cetearyl alcohol	3.0
	Cetyl palmitate	1.0
	Myristyl myristate	1.0
	Dimethicone 50 cSt	2.0
	Decyl oleate	4.0
	Mineral oil	5.0
B	Deionized water	69.0
	Glycerin	10.0
	Xanthan gum	0.2
	Magnesium aluminum silicate	0.6
C	Preservatives	*q.s.*
	Color	*q.s.*
	Fragrance	*q.s.*

Procedure: Heat parts A and B separately to 70°C. Add part A to part B with good agitation. Mix to 35°C and add part C. Continue mixing until dispersion is complete.

3.6.4 *Facial moisturizer lotions*

In terms of esthetics, facial moisturizer lotions are very different from hand and body lotions. They contain occlusive agents and humectants at the lower concentration levels of hand and body lotion formulations. Great attention is paid to quick-breaking, instant-absorbing products with no greasy afterfeel. These effects are obtained by the use of emulsions that break down quickly, dry to a matte finish, and utilize light emollients such as esters and fatty alcohols and, where appropriate, additives such as volatile silicones. Performance characteristics are good moisturizing capability, the ability to reduce MVTR rates, and also an immediate and a short-duration effect on superficial facial lines. Visual performance effects and esthetics are key factors in successful products.

Typical formulation of a facial moisturizing lotion

Part	*Ingredient*	*Wt%*
A	Deionized Water	63.90
	Methylparaben	0.20
	Sorbitol, 70% solution	2.00
	Propylene glycol	3.00
	Carbomer 940 (2% dispersion)	15.0
	Triethanolamine	0.60
B	Glyceryl monostearate	4.00
	Mineral oil	3.00
	Caprylic/capric triglycerides	4.00
	Hydrogenated vegetable oils	1.50
	Stearic acid	2.00
	Laureth-23	0.80
C	Fragrance and preservatives	*q.s.*

Procedure: Mix all of part A except the triethanolamine. After all of part A is dispersed, add TEA. Heat parts A and B to 70°C. Add part B to part A with good agitation. Mix to 35°C and add part C. Continue mixing until dispersion is complete.

3.6.5 *Facial moisturizer creams: night creams*

Facial moisturizer creams are often sold as night creams or heavier duty moisturizer products than their lotion counterparts. They generally have higher levels of occlusive and humectant materials than facial moisturizer lotions, but are formulated (with esthetics as a principal factor) using lighter emollients, quick-breaking emulsions, and controlled levels of humectant materials. Additives such as volatile silicones are often added to improve rub-in, break and a non-greasy afterfeel which, together with a matte finish, are key esthetic features. Performance characteristics are excellent moisturizing capabilities and a good effect on superficial facial lines.

Typical formulation of a night cream

Part	*Ingredient*	*Wt%*
A	Ceteareth-20 and cetearyl alcohol	3.0
	Stearyl alcohol	2.0
	Glyceryl monostearate	2.0
	Diisopropyl dimerate	5.0
	Octyl palmitate	5.0
	Petrolatum USP	3.0
	Dimethicone 350 cSt	1.0

B	Deionized water	70.0
	Hydroxypropyl guar gum	0.3
	Glycerin	7.0
	Magnesium aluminum silicate	0.9
C	Preservative and fragrance	*q.s.*

Procedure: Heat parts A and B separately to 70°C. Add part A to part B with good agitation. Mix to 35°C and add part C. Continue mixing until dispersion is complete.

3.7 Anti-ageing products

The emergence of new technology with proven performance to reverse damage from photoageing necessitates a separate section on anti-ageing products. Many such products are sold but, prior to discovery by Kligman [51] of the effect of retinoic acid on photodamage, and patents issued to Van Scott and Yu [52] on α-hydroxy acids, these products had primarily cosmetic benefits; that is, complete hydration of the skin, a return of suppleness, the reduction of superficial facial lines and, at times, a masking effect, giving lines a more youthful, albeit transient, improved appearance.

Ageing is well understood and, although much can be done in the laboratory to affect cellular vitality, or to affect the overall longevity of laboratory animals, the human species has not materially advanced in its maximum age as proposed in the original theories of Hayflick and Moorehead [53]. A more optimistic review by Pugliese [54], both of ageing and of events within overall cellular mortality, is of interest and suggests that a proper understanding of the entire ageing process is the key to effecting changes in it. This review offers many areas of interest to the cosmetic chemist. Ageing results in subtle and not so subtle changes in the appearance of the skin. It is probable that many of the effects seen on the skin with ageing are an indirect result of the decrease in the blood supply to the dermis [55]. The number of sebaceous ducts and the level of secretion from them also decreases, however, many of the remaining ducts increase in size [56]. An excellent review of these changes in physiology and pathophysiology by Gilchrest [57] is worth reading in order to understand the substrate in question.

Manifestations of ageing that are of primary concern to the cosmetic chemist are wrinkles. These lines, which become more pronounced with age, especially on the face, are caused by a shrinking of the superficial muscles, which have their points of insertion in the dermis. The facial expression muscles are the first areas to change with age. They also develop the deepest wrinkles. As they lose superficial mass, thinning of the epidermis and a loss of collagen and elastin are apparent. These all contribute to the visible process called ageing. As yet, it has not been possible to affect this frank, deep cellular ageing. One reason is that there is nothing histological to differentiate the cellular structure of a wrinkle from that of a non-wrinkled epidermal area. An excellent review on this entire subject is provided by Wright and Shellow [58], who studied the histology of deep wrinkles and adjacent non-wrinkle areas.

Advances are continually being made to show improvements in ageing of photodamaged skin, due to retinoid and retinoid-like effects, more recently demonstrated by α-hydroxy acids and related compounds. Several natural physiological vitamin A or retinoid effects that are known to be ongoing in the dermis are dermal development, collagen synthesis, hyaluronic acid synthesis, DNA content and epidermal protein expression. Since these factors are, in part, controlled by vitamin A and its derivatives in healthy skin, it has proven worthwhile to seek profound effects such as those demonstrated by topical treatment of retinoic acid. Achieving a retinoid effect is not possible by simply adding vitamin A or another retinoid to a cosmetic formula. Following topical application, the effective drug must be available to the dermis and epidermis, without untoward side effects. Recent work [59] with the hairless mouse model has correlated a dose response to increasing levels of retinyl palmitate in a suitable cosmetic preparation. This response from topical treatment resulted in an increase in protein and collagen synthesis, and in DNA content, together with a thickening of the epidermis. Prior work with a topically applied retinoid has shown accelerated dermal repair to UV damaged dermis, due to increasing regeneration of connective tissue [60]. Emerging work has shown quantifiable effects on human photodamaged skin, and significant biochemical changes in the hairless mouse model, due to topical application of a common cosmetic retinoid. This, coupled with the ability to increase cell turnover from use of a cosmetic product, the reduction of superficial facial lines through new and improved materials and mechanisms, and the 'plumping' of the epidermis by occlusive or humectant agents can, in theory, be combined to give excellent visual anti-ageing results.

Both Kligman and Van Scott [51, 52] discuss the effect of both retinoids and α-hydroxy acids on photodamage. That photodamage has a profound effect on the appearance of ageing of the skin, is now well understood and discussed under section 3.8.1 of this chapter. Damage that can manifest itself as ageing, stems from the effect on the elastic fibers that provide smoothness and suppleness [61, 62]. Evidence that these changes can be reversed was demonstrated by Kligman *et al.* [63].

Recently, α-hydroxy acids, primarily lactic and glycolic, have entered the marketplace in many skin-care preparations. Smith [64] has demonstrated various effects of these materials and has compared their efficacy. Derivatives of α-hydroxy acids have also been marketed [65].

A growing problem is whether or not retinoid-like materials are drugs. In the US, and increasingly throughout the world, courts and regulatory bodies are telling us that if a product *contains* a 'drug', it clearly *is* a 'drug'. Even if it does not contain what has been previously classified as a drug, but alters the structure or function of the human body, it may well still be regarded as a drug. In the US, products containing retinoic acid are classified as drugs and are only available on prescription allowed to be used under the supervision of a physician. At the present time, such products are cleared for use only in the treatment of acne, but are being mis-prescribed by many physicians for treatment of photoageing and its associated dermal damage. They are not without side effects, including significant levels of irritation in some patients.

The status of products claimed to possess anti-ageing, and other properties has recently come under scrutiny by the US FDA. The FDA has issued several regulatory letters to companies, requesting them to stop making drug claims, i.e. "…altering the structure or function of the body." Most of these regulatory letters have been resolved by the companies modifying their claims following negotiations with the FDA. One company has, however, initiated suit over this issue, which will no doubt be decided in the courts. A review of performance claims for skin-care cosmetics, and the basis of US regulations, was written by McNamara [66]. In this review, the legal basis for the distinction between drugs and cosmetics, based on both claims and ingredients, is discussed.

Anti-ageing products have been marketed by the cosmetic industry for many years. Successful new products were often launched by interesting claims such as "penetrates 21 cell layers", "speeds up cell turnover" or "enhances overnight repair at a cellular level". However, although these products did have performance characteristics, proven by new biophysical tests that were optimized in the product, they did not truly change the ageing process. The proven ability of retinoic acid to reverse photodamage has now altered this situation, and true pharmacological effect from topically applied products is now a reality. The successful formulator and marketer will have to keep up with this rapidly evolving and most exciting area of science.

3.8 Sunscreen products

Sunscreen products form an important subclass of skin-care products. In order to have a good understanding of these products, the chemist must first understand the physics and pharmacology of solar radiation, and its effect on the body.

3.8.1 *Solar radiation*

Skin damage, and the development of products that help prevent it, require an understanding of the ultraviolet spectrum of light. The ultraviolet spectrum of concern is between 200 and 400 nanometers (nm) and, in practice between

280 and 400 nm. The ozone layer around the earth absorbs the shorter wavelengths [67], therefore the important wavelengths are UV-B (280–320 nm) and UV-A (320–400 nm). The penetration of the skin by ultraviolet radiation depends on wavelength. As the wavelength increases from 200 to 400 nm, the penetration also increases. Consequently, UV-B penetrates (depending on the thickness of the skin in the area of the body irradiated) into the upper layers of the dermis, whereas the longer wavelengths from UV-A penetrate deeply into the dermis. Due to the variable thickness of the skin on various areas of the body, some areas are more susceptible to sunburn and photodamage (ageing) than other areas.

The immediate symptom of excessive UV-B exposure is sunburn. This is caused by a reaction of UV-B radiation with an absorbing material within cells. This absorbing material has been shown to be DNA, which is then depressed prior to cellular propagation [68]. The ability to penetrate varies greatly according to wavelength [69], with small percentages of UV-B penetrating to the capillary system in the dermis at lower UV-B wavelengths, compared to roughly 50% at its greatest wavelength (320 nm). UV-B, however, also stimulates the dermis to form melanin by a complex series of reactions that stimulate the melanocytes to trigger tyrosine in a multi-step reaction to form melanin. For this reaction to occur, the skin must be stimulated by a dose of UV-B, close to the dose that causes redness. On excessive exposure to UV-B (beyond one minimal erythemal dose (MED)), the skin becomes pink after a delay of two to four hours. Redness occurs, which, depending on the severity of the burn, will begin to fade from overnight to within three to four days. In severe cases, the redness may result in separation of the stratum corneum by edema and 'blistering'. The wavelength of maximum redness within the UV-B range is near the peak of UV-B or 305 nm [70].

UV-A is not directly associated with erythema, and has a redness-producing potential compared to UV-B of about 1/1000. However, it has been shown to damage the skin by penetration to connective tissue, and is also capable of producing tumors. In addition to damage by the deeper penetration of UV-A, photosensitizing or allergic response due to a causative chemical and UV appears to be primarily caused by UV-A [71]. UV-A is able to produce melanin, but the tan is more unstable than the melanin created by UV-B exposure. UV-A is used in tanning salons because of its lack of erythema, and its relatively quick tanning. It is doubtful, however, whether tans produced by exposure to UV-A will provide resistance to significant amounts of UV-B [72]. UV-A exposure in tanning salons is not without its hazards, especially to the deeper underlying structure of the dermis.

Due to the damaging potential of UV-B, both near term (sunburn) and long term (skin damage), and the deeper penetration and potentially greater damage from UV-A, proper protection from UV-A and UV-B must be exercised. Exposures to both UV-A and UV-B have been implicated in increased numbers of skin cancers.

The amount of solar radiation varies depending on the season, the latitude, the time of day, and the altitude. The closer to the equator, the greater the total radiation from ultraviolet. The same is true for any location during the summer, when the sun is at its closest point. From 9 a.m. to 3 p.m. represents the period of greatest exposure, while increasingly higher altitudes add to exposure levels. In addition, when directly in the sun, there is from 10 to 100 times more UV-A exposure than UV-B exposure. Consequently, both bands contribute significantly to harmful effects. An additional concern is that the vast majority of UV emitted by the sun is absorbed by stratospheric ozone. There is recent evidence that this ozone layer is decreasing, which will allow more UV-B to penetrate. A decrease in the ozone layer of 10% is estimated to increase melanoma by 10–20%, basal cell carcinoma by 25–35%, squamous cell carcinoma by 50–60%, and cataracts by 6–10% [73].

Sensitivity to UV irradiation depends on the amount of melanin, which absorbs UV, in the skin. Fitzpatrick and Pathak [74] have developed a classification for skin types, depending on their response to UV irradiation (Table 3.2). This response is generally due to the level of melanin pigment in the skin. It is also typical of various racial or ethnic groups. Those groups who have a long history of sun exposure, e.g. black Africans, tend to be classification V or VI, while those with little history of sun exposure, e.g. people of Irish or Celtic heritage, tend to be category I or II. The normal reaction following exposure to meaningful amounts of UV is for the skin not only to produce more melanin, but also to thicken [72]. The most serious effects are photodamage and cancer. Prolonged exposure can lead to profound changes in the fibrous component of the dermis, with a decrease in collagen, alteration in keratinocytes and melanocytes and increased deep wrinkling and development of age spots. Continuous exposure leads to development on exposed areas of a wrinkled leather-like appearance or 'red

Table 3.2 Skin classification [74]

Skin type	Classification	Reaction to sun
I	Sensitive	Burns easily Never tans
II	Sensitive	Burns easily Tans minimally
III	Normal	Burns moderately Tans gradually
IV	Normal	Burns minimally Always tans well
V	Insensitive	Rarely burns Tans profusely
VI	Insensitive	Never burns Deeply pigmented

neck'. Skin cancers have been shown to be the direct result of over-exposure to the sun, and the propensity of the individual to erythema and damage.

Australia, which has a high concentration of northern Europeans with Type I or Type II skin, and a location close to the equator giving maximized ultraviolet radiation exposure, has the highest reported incidence of skin cancer for a developed population. The most prominent form of skin cancer is basal cell carcinoma. This cancer is typically preceded by a pre-cancerous lesion, a solar keratosis. This solar keratosis is easily removed by a dermatologist. The most serious form of skin cancer is a malignant melanoma, which usually arises from a mole and, if not treated early, is often fatal. Given the potential sun-damage to the skin, the need for sunscreen products is obvious. Some data [75] suggest that 50% of a person's lifetime solar exposure occurs by the age of 18, therefore early usage is critical.

3.8.2 *Sunscreen chemicals*

Sunscreen products are simply vehicles for materials that prevent meaningful amounts of UV radiation from reaching the skin. These materials are usually classified as chemical absorbers, or physical blockers and scatterers of UV radiation. In the US, they are recognized by the FDA 'over-the-counter drug regulations', which list various materials as category I (safe and effective) sunscreens. In Europe, they form part of the EC and European Federation of the Perfume, Cosmetics and Toiletries Industry (COLIPA) positive list. An excellent review concerning the use of UV absorbers in sunscreen products has been written by Shaath [76]. The materials can be grouped by name, chemical type, the wavelength at which they have maximum absorbance, allowed percentage usage in the US and EC, and by EC and COLIPA list numbers (Table 3.3). The vast majority of the products listed in Table 3.3 are UV-B absorbers. The benzephones and butyl methoxy dibenzoyl methane are UV-A absorbers. Titanium dioxide is a broad spectrum physical block.

3.8.3 *Testing*

Sunscreen products formulated with the sunscreen materials shown in Table 3.3, at allowed or appropriate amounts, must be tested for performance. A number of *in vitro* performance tests have been developed and advocated, but the only currently accepted tests are *in vivo* tests. The standard tests vary from country to country, but are all similar. They are based on exposure to the sun or, more usually, to lamps that give carefully controlled amounts of radiation of very specific wavelengths. A useful value has evolved to identify the efficacy of sunscreens. This is the sun protection factor or SPF, which is based on production of redness on a test subject with a minimal erythemal dose (MED), representing the amount of exposure required to produce

Table 3.3 UV absorbers for sunscreen chemicals

CTFA name [Common or chemical name]	Chemical type	Maximum absorbance (nm)		FDA Category I	% allowed		EC No.	COLIPA No.
		Alcohol	Mineral oil		FDA	EC(max)		
Benzophenone-3 [Oxybenzone]	Benzophenone	288/325	288/329	Yes	2–6	10	1.4	S 38
Benzophenone-4 [Sulisobenzone]	Benzophenone	286/324	N/A	Yes	5–10	5	2.17	S 40
Benzophenone-8 [Dioxybenzone]	Benzophenone	284/327	296/352	Yes	3	—	—	—
DEA methoxycinnamate [4-Methoxycinnamic acid salts]	Cinnamate	290	N/A	Yes	8–10	8	2.9	S 24
Ethyl dihydroxypropyl *p*-Amino benzoic acid (PABA) [Ethyl-4-*bis*(hydroxypropyl) amino benzoate]	PABA derivative	312	N/A	Yes	1–5	5	2.1	S 2
Glyceryl PABA [Glyceryl 1-(4 amino-benzoate)]	PABA derivative	297	N/A	Yes	2–3	5	2.4	S 6
Homosalate [Homomenthyl salicylate]	Salicylate	306	308	Yes	4–15	10	1.3	S 12
Menthyl anthranilate [Menthyl-*o*-aminobenzoate]	Anthranilate	336	334	Yes	3.5–5	N/A	N/A	N/A

Octo crylene [2-Ethylhexyl-2-cyano-3, 3-diphenyl acrylate]	Diphenyl acrylate	303	N/A	Yes	7–10	N/A	N/A	N/A
Octyl dimethyl PABA [2-Ethylhexyl 4-dimethyl aminobenzoate]	PABA derivative	311	300	Yes	1.4–8	8	2.5	S 8
Octyl methoxycinnamate [2-Ethylhexyl-*p*-methoxy cinnamate]	Cinnamate	311	289	Yes	2–7.5	10	2.13	S 28
Octyl salicylate [2-Ethylhexyl salicylate]	Salicylate	307	310	Yes	3–5	5	2.6	S 13
PABA [4-amino benzoic acid]	PABA	283	N/A	Yes	5–15	5	1.1	S 1
2-Phenyl-benzimidazole-5-sulphonic acid [Novantisol]	Sulphonic acid derivative	Alkaline Solution	310	Yes	2–6	8	1.6	S 45
TEA-salicylate	Salicylate	298	N/A	Yes	5–12	2	2.11	S 9
3-(4-methylbenzylidene)-camphor	Miscellaneous	300	N/A	No	N/A	6	2.25	S 60
4-Isopropyl dibenzoyl methane [1-*p*-cumenyl-3-phenylpropane-1,3-dione]	Miscellaneous	345	N/A	No	N/A	5	2.28	S 64
Butyl methoxydibenzoyl methane	Avobenzone	358	N/A	No	N/A	5	2.31	S 66
Titanium dioxide		Broad spectrum		Yes	2–25			

redness. The SPF is then the protection factor achieved with a specified layer of sunscreen applied to test subjects. For instance, if individuals on average show a redness (MED) at 20 min and, after the application of a fixed amount of sunscreen product, show redness at 80 min, the product would have an SPF of 4. Much experimentation has been done to evolve alternative testing methods, and these are discussed in an excellent overall review of the subject of light protection by Groves [77].

The thickness of the applied film and lamps specified in sunscreen testing vary from country to country. The US and Europe differ significantly. Australia has separate standards and Japan is promulgating guidelines. Since these methods are under review and change, it is best to seek advice on acceptable testing from local trade associations, testing companies, sunscreen suppliers or regulatory bodies.

Another area of testing is that of waterproofness, which also varies from location to location. For instance, to support claims of waterproofness, the US FDA requires an *in vivo* test in which the subject is immersed in water for 80 min after application of product and then tested using the standard protocol. The claimed SPF must not be below the value achieved after immersion. Tests are now evolving and being advocated for testing protection versus UV-A exposure. At this point in time, no one test is yet accepted as standard.

3.8.4 *Sunscreen formulations*

Products that contain active sunscreens must be carefully formulated in order to ensure that the sunscreen material is in its active form, is available when used, does not contain materials that adversely shift or diminish its absorbance, meets the SPF claimed, is waterproof if required, and is cosmetically elegant. There are many forms available, including the common lotions (generally O/W emulsions), oils (solutions of sunscreens in mineral oil, vegetable oil, volatile silicones or esters), gels of a water or water/alcohol character, sprays, and sticks for special applications. Many modern sunscreens are formulated with combinations of raw materials that have absorbance in both the UV-A and UV-B range, as well as physical blocks to augment protection.

The esthetic properties of products are also important, as the product must go on easily, spread uniformly and, if necessary, depending upon usage and claims, establish a waterproof film. In general, there has been a steady increase in SPF value, the spectrum absorbed and blocked, and the requirement for formulas to be waterproof. The US FDA has established regulations and, for instance, does not allow products to be marketed as sunblocks unless they have a proven SPF of not less than 15. It is now common for products to have SPFs greater than 15, with some having SPFs approaching 50. Products with high SPFs require considerable formulation skills to maximize the effect of the sunscreens used, and to make an elegant and often waterproof product with actives that may be used at levels greater than 20%. Increasing levels of sunscreen can also lead to problems of product irritation.

Due to the variation in sunscreen formulations, and the multitude of active ingredients and product forms, a careful study of available literature and the marketplace is essential before meaningful exploratory work with formulation can be initiated. Although there are relatively few manufacturers of sunscreen chemicals, their help can be invaluable.

3.8.4.1 *Suntan lotions* These are generally O/W emulsions containing levels of UV absorber or physical blocks to give a desired SPF. Their formulation may differ from that of conventional moisturizing lotions, in that significant amounts of oil phase are replaced by oil-based UV absorbers. In addition, if suntan lotions are required to be waterproof, they require addition of a water-repellent film former or some other means of ensuring adherence of the active to the skin. The following formula is that of a low SPF non-waterproof suntan lotion.

Typical formulation of an SPF 4 non-waterproof suntan lotion

Part	*Ingredient*	*Wt%*
A	Ethylhexyl *p*-methoxycinnamate	2.0
	Oxybenzone	1.0
	Isopropyl palmitate	6.0
	Stearic acid	3.0
	Cetyl alcohol	1.0
	Glycerol monostearate	1.0
B	Deionized water	71.0
	Sorbitol	3.0
	Carbomer 934 (2% dispersion)	10.0
	Triethanolamine, 99%	1.2
C	Fragrance and preservative	*q.s.*

Procedure: Heat parts A and B separately to 70°C. Add part A to part B with good agitation. Mix to 35°C and add part C. Continue mixing until dispersion is complete.

3.8.4.2 *Waterproof sunblock creams* These formulas incorporate significantly higher levels of sunscreens, and also contain waterproofing agents. The following formula is for an SPF 15 waterproof O/W lotion.

Typical formulation of a waterproof sunblock cream SPF 34 (US)

Part	*Ingredient*	*Wt%*
A	Stearic acid	4.0
	Cetyl alcohol	1.0
	DEA cetyl phosphate	2.0
	PVP eiocosene copolymer	3.0

	Dimethicone	0.5
	Ethylhexyl *p*-methoxycinnamate	7.5
	Oxybenzone	6.0
	Octyl salicylate	5.0
	Octyldodecyl neopentanoate	10.0
B	Deionized water	53.9
	Glycerin	5.0
	Carbopol 940	0.1
C	Deionized water	0.9
	Triethanolamine, 99%	0.1
D	Fragrance and preservative	*q.s.*

Procedure: Add part C to part B and mix until uniform. Add ingredients of part A and mix to dissolve evenly. Hold at 85°C. To form the emulsion, add part A to the mixed parts of B and C at 85°C. Mix and cool to 35°C. Add part D.

3.8.4.3 *Suntan oils* Suntan oils are typically used by individuals who are seeking a tan, rather than by individuals who are seeking protection. They usually have a low SPF and are used not only to afford some protection for a longer exposure, but also to give a glistening appearance to pigmented skins. They are seldom waterproof and incorporate oil-soluble sunscreens into mineral oil, vegetable oils and derivatives, fatty esters, and combinations of the above, at times with volatile silicone added to decrease oiliness. Products should be formulated to give hedonistic values upon application.

Typical formulation of a suntan oil

Ingredient	*Wt%*
Cyclomethicone	64.9
Octyl dimethyl PABA	6.0
Capric/caprylic triglyceride	5.0
Fragrance	0.1
Isopropyl palmitate	24.0

Procedure: Blend all ingredients at room temperature.

3.8.4.4 *Sun gels* These products are often alcoholic/water solutions of alcohol-soluble sunscreens, thickened with gums or polymers. They are cool to apply and represent a small but loyal sub-segment of the category.

Typical formulation of a sunblock gel SPF 15

Ingredient	*Wt%*
Alcohol, 190 proof	75.0
Water	10.4
Octyl dimethyl PABA	7.0

Octyl salicylate	3.0
Oxybenzone	2.0
PPG-15 stearyl ether	1.0
Carbomer 940	0.6
Triethanolamine, 99%	1.0
Fragrance	*q.s.*

Procedure: Blend in order at room temperature. Ensure thorough dispersion of carbomer before adding triethanolamine.

3.8.4.5 *Sprays* Most sprays are solutions of sunscreens in alcohol with or without volatile silicone, or alcohol-soluble emollients. They can either be pump sprays or aerosol sprays, usually propelled by hydrocarbon propellants. Such products are, by nature, light and easily spread on the skin. Since they are in volatile bases, they tend to leave mainly sunscreen actives on the skin. Great care must be taken to ensure spreadability, and enough dispersion to ensure performance as well as maintenance of good esthetics.

Typical formulation of a spray-pump SPF 15

Ingredient	*Wt%*
Cyclomethicone	20.0
Ethyl alcohol, 200 proof	25.0
Diisopropyl adipate	20.0
Isopropyl palmitate	13.5
Octyl dimethyl PABA	7.5
Octyl salicylate	5.0
Oxybenzone	6.0
PVP/hexadecene copolymer	3.0
Fragrance	*q.s.*

Procedure: Blend in order at room temperature. Ensure solution before addition of the next ingredient. Filter well before packaging.

Typical formulation of an aerosol-spray SPF 15

Part	*Ingredient*	*Wt%*
A	Cyclomethicone	20
	Ethyl alcohol, 200 proof	25
	Diisopropyl adipate	20
	Isopropyl palmitate	13.5
	Octyl dimethyl PABA	7.5
	Octyl salicylate	5.0
	Oxybenzone	6.0
	PVP/hexadecene copolymer	3.0
	Fragrance	*q.s.*
B	Propellant	20

Procedure: Blend in order at room temperature. Ensure solution before addition of the next ingredient. Filter well before packaging. Part A is a 100% formula for the concentrate. Fill into an aerosol container and charge with a suitable propellant.

3.8.4.6 *Sticks* Sticks are usually for special application. They are of two general types: (i) for lips; and (ii) for areas that burn easily, such as the nose, ears and cheekbones. The formulas for lips are similar in nature to lip balm or lip-stick formulas, and must be made of cosmetic raw materials that are edible. The products for areas that burn easily are often formulated with pigments that can act as total blocks. Zinc oxide, titanium dioxide, and talc are frequently used in a waxy matrix, not altogether different from the lip-balm products.

Typical formulation of a lip-balm stick SPF 15

Ingredient	*Wt%*
Petrolatum USP	72.5
Paraffin wax	17.0
Octyl dimethyl PABA	7.5
Oxybenzone	3.0
Color	*q.s.*
Fragrance	*q.s.*

Procedure: Melt petrolatum and paraffin wax. Add other materials. Pour into molds or swivel sticks.

Typical formulation of a burn-block pigmented stick

Ingredient	*Wt%*
Titanium dioxide (oil dispersible)	10.0
Petrolatum USP	74
Paraffin wax	16
Color	*q.s.*
Fragrance	*q.s.*

Procedure: Melt petrolatum and paraffin wax. Add other materials. Pour into molds or swivel sticks.

3.8.4.7 *Everyday cosmetics with UV screens* Many everyday cosmetic products, including color cosmetics, are now being marketed with sunscreen actives in their formulation. Recently, protective UV variants of the most common hand and body moisturizer, as well as the most common facial moisturizer, were launched. These are products that contain sunscreens with an SPF of 4 and 15, respectively. It appears that many skin-care products for everyday usage will contain UV screens. It is critical for these products that the formulator takes extra care to ensure that esthetics do not suffer, that the active is stable, and that the products do not cause irritations with the mass of consumers who may be exposed to them on an everyday basis.

3.9 Acne

Acne vulgaris is a disease that can range from occasional blemishes to a devastating, continuing episode leading to permanent scarring and much anguish. This section will discuss the various forms of acne and their treatment, except the treatment of cystic acne, which should only be treated professionally.

There are many misconceptions about acne. It is not caused by diet, lack of cleansing, sleep or social habits [78], but by hormonal control. Both males and females secrete the male sex hormone testosterone. These secretions stimulate the sebaceous glands to secrete sebum [3]. Acne is a disease that often develops due to genetic predisposition during puberty, and typically decreases during adulthood. The years of greatest severity are from 16 to 18 for women and from 18 to 19 for men. In white women, the incidence of clinical acne during the years of greatest severity is 40%; in men, the incidence is 35%. By the age of 40, 1% of men and 5% of women have lesions. The incidence among orientals and blacks is lower [79]. The location of acne lesions is generally on the face, neck, back and chest.

As discussed previously (section 3.2.2.4), the pilosebaceous system contains sebaceous glands, and their associated hair follicles, both of which secrete sebum. Acne generally occurs in sebaceous glands that are either associated with fine or vellous hairs, or not associated with any hair. Humans are the only mammals that routinely develop acne; this is because they have lost their need for a hairy coating over the body, but still have many sebaceous glands without terminal hairs. In the presence of a coat, these glands would serve to wick sebum out on to the hair for the purpose of lubricating and waterproofing terminal hairs, or for the purpose of providing an identifying odor.

3.9.1 *Sebaceous gland*

As indicated previously (section 3.2.2), the sebaceous gland, with or without a hair, has its origin within the dermis. The follicle is, however, lined with stratum corneum, creating the follicle wall. This stratum corneum sheds flattened dead cells into the follicle channel. This, combined with sebum, produces a mixture which, if it does not flow freely to the skin's surface, has the opportunity to plug, get infected and, through the enzymatic action of bacteria, break down certain components of sebum. There are between 400 and 800 sebaceous glands per square centimeter, and it takes only one to become inflamed to cause an acne lesion. All of these are factors in the disease called acne, but none alone is the direct, single cause. All individuals have sebaceous glands, containing sebum and bacteria. Some individuals have severe acne, while others have little or no acne, sometimes only during adolescence. Individuals who develop severe acne are differentiated by the speed at which the epidermis lining the sebaceous gland produces and sheds cells. This is called retention hyperkeratosis. The increase of cells into an area rich with bacteria and lipid materials, including fatty acids, can cause an acne lesion. The acne lesion can be classified, and the acne graded by severity.

3.9.2 *Acne nomenclature*

Closed comedones or whiteheads are impactions of cellular debris, sebum, etc., that build up under the skin in the shape of an inverted 'U' with a nearly closed opening. The development of a comedone takes weeks and can lead to other stages, some of which are more serious. If the comedone matures without irritation, it becomes an open comedone or blackhead. The dark color is due to the accumulation of dead cells and oxidized sebum, or collected melanin, not to dirt. If the comedone, with its rich supply of cellular debris, sebum and bacteria, becomes inflamed it is then a closed, red lump called a papule (commonly a pimple, a spot, or a zit). As the process of inflammation continues, the papule can build up white blood cells or pus, which arise from the body's attempt to fight the irritation. This is known as a pustule, or fluid-filled pimple. If the inflammation continues, the follicle wall ruptures, and the inflammation spreads under the epidermal surface and into the dermis, creating a nodule which is firm, red, but much larger and deeper than a pimple. In severe acne, several follicles become inflamed in an area and form a very large, merged, multiple pustule known as a cyst. Cystic acne is deep and can lead to much loss of tissue and scarring. The entire process, starting from an impacted sebaceous gland, and leading to cystic formation, can take months from start to finish. In severe cystic acne, there is overlap of one episode to another.

3.9.3 *Acne grades*

Acne is identified by four severity grades.

Grade I is a mild acne, which has both open and closed comedones, or whiteheads and blackheads. This level is without redness, and does not spread to adjacent tissue, or form pus. It is the first stage of the disease. Depending on the individual, it may develop into more severe acne, remain at this stage because of treatment, or may actually recede because of treatment or the natural course of the disease.

Grade II consists of closed comedones or whiteheads that are present in a large number. They cause a bumpy surface and do not develop into open comedones or blackheads. This bumpy surface is generally without inflammation but, if left untreated, can lead to papules and pustules.

Grade III consists of both open and closed comedones, as well as comedones that have become inflamed and have progressed to papules and pustules. This grade of acne is severe, involves regions of redness, and can lead to light scarring.

Grade IV, or cystic acne, is most severe. Individuals in Grade IV not only have open and closed comedones, but papules and pustules that have merged together with much irritation, inflammation and eventual loss of tissue with scarring.

3.9.4 *Treatment*

Acne of all but the most mild and transient forms should be treated professionally. Treatment can be either topical or systemic. Various treatments are available, either at retail or on a prescription from a physician, depending on their nature. Topical treatments sold at retail consist of benzoyl peroxide, salicylic acid, sulfur, resorcinol, and a combination of ethyl lactate and lactic acid. Topical preparations available on prescription are retinoic acid, and various antibiotics such as clindamycin and tetracycline. Oral therapy consists of antibiotic therapy with tetracycline or erythromycin and 13-*cis*-retinoic acid. The retinoids have revolutionized the treatment of severe acne. Topical retinoic acid [80] decreases the cohesiveness of the pre-acne follicular epithelium, which leads to decreased microcomedo formation. In addition, it stimulates mitotic activity and causes extrusion of existing comedones. Systemic or oral administration of isotretinoin is used for severe or cystic acne. It has been shown to affect sebum secretion, to have an inflammatory action, and to stimulate the immune system. Its primary function is to reduce over-active sebaceous glands to epithelial buds. It often results in a complete and prolonged remission of cystic acne [81]. This oral therapy has severe side effects, including fetal abnormalities, extreme dry skin, and peeling, especially of the lips. Consequently, it must be carefully controlled and monitored, and is contra-indicated in female patients of child-bearing age, unless adequate steps to prevent pregnancy are taken. Treatment for Grades I, II or III acne is often a combination of oral and topical treatment. Isotretinoin is reserved for Grade IV acne that has been shown to be unresponsive to other treatments.

A wide variety of cleansing products are sold to acne-prone individuals. This includes astringents, exfoliating scrubs, and buffing pads that are used to get the face very clean and to remove surface sebum. The products sold are satisfactory as cleansers, but do not provide adequate therapy for acne. This is because acne is caused by a combination of factors, all of which are operating in the sebaceous gland below the surface of the epidermis and, are not therefore, vulnerable to topical or surface cleansers, be they solvents (alcohol), exfoliants (polyethylene beads) or scrubbing pads of rough synthetic material.

As already stated, acne is a disease that can have devastating results on the appearance and psyche of individuals affected. In all but the most mild and transient cases, it is advisable to seek competent professional advice.

3.10 Liposomes

The use of liposomes in cosmetic products has increased during the 1980s and 1990s. The attractiveness of liposomes to both the cosmetic chemist and marketeer comes from several factors. Liposomes are very small spheres consisting of a lipid bilayer. It is interesting to note that the skin's lipids are also organized into a similar bilayer structure [82]. Liposomes range from 25–500 nm and vary with the lipids used and method of preparation. Liposomes are often formed with phospholipids, from lecithin of either vegetable (soybean) or animal (egg yolk) source [83]. However, liposomes can also be formed with non-ionic surfactant vesicles and other materials [84]. Liposomes have been shown to be effective carriers for a wide variety of materials, both active and cosmetic.

Liposomes formed from phospholipids are notoriously unstable in finished formulations and may be stabilized by the addition of cholesterol to increase bilayer viscosity. The liposome may also be given a positive or negative charge by the addition of dicetyl phosphate or stearylamine [85]. Liposomes have been shown to be capable of carrying drug materials within the sphere. Many drugs are made more available to the body by the use of liposomes [86]. Ghyczy *et al.* [87] have shown that even so-called empty liposomes formed from vegetable source phospholipids are able to increase the moisture content of the skin and improve roughness in comparison with a typical O/W emulsion.

Formulating with liposomes requires care in order to maintain the stability of the liposome. Rules advanced by Brooks and McManus [83] are:

- Avoid surfactants or materials that can act as a co-solvent for the phospholipid.
- Avoid materials with high ionic strength such as salts.
- Avoid solvents such as alcohol that can remove the water that stabilizes the surface of the bilayer.
- Avoid excessive shear or temperature when adding the liposome to your final formula.

Typical formulation of a liposome-containing facial lotion

Part	*Ingredient*	*Wt%*
A	Deionized water	67.7
	Magnesium aluminium silicate	0.3
	Carbomer 941 (2% dispersion)	7.5
	Tetrasodium EDTA	0.1
	Glycerin	3.0

B	Ceteryl alcohol and ceteareth-20	0.8
	Sorbitan stearate	0.5
	Stearic acid	0.5
	Glyceryl stearate	1.0
	Cetearyl alcohol	1.4
	Cetyl alcohol and acetylated lanolin alcohol	0.5
	C 12-15 alcohols benzoate	0.4
	PPG-12 stearyl ether	0.4
	Dimethicone	0.2
	Mineral oil 70 vis	3.0
C	Potassium hydroxide (10% soln)	1.5
D	Soy lecithin and tissue respiratory factors	10.0
E	Preservatives	q.s.
	Fragrance	q.s.

Procedure: Heat parts A and B separately to 75°C. Add part A to part B with good agitation and mix until uniform and creamy. Add part C and begin cooling with mixing. Add part E, then part D (liposome) and mix gently to 30°C.

References

1. Marzalli, F.N. and Maibach, H.I. (1984) Permeability and reactivity of skin as related to age. *J. Soc. Cosmet. Chem.* **35** 95–102.
2. Montagna, W. and Parakkal, P.F. (1974) *The Structure and Function of Skin*, 3rd edn. Academic Press, New York.
3. Hamilton, J.B. (1941) Male hormone substance: a prime factor in acne. *J. Clin. Endocrinol* **1** 570–592.
4. Kligman, A.M. and Shelley (1958) An investigation of the biology of the human sebaceous gland. *J. Invest. Dermatol.* **30** 99–125.
5. Strauss, J.S. and Kligman, A.M. (1958) Pathologic patterns of the sebaceous glands. *J. Invest. Dermatol.* **30** 51–61.
6. Uttley, M. (1972) Measurement and control of perspiration. *J. Soc. Cosmet. Chem.* **23** 23–43.
7. Froebe, *et al.* (1990) Auxillary malodor production: A new mechanism. *J. Soc. Cosmet. Chem.* **41** 173–185.
8. Blank, I.H. (1952) Factors which influence the water content of the stratum corneum. *J. Invest. Dermatol.* **18** 433–440.
9. Kligman, A.M. (1978) Regression method for assessing the efficacy of moisturizing. *Cosmet. Toilet.* **93** 27–35.
10. Boisits, E.K. *et al.* (1989) The refined regression method. *J. Cutaneous Aging and Cosmetic. Dermat.* **1** 155–163.
11. Packman, E.W. and Gans, E.H. (1978) Topical moisturizers: quantification of their effect on superficial facial lines. *J. Soc. Cosmet. Chem.* **29** 79–90.
12. Packman, E.W. and Gans, E.H. (1978) The panel study as a scientifically controlled investigation: moisturizers and superficial facial lines. *J. Soc. Cosmet. Chem.* **29** 91–98.
13. Jansen, L.H. *et al.* (1974) Improved fluoresence staining technique for estimating turnover of the human stratum corneum. *Brit. J. Dermatol.* **90** 9–12.
14. Cook, T.H. (1980) Profilometry of skin: a useful tool for the substantiation of cosmetic efficacy. *J. Soc. Cosmet. Chem.* **31** 339–359.

15. Dorogi, P.L. and Zielinski, M. (1989) Assessment of skin conditions using profilometry. *Cosmet. Toilet.* **104** 39–44.
16. Hargens, C.W. (1981) The gas-bearing electrodynamometer (GBE) applied to measuring mechanical changes in skin and other tissues. *Bioeng. Skin* **1** 133–122.
17. Christensen, M.S. *et al.* (1977) Viscoelastic properties of intact skin: instrumentation, hydration effects and the contribution of stratum corneum. *J. Invest. Dermatol.* **69** 282–286.
18. Nucci, J. (1989) Measurement of skin surface viscoelasticity for claim support and product development. *Cosmet. Toilet.* **104** 47–51.
19. Aubert, L. *et al.* (1985) An *in vivo* assessment of the biomechanical properties of human skin modifications under the influence of cosmetic products. *Int. J. Cosmet. Sci.* **7** 51–59.
20. Berube, G.R. *et al.* (1971) Measurement *in vivo* of transepidermal moisture loss. *J. Soc. Cosmet. Chem.* **22** 361–368.
21. Berube, G.R. and Berdick, M. (1974) Transepidermal moisture loss II. The significance of the use thickness of topical substances. *J. Soc. Cosmet. Chem.* **25** 397–406.
22. Berube, G.R. and Tranner, F. (1979) Transepidermal moisture loss III. An *in vitro* approach. *J. Soc. Cosmet. Chem.* **30** 181–190.
23. Steitz, J.C. and Spencer, T.S. (1982) The use of capacitative evaporimetry to measure effects of topical ingredients and transepidermal water loss (TEWL). *J. Invest. Dermatol.* **78** 351–356.
24. Potts, R.O. (1985) *In vivo* measurement of water content of the stratum corneum using infrared spectroscopy: a review. *Cosmet. Toilet.* **100** (10) 27–31.
25. Salter, D.C. (1987) Instrumental methods of assessing skin moisturization. *Cosmet. Toilet.* **102** 103–109.
26. Rieger, M.M. and Deem, D.E. (1974) Skin moisturizers, methods for measuring water regain, mechanical properties, and transepidermal moisture loss of stratum corneum. *J. Soc. Cosmet. Chem.* **25** 239–252.
27. Nishiyama, S. *et al.* (1983) A study on skin hydration with cream, influence of its components on skin hydration. *J. Soc. Cosmet. Chem. Japan* **16**(2) 136–143.
28. Nishiyama, S. *et al.* (1983) A study on skin hydration with W/O type cream. *J. Soc. Cosmet. Chem. Japan* **17**(2) 116–120.
29. Quattrone, A.J. and Laden, K. (1976) Physical techniques for assessing skin moisturization. *J. Soc. Cosmet. Chem.* **27** 607–623.
30. Kligman, A.M. *et al.* (1986) Sebutape: A device for visualizing and measuring human sebaceous secretion. *J. Soc. Cosmet. Chem.* **37** 369–374.
31. Frosch, P.J. and Kligman, A.M. (1979) The soap chamber test: A new method for assessing the irritancy of soaps. *J. Am. Acad. Dermatol.* **1** 35–41.
32. Murahata, R.I. *et al.* (1986) The use of transepidermal water loss to measure and predict the irritation response to surfactants. *Int. J. Cosmet. Sci.* **5** 225–232.
33. Lukacovic, M.F. *et al.* (1988) Forearm wash test to evaluate the clinical mildness of cleansing products. *J. Soc. Cosmet. Chem.* **39** 355–366.
34. Strube, D.D. *et. al.* (1989) The flex wash test: A method for evaluating the mildness of personal wash products. *J. Soc. Cosmet. Chem.* **40** 297–306.
35. Frosch, P.J. (1982) Irritancy of soaps and detergent bars. In *Principles of Cosmetics for the Dermatologist* Frosch, P. and Horowitz, S.N. (Eds). Mosby, St. Louis, pp. 5–12.
36. CTFA (1991) *International Cosmetic Ingredient Dictionary*, 4th Edn. CTFA Inc., Washington, D.C.
37. US Phamacopea XXI (1970) Mack Publishing Co., Easton, PA, p. 143.
38. US Patents 2, 894, 912; 3, 376, 229; 3, 394, 155; 3, 420, 858.
39. US Patents 4, 132, 679; 4, 139, 485; 4, 369, 134; 4, 536, 519; 4, 758, 376.
40. Bettley, F.R. (1960) Some effects of soap on the skin. *Brit. Med. J.* **5187** 1675–1679.
41. Prottey, C. *et al.* (1972) The effect of soap upon certain aspects of skin biochemistry. *J. Soc. Cosmet. Chem.* **24** 472–492.
42. Loden, M. and Bengtsson, A. (1990) Mechanical removal of the superficial portion of the stratum corneum by a scrub cream: methods for the objective assessment of the effects. *J. Soc. Cosmet. Chem.* **41** 111–121.
43. Komp, B. (1987) Skin compatibility tests—Importance in skin cleansing product development. *Cosmet. Toilet.* **102** 89–98.
44. Middleton, J.D. and Allen, B.M. (1973) The influence of temperature and humidity on stratum corneum and its relation to skin chapping. *J. Soc. Cosmet. Chem.* **24** 239–243.

45. Idson, B. (1973) Water and skin. *J. Soc. Cosmet. Chem.* **24** 197–212.
46. Blank, I.H. (1953) Further observation on factors which influence the water content of the stratum corneum. *J. Invest. Dermatol.* **21** 259–271.
47. Laden, K. and Spitzer, R. (1967) Identification of a natural moisturizing agent. *J. Soc. Cosmet. Chem.* **18** 351–360.
48. Rieger, M.M. and Deem, D.E. (1974) Skin moisturizers II. The effects of cosmetic ingredients on human stratum corneum. *J. Soc. Cosmet. Chem.* **25** 253–262.
49. Bissett, D.L. and McBride, J.F. (1984) Skin conditioning with glycerol. *J. Soc. Cosmet. Chem.* **35** 345–350.
50. Middleton, J.D. (1974) Development of a skin cream designed to reduce dry and flaky skin. *J. Soc. Cosmet. Chem.* **25** 519–534.
51. US Patents 4, 603, 146, and 4, 888, 342.
52. US Patents 3, 879, 537; 3, 920, 835; 4, 105, 783; 4, 234, 599; 4, 363, 815; 4, 380, 549; 4, 518, 789; 5, 091, 171; 5, 093, 360; 5, 258, 391; 5, 385, 938.
53. Hayflick, L. and Moorehead, P.S. (1961) The serial cultivation of human diploid cell strains. *Exptl. Cell Res.* **25** 585–621.
54. Pugliese, P.T. (1987) Concepts in aging and the skin. *Cosmet. Toilet.* **102** 19–44.
55. Ryan, J.C. (1966) The microcirculation of the skin in old age. *Gerontol. Clin.* **8** 327–337.
56. Plewig, G. and Kligman, A.M. (1972) Follikulare pusteln im kaliumjodid-epicutantest. *Arch. Dermatol. Forsch.* **242** 137–152.
57. Gilchrest, B.A. (1991) *Physiology and Pathophysiology of Ageing Skin in Physiology, Biochemistry and Molecular Biology of the Skin.* Goldsmith, L.A. (Ed.). Oxford University Press, pp. 1425–1444.
58. Wright, E.T. and Shellow, W.V. (1973) The histopathology of wrinkles. *J. Soc. Cosmet. Chem.* **24** 81–85.
59. Counts, D.F. *et al.* (1988) The effect of retinyl palmitate on skin composition and morphometry. *J. Soc. Cosmet. Chem.* **39** 235–240.
60. Kligman, L. (1986) Effects of all trans-retinoic acid on the dermis of hairless mice. *J. Am. Acad. Dermatol.* **15** 779–785.
61. Braverman, I.M. and Fonferko, E. (1982) Studies in cutaneous ageing I: the elastic fiber network. *J. Invest. Dermatol.* **78** 434–443.
62. Smith, J.G. *et al.* (1962) Alterations in human dermal connective tissue with age and chronic sun damage. *J. Invest. Dermatol.* **39** 347–350.
63. Kligman, L.H. *et al.* (1983) Sunscreens promote repair of UV radiation induced dermal damage. *J. Invest. Dermatol.* **81** 94–102.
64. Smith, W.P. (1994) Hydroxy acids and skin aging. *Cosmet. Toilet.* **109** 41–48.
65. Abamba, G. (1993) Skin preparation of Poucher's Perfumery, *Cosmetics and Soaps.* Volume 3, Poucher, W.A. and Butler, H. (Eds). pp. 335–392.
66. McNamara, S.H. (1987) Performance claims for skin care cosmetics. *Cosmet. Toilet.* **102** 81–88.
67. Koller, L.R. (1965) *Ultraviolet Radiation*, 2nd edn. John Wiley and Sons Inc., New York, p. 107.
68. Wier, K.A. *et. al.* (1971) Nuclear changes during ultraviolet light-induced depression of ribonucleic acid and protein synthesis in human epidermis. *Lab. Invest.* **25** 451–456.
69. Everett, M.A. *et al.* (1966) Penetration of epidermis by ultraviolet rays. *Photochem. Photobiol.* **5** 533–542.
70. Olson, R.L. *et al.* (1966) Effect of anatomic location and time on ultraviolet erythema. *Arch. Derm.* **93** 211–215.
71. Ibson, A. and Blank, H. (1967) Testing drug phototoxicity in mice. *J. Invest. Dermatol.* **49** 508–511.
72. Kaidby, K.H. and Kligman, A.M. (1978) Sunburn protection by longwave ultraviolet radiation induced pigmentation. *Arch. Derm.* **114** 46–48.
73. Editorial (1991) Protecting man from UV exposure. *The Lancet* **337** 1258–1259.
74. Fitzpatrick, T.B. and Pathak, M.A. (Ed.) (1974) In *Sunlight and Man.* University of Tokyo Press, Tokyo.
75. National Institutes of Health (1989) *Sunlight, Ultraviolet Radiation and the Skin.* NIH Consensus Development Conference Statement **7,** No. 8, pp 1–10. NIH, Bethesda, MD.
76. Shaath, N.A. (1987) Encyclopedia of UV absorbers for sunscreen products. *Cosmet. Toilet.* **102** 21–39.

77. Groves, G.A. (1980) Sunburn and its prevention. *Aust. J. Dermatol.* **21** 115–141.
78. Fulton, J.E. and Black, E. (1983) *Dr. Fulton's Step-By-Step Program for Clearing Acne.* Harper and Row, New York.
79. Rook, A. *et al.* (1986) *Textbook of Dermatology.* Blackwell Scientific, Oxford, England.
80. Frank, S.B. (1971) *Acne Vulgaris.* C.B. Thomas, Springfield, Ill.
81. *Physicians Desk Reference* (1986) Medical Economics, Oradell, N.J.
82. Elias, P.M. (1991) Epidermal barrier function: intercellular lamellar lipid structures, origin, composition and metabolism. *J. Contr. Release* **15** 199–208.
83. Brooks, G.J. and McManus, R.C. (1990) Finding new uses for liposomes in cosmetics. *Inform* **1** 891–896.
84. Abate, K. (1993) Non-phospholipid liposomes in cosmetics. *Soap/Cosmet/Chem Specialities* **69** (S) 37–40.
85. Perrier, P. and Redziniak, G. (1989) Liposomes in cosmetics. *Soap, Perfumery Cosmet.* **62** (2) 29–30.
86. Talsma, H. and Crommelin, D.J.A. (1992) Liposomes as drug delivery systems, part 1: preparation. *Pharm. Tech.* **16** 96–106.
87. Ghyczy, M. *et al.* (1992) Effect and production of liposomes from vegetable phosphatidylcholine (lecithin) in cosmetics. *Society of Cosmetic Chemists Annual Meeting*, Dec. 3–4, New York.

4 Color cosmetics

J. CUNNINGHAM

4.1 Introduction

The primary reason a pigmented product is purchased is for the color it imparts. Other beneficial characteristics will be appreciated by the consumer, but short of total discomfort, these will be sacrificed if the color is 'just right'. This color selection will be based on the purchaser's general complexion, and on what he or she perceives as making him or her look attractive, much the same criteria that dictate the purchase of clothing and other fashion accessories. In a nutshell, then, when speaking of colored cosmetics, it is fashion first and science second.

The role of cosmetic formulators is important. Product safety must be ensured, and creative work on the secondary benefits such as application and texture can often mean the difference between the success or failure of the product in a competitive environment. For this reason, each product category treated in this chapter is prefaced with a list of consumer wishes after the shade has been chosen.

4.2 Lip color

Lip products fall into three main categories; (i) lipsticks; (ii) glosses; and (iii) liners. Lipsticks, as the name suggests, are molded in the form of a crayon and are dispersed from a swivel-up case (Figure 4.1). Lip glosses, too, can be marketed in stick form, but the majority of products in this category are semi-solid and dispensed from a tube with a sponge-type applicator. The majority of liners are sold in pencil form. The share of lip make-up (US dollars) is shown in Figure 4.2. Total sales in the US and Europe are estimated to be around 1.5 billion dollars per year.

4.2.1 *Lipsticks*

4.2.1.1 *Consumer expectations* A lipstick should apply easily, give good color coverage, yet look natural. It should feel moist, not dry, and should not bleed (flow) into the lines around the mouth. It should not change color

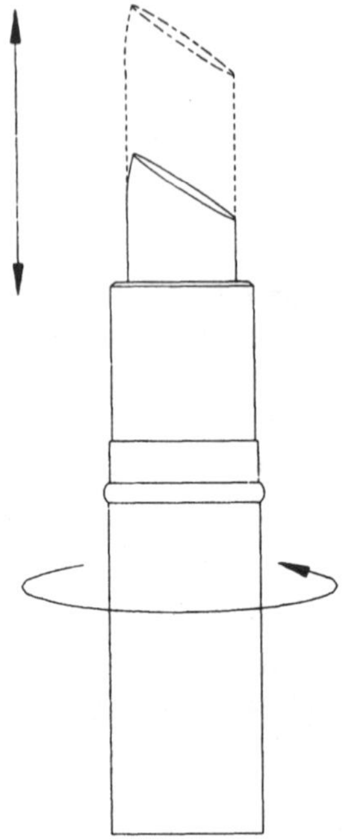

Figure 4.1 Standard lipstick case.

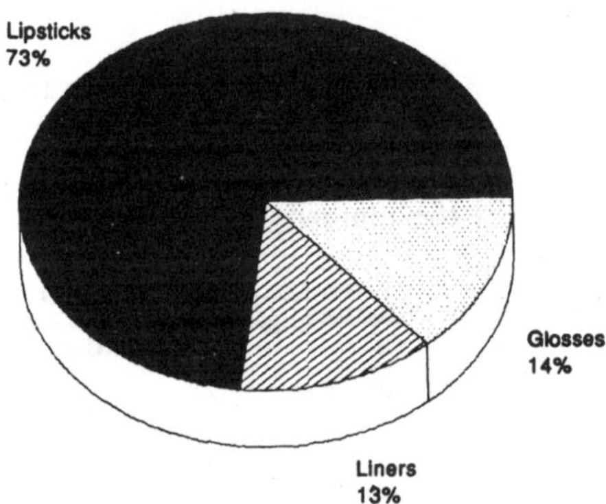

Figure 4.2 Share of lip make-up sales (dollars) in the US and Europe during 1990.

during wear and should have an acceptable flavor fragrance. It should keep lips from cracking and peeling, and last a minimum of three to four hours. A wide range of colors, in both cream and frost (or pearl), with varied shine levels from matte (dull) to cream (soft shine) to glossy (high shine) is essential.

4.2.1.2 *Formulation* Lipsticks are mixtures of waxes, oils and pigments. Varying the ratio of these ingredients determines the final product characteristics. In the case of lipstick, a high wax, low oil, high pigment formulation generally results in a long-wearing product that sacrifices a varying degree of gloss and texture. Meanwhile a lower wax, higher oil base will apply more smoothly, have greater shine, but will not wear as long. More recently several lipsticks have been introduced which incorporate a volatile silicone or hydrocarbon. These formulations apply smoothly but soon dry down to leave a matte and rather dry film on the lip. These newer products, wear longer and are more resistant to the common bane of the lipstick consumer 'transfer of the lipstick to the coffee cup'. Consequently, the claims that are to be made for a product must be known before embarking on the formulation process. In addition, if it is decided to incorporate a volatile into the lipstick it is important that an airtight package be sourced, and it is just as critical that the packaging materials be occlusive to the passage of the chosen solvent. The following is an example of a basic lipstick formulation. Other examples are shown in Table 4.1.

Trade name	*Manufacturer:*	*Wt%*
Indopol H-100	Amono Chemical	6.00
Eutanol G	Henkel, Inc.	7.50
Candelilla wax	Strahl & Pitsch	9.00
Multiwax ML-445	Sonneborn	3.00
Satulan	Croda	7.50
Pigments phase (iron oxides and lakes)		8.00
Pearlescent pigments		6.00
Castor oil		52.00
Fragrance (if desired)		
Preservatives and antioxidants		1.00

Table 4.2 gives examples of formulations which incorporate volatile solvents. Recent estimates suggest that the market share of such formulations exceeds 10% in the US.

Candelilla wax Candelilla serves two purposes in this formulation. Primarily, it is a hard wax which gives the stick rigidity. Its high melting point aids the overall high temperature stability of the finished stick. In most formulations, carnauba wax can be used either as a substitute, or in combination with candelilla.

Multiwax ML455 (microcrystalline wax) Microcrystalline wax is used to give the stick structural integrity. Like candelilla, it has a high melting point (approx. 80°C), but is softer and less brittle. This improves the overall tensile strength of the stick. Ozokerite and cerasynt waxes are often used to achieve the same end.

Indopol H-100 (polybutene) Indopol is a viscous semi-solid and is primarily used to increase the gloss of the stick and the lay down (imparting the color film). It also serves to increase the viscosity of the oil phase, which

Table 4.1 Lipstick formulas (wt%) by Mearl Corporation

	Soft		Medium		Firm	
	A	B	A	B	A	B
Carnauba wax	12.0	11.6	13.7	13.3		
Candelilla wax					11.5	11.5
Beeswax			8.5	8.5	15.0	15.0
Hydrogenated lanolin (Lipolan)	13.0	13.0				
Pur-cellin wax	15.0	15.0				
Isopropyl lanolate (Amerlate P)					15.0	15.0
Castor oil			32.8	10.7	16.0	
Isostearyl alcohol (Adol 66)					14.0	14.0
Lauryl lactate (Ceraphyl 31)	16.0	8.0				
Diisopropyl adipate (Ceraphyl 230)					13.5	7.0
Decyl oleate (Ceraphyl 140-T)	15.0	7.5				
Butyl stearate	14.0	7.4				
Isopropyl myristate			30.0	30.0		
Flamenco Superpearl 100[a]	15.0		15.0		15.0	
Flamenco Superpearl CC[a]		37.5		37.5		37.5
Antioxidant	*q.s.*	*q.s.*	*q.s.*	*q.s.*	*q.s.*	*q.s.*
Antimicrobial	*q.s.*	*q.s.*	*q.s.*	*q.s.*	*q.s.*	*q.s.*
Perfume/flavor	*q.s.*	*q.s.*	*q.s.*	*q.s.*	*q.s.*	*q.s.*

[a] Mearl Corporation

Table 4.2 Long wearing lipstick formulas (wt%)

Trade name	Manufacturer	A	B
Liponate TDTM	Lipo	12.4	13.0
Synthetic Wax	Dura	12.3	13.0
Paraffin	Ross	6.5	6.5
Ozokerite	Strahl & Pitsch	5.0	5.0
Emerest 2452	Henkel	4.7	5.0
Silicone Fluid	G.E.	6.7	5.0
Abil Wax 9800	Goldschmidt	5.3	5.0
PEG-4	Lipo	1.4	2.0
Indopol H-100	Chem Central	4.6	–
Masil SFVV	Mazer	30.0	20.0
Permethyl 99A	Presperse		10.0
Pigments		10.0	12.0
Preservatives and fragrance		*q.s.*	*q.s.*

in turn reduces the potential of color bleeding into the fine lines around the mouth.

Eutanol and satulan (*octyldodecanol and hydrogenated lanolin*) Both these items are used to increase the slip and moisturizing feel on the lips. They can be replaced by a myriad of similar emollients, such as mineral oil, isopropyl myristate, etc.

Castor oil Castor oil is the backbone of most lipstick formulations. It is an excellent pigment dispersing aid and imparts a superb creamy, moisturizing feel on the lips.

Pigments In lipsticks, color concentrations range between 2 and 10%. Bismuth oxychloride and titanium-coated micas are used to impart a pearlescent look (frost).

Fragrance Fragrance/flavor is often added, usually at less than 1%, to improve the taste of the product on the lips.

4.2.1.3 *Manufacture* Lipsticks are generally manufactured in four stages:

(i) Pigment milling
(ii) Combination of pigment phase into the base
(iii) Molding
(iv) Flaming

Milling Milling is required to break up pigment agglomeration rather than to reduce particle size. There are no hard and fast rules on how this can be accomplished, but today most people opt for the three-roll mill. Carborundum and colloid mills are also used. The pigment must first be wetted with one of the liquid constituents of the base. It is wise, therefore, to incorporate a good milling media into the formula recipe. The majority of formulations on the market today use castor or lanolin oils for this purpose. To determine the pigment/oil ratio best for milling, a good rule of thumb is to use two parts of oil for one part of pigment, but a formulation particularly high in organic pigments may necessitate an increase in the oil level. The pigment/oil mixture is usually blended together in a Cowles Dissolver or any high-shear mixer that can handle high viscosity materials. The resultant paste can then be passed through the mill. The blend is milled until it reaches a satisfactory particle size (usually 20 μm) on a standard paint gauge. Often two passes through the roller mill are enough to accomplish this but, depending on the shade, three passes are sometimes required.

Combination of pigment phase into the base The batching of the lipstick itself is a simple procedure. The wax and oil phase are usually melted in a steam-jacketed kettle equipped with a single propeller agitator. Following this, the milled pigment phase is added.

Two points should be considered during manufacturing. The first is that a conscious effort should be made to minimize the incorporation of air into the mass. Air in the bulk is difficult to remove and will lead to unsightly pinholes in the stick during molding, resulting in slow production and increase rejection rate. The second point is that it is often wise to complete a small trial batch (as little as a 100 g) on the laboratory bench using the pigment shade phase before preparing the wax/oil/pigment blend on full scale. The reason for this is that color mismatch will show up on the laboratory trial batch and, if major color adjustments are needed, it is better to make the changes to the pigment/oil shade blend rather than the finished batch.

It should be noted that in the case of the formulations containing cyclomethicone (Masil SFVV) or Permethyl 99A (Isododecane) several modifications to this procedure will be necessary. First of all, because of the fugitive nature of the volatiles it is best to add these materials at the last possible moment. Some manufacturers will add the volatiles just before the finished batch is dropped into storage, while others delay the addition until the molding process itself. Then again, and of course, this will vary with the specific formulation, since owing to the lack of a sufficient quantity of a good milling media, it probably will be necessary to mill the pigment into the total base (minus the volatiles). Ultimately manufacturers prefer to fill these formulations directly into the case rather than going through the intermediate block molding routine.

Molding The most common method for molding lipsticks is by use of verticle split molds. The split itself runs down the center of each row of sticks (Figure 4.3). Most lipstick formulations mold well between 75 and 85°C. Pre-heating the mold to around 35°C avoids the formation of 'cold marks' on the stick, and holding the mold at a slight angle to the verticle is a technique often used to avoid air entrapment. The mass should not be poured directly into the cavities for the same reason. Rapid cooling after the mass is poured into the mold is important since it leads to a smaller, more uniform crystalline structure, which, in turn, leads to better stability and gloss. Once cooled, the molds can be split open and the sticks ejected on to trays or some other suitable storage container until they are ready for flaming.

Flaming Flaming is a procedure for passing the molded stick through the flame of a gas burner, or series of burners, to increase the surface gloss.

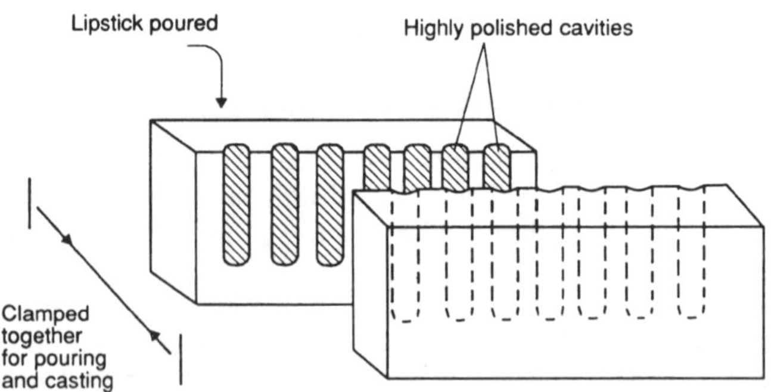

Figure 4.3 Lipstick split mold.

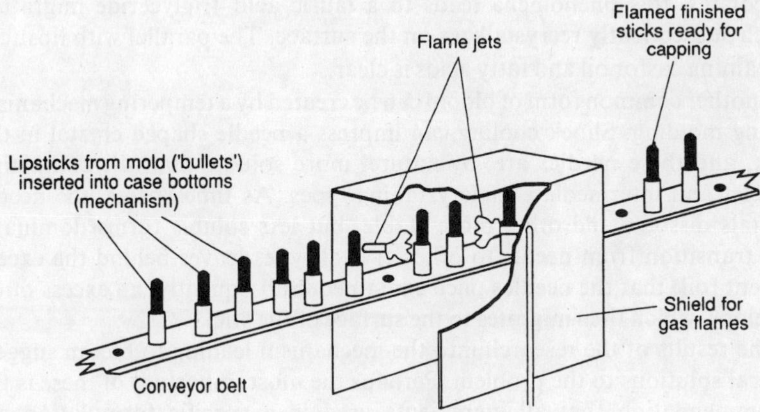

Figure 4.4 Lipstick flame line.

The flames are adjusted to a level just hot enough to just melt the surface of the stick (Figure 4.4).

Although only one method of lipstick manufacture has been described here, several high-speed, automatic molding machines, which produce high quality lipsticks are available. For more information, the reader is referred to the manufacturers.

4.2.1.4 *Blooming* Under certain conditions, particularly high ambient temperatures, certain liquid fractions in the formulation, including the higher melting point portions in solution, can migrate through the lipstick manifesting as unsightly oil droplets on the surface. This phenomenon is known as 'blooming' or 'syneresis'.

What happens next depends on the subsequent environmental conditions in which the lipstick is stored. Under some conditions the oil can be reabsorbed into the stick leaving the product with an unesthetic blotchy appearance, while on other occasions the oil exudate and its solute can recrystallize giving the composition a mottled salty look. These problems are shared with the manufacturers of chocolate, and the confectionery industry has contributed much research into the cause of this manifestation.

One result of this research indicates that the root cause of bloom can be traced to an incompatibility between individual ingredients in the formulation. The studies also show that bloom can be exacerbated by processing and, ultimately, the storage conditions of the finished product. Although these statements may seem to be stating the obvious, the real story is extremely complex. For example, many waxes, oils and fatty acids commonly used in lipsticks are completely miscible in the liquid phase, tolerate each other in the solid phase, but are incompatible in the eutectic state. In chocolates, this phenomena leads to a lauric acid–triglyceride migration which subsequently recrystallizes on the surface. The parallel with lipsticks containing castor oil and fatty acids is clear.

Another common form of bloom can be created by a tempering mechanism during molding. Shock cooling can impress a needle shaped crystal in the stick, and these needles are, by nature, more soluble than the competing platelet and intermediate malcrystalline types. As time passes, the needle crystals dissolve and other more stable, but less soluble forms dominate. The transition from needle to other crystal types leaves behind the excess solvent (oil) that the needles once contained. Subsequently, an excess of oil develops, which then migrates to the surface of the stick.

The results of the research into the mechanism leading to bloom suggest several solutions to the problem. Perhaps the most important of these is the recommendation that all ingredients used in a specific formulation are compatible at the ratios in which they are incorporated. It is suggested that no assumptions be made in this regard. Most formulations are developed by trial and error with aesthetic considerations being the primary goal and stability being considered only after this has been achieved. The rules can be ignored if required and sometimes the aesthetic qualities might be worth a minor stability problem. However, it is better that this be a choice rather than an accident.

Another way to minimize the bloom is to incorporate anti-blooming agents into the formulation. The chocolate industry has found that sorbitan tristearate, at levels up to 2.5% has yielded excellent results. Lactic acid esters of monoglycerides have also shown promise. These compounds have also been employed successfully in cosmetics. Meanwhile, ascorbic palmitate, at levels of around 1% and octyldodecanol, at 5% and above, have also been found to reduce the propensity to bloom.

This discussion on blooming only touches upon some of the more important issues involved in this most complex subject. As mentioned, most of the research work on this area has taken place in the confectionery industry, and although parallels can be made with cosmetics, it should be pointed out that chocolate formulations are far simpler than the multi-waxed lipstick formulation. In addition, no attempt has been made here to discuss 'super cooling', which maintains waxes in liquid form well below their normal solidification point, thus creating the eutectic effect mentioned earlier.

Finally, literature is available on the intercompatibility of wax systems and can be obtained from the raw material suppliers or from several publications on industrial waxes [1–3].

4.2.2 *Lip glosses*

4.2.2.1 *Consumer expectations* A lip gloss should apply easily and provide a wet, shiny look. It should have a transparent color coverage (sheer) and feel very moist, never dry, on the lips. The fragrance/flavor may be higher than in lipstick. As with lipstick, lip gloss should come in a wide variety of colors, both cream and frost (pearl).

4.2.2.2 *Formulation* Like lipstick, lip gloss consists of a mixture of waxes, oils and pigments. The major difference is that gloss and transparent coverage are the key properties required and, consequently, the oil to wax ratio is higher and the pigment level is lower. A typical lip gloss formulation is shown below.

Trade name	*Manufacturer:*	*Wt%*
Deodorized lanolin	Amerchol	44.40
Indopol H-100	Amoco	32.00
Beeswax	Strahl & Pitsch	6.00
Amerlate P	Amerchol	23.00
D&C Red #7	W, C, D	3.00
D&C Red #27	Sun Chemical	1.50
Yellow iron oxides	W, C, D	0.10
Fragrance (to taste)		
Preservatives		

Deodorized lanolin Lanolin, like castor oil, is an excellent pigment dispersant. It also has an excellent feel on the lips, and imparts an excellent gloss.

Indopol H-100 Details of this are given in section 4.2.1.2.

Beeswax Beeswax is used to build viscosity and lengthen the wear.

Amerlate P (*lanolin oil*) Imparts excellent slip and moisturizing feel to the lips.

Pigments Pigments range from 0 to 5% in most lip glosses. As there is desire for transparency, inorganic pigments are usually used at low levels.

4.2.2.3 *Manufacture* Lip gloss manufacture is identical to that of lipsticks, but, if the product is not in stick form, the molding and flaming steps are omitted.

4.2.3 *Lip liners*

4.2.3.1 *Consumer expectations* Liners should have high pigment coverage to accent the line of the lip, and should be firm enough not to run into the lines around the lip. A wide range of colors is desirable, and the products must be suitable for drawing a thin, clearly defined line around the periphery of the lips.

4.2.3.2 *Formulation* As with the other lip products, lipliners are mixtures of waxes, oils, and pigments. Emolliency is not the necessity it is with lipsticks and lip glosses, but pigment cover is essential. For this reason, the wax/pigment level is elevated and the oil level lowered.

4.2.3.3 *Manufacture* Although a small segment of the lipliner market uses small diameter molded crayons, the majority of products sold are in the form of woodcased pencils, most of which have extruded 'leads.' Bulk manufacturing usually takes place in a steam-jacketed kettle, but, due to the heavier pigment/wax concentration the bulk is more pasty and requires a kneading-type action to ensure proper mixing. Due to the high pigment concentration, pre-milling of pigment is often impractical and it is best to roller mill the whole batch as the final step. Roller milling helps to remove any entrapped air.

Examples of extruded bulk lipliner formulations

Trade name	*Manufacturer:*	*Wt%*
Japan wax	Robeco Chemical	43.41
S70-XXH	Durkee Foods	7.66
EE hard fat flakes	Swift & Company	3.91
Lanolin, anhydrous	Croda	8.81
Bentone 34	N.L. Industries	8.26
D&C Red #27	W, C, D	8.00
D&C Red #6	W, C, D	5.81
D&C Red #7	W, C, D	3.07
Titanium dioxide	W, C, D	10.63
Preservatives		0.44

Trade name	*Manufacturer:*	*Wt%*
Paramount X	Durkee	46.46
EE hard fat flakes	Swift & Company	2.70
Japan wax	Robeco	14.70
'B' wax	Int'l Wax Refining	7.70
Bentone 34	N.L. Industries	2.80
D&C Red #7	Sun Chemical	2.22
Flamenco Superpearl 100	Mearl Corporation	21.16
D&C Yellow #6	W, C, D	1.69
Cosmetic brown C33–115	Sun Chemical	0.11
Manganese violet 7101	W, C, D	0.02
Preservatives		0.44

Extrusion The liner bulk is loaded into a cylinder and forced by a piston through a narrow opening of desired shape and size of the crayon (usually round). At this point, the crayon is generally soft and flexible, and must undergo a tempering phase of up to five days before it will reach its optimum crystalline form. The sticks are then placed in wooden slats, and shaped, pointed, painted and sharpened.

4.3 Nail polish

4.3.1 *Consumer expectations*

Nail polish should be easy to apply, giving a good cover of the nail with two coats, and should not streak or apply unevenly. The polish should dry quickly to the touch and leave a high-shine finish film that will not stain the nail when removed. The film should be sufficiently hard so it will not mar easily and, during the wear cycle should not chip or peel. Durability and wear should last for 4 to 5 days. It is particularly important that nail polish is available in a wide range of colors, cream and frost (pearl), transparent and opaque. The polish is mostly dispensed from a bottle with a brush applicator (Figure 4.5).

Sales of nail polish in 1990 were approximately \$660 million in the US and \$365 million in Europe.

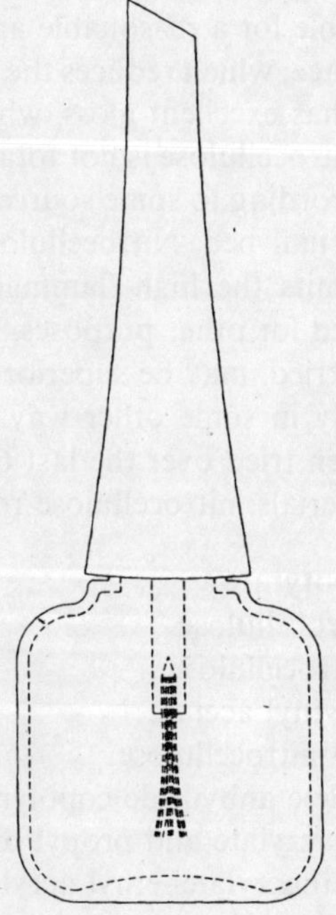

Figure 4.5 Nail polish bottle.

4.3.2 *Formulation*

Nail polish, simply stated, consists of pigments suspended in a volatile solvent in which a film former or film formers have been dissolved. On application, the solvent evaporates leaving the color and film former on the nail. A standard nail polish formulation is as follows:

	Wt%
Nitrocellulose 1/2 sec	15.00
IPA	4.50
Polyester resin	8.00
Ethyl acetate	28.00
Butyl acetate	40.00
Bentone	1.00
Camphor	3.00
Color	0.50

Nitrocellulose Although the choice of film former is not absolutely universal, the vast majority of formulations use nitrocellulose. This is not accidental for, although it is not perfect, nitrocellulose has a blend of all the qualities necessary to make a successful product: (i) nitrocellulose films are quite hard yet remain flexible for a reasonable amount of time; (ii) the films adhere well to the nail surface, which reduces the occurrence of chipping and peeling; (iii) nitrocellulose has excellent gloss, which is extremely important to the consumer; and (iv) nitrocellulose is not totally occlusive to the passage of water and air, which, according to some sources, eliminates the possibility of fungal infections in the nail bed. Nitrocellulose is usually supplied as a solution (lacquer). This limits the high flammability risk of the so-called dampened material supplied for other purposes. Other polymeric materials, of which many have been tried, may be superior in one specific attribute of nitrocellulose but fail badly in some other way. Listed below are some of the materials that have been tried over the last 60 years. Despite optimistic claims made for these materials, nitrocellulose remains supreme.

- Cellulose acetate isobutyrate
- Methacrylate and nitrocellulose
- Ethylcellulose and nitrocellulose
- Nitrostarch and polyvinyl acetate
- Polyvinyl acetate and nitrocellulose
- Alkyl vinyl ether–maleic anhydride copolymer
- Methyl or ethyl methacrylate and propyl or butyl methacrylate
- Acrylamide–butyl methacrylate–ethyl acrylate copolymer
- Polyvinyl butyral
- *cis*-4 Cyclobexane, 1,2 anhydrodicarboxylic acid–2,2 dimethyl 1-3 propane diol copolymer
- Siloxanylalkyl ester–acrylate–methacrylate copolymer
- Polyacrylic and polyethyleneimine crosslinked with polyepoxide
- Cellulose acetate
- Cellulose acetate propionate

Isopropyl alcohol (IPA) Nitrocellulose is generally shipped 70% active in IPA, therefore IPA is present incidently.

Butyl and ethyl acetate Butyl and ethyl acetates are solubilizers for the nitrocellulose. By modifying the ratio of the two solvents, the drying time of the film on the nail can also be modified. Since ethyl acetate is approximately four times more volatile than butyl, the higher the level of ethyl, the faster the film will dry. This is often a delicate balance in nail polish formulations—if the film dries too fast it can cause streaky application and low gloss on the nail, while if it dries too slowly it is an obvious inconvenience to the consumer.

Camphor Camphor is a plasticizer for nitrocellulose. Its inclusion increases the flexibility of the film, reducing brittleness, thus reducing the potential for film chipping. Dibutyl phthalate is also commonly used as a plasticizer.

Bentone (*Quaternium-18 bentonite*) Bentone is used in modern formulations as a suspending agent. Being thixotropic, its 'at rest' viscosity is relatively high, making it ideal for suspending the pigments. On application, through the action of a brush or when the bottle is shaken, the viscosity drops, allowing the user to apply an even film to the nail.

Polyester resin Polyester resin is used as an auxillary resin to nitrocellulose. Its addition improves hardness and increases the gloss of the film. The following resins have also been used as adjuncts to nitrocellulose.

- Santolite MHP
- Dammar
- Sandarac
- Pontianac
- Resin and esters
- Ester gum
- Resin and esters
- Ester gum
- Hexyl methacrylate–methylmethacrylate copolymer
- Acrylic ester oligomers
- Alkyd resin
- Polytetrahydrofuran
- Polyamides
- Laurolactam–caprolactam–hexamethylene diamine adipate terpolymer
- Helamine resin

4.3.3 *Manufacture*

The manufacture of nail polish involves the high-shear mixing of highly flammable and volatile materials and, because of this, involves high risk. Precautions must be taken to ensure that the equipment and premises used are adequately protected. Both buildings and manufacturing equipment must be flame-proof and carefully monitored for harmful vapours.

As with all color cosmetic manufacture, pigment preparation is the most important step. This is particularly true of nail polish since the finer the pigment is ground, the higher the gloss that will be achieved, and the more stable the finished product. Pigments for nail polish are usually prepared in 'chip' form. The required pigment is blended with nitrocellulose in a mix of bentone solution and plasticizer. The resultant mixture is then ground through a triple roll mill, dried and 'chipped' (i.e. split up into solid fragments). Although the final product is stable, the process involved is very hazardous, and explosions are not unknown. As a result, most nail polish manufacturers prefer to buy the 'chips' already made.

The color chips are blended in the desired shades and dissolved in the nitrocellulose solution (lacquer) using a high-shear mixing blade (Cowles) under flame-proof conditions. The temperature must be carefully monitored to avoid a large increase. The other solvents and additives are added when a uniform color has been achieved. Following this, the bentone is added and the viscosity adjusted by addition of lacquer or thinners. Special viscosity modifying additives are also required. For this mixing, a normal propeller-type blade may be used.

For convenience, all the materials used may be obtained as solutions in the mixed solvent base. The operation consequently becomes a single liquid-blending operation to obtain the correct shade, with appropriate viscosity adjustment as required. To ensure that the correct shades are achieved and that the product will remain stable on-shelf, careful control over the thixotropy and viscosity of the finished product is essential. The quality of the incoming raw materials must also be carefully controlled. The viscosity of nail polish is sometimes adjusted by adding small quantities of phosphoric acid or organic acids to modify the gelling system. This is a delicate operation, which requires very careful monitoring and control shade-by-shade to avoid unwanted overgelling or loss of viscosity with time.

4.4 Face make-up

Face make-up can be divided into three segments: (i) face powders; (ii) liquid foundation, and (iii) blushers. In 1990, total sales in the US and Europe were estimated to be in the area of 3.5 billion dollars. Of this, 46% was for liquid foundation, 29% was for blush and, 25% was for face powder (Figure 4.6).

4.4.1 *Consumer expectations*

Face make-up should apply easily to give smooth, even coverage, but a natural look. The product should blend easily and not settle into the lines and pores on the face. The shades for foundation and powders should be

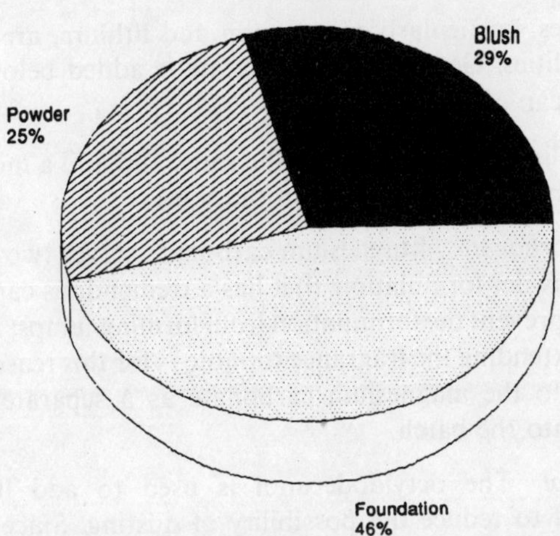

Figure 4.6 Share of face make-up sales (dollars) in the US and Europe during 1990.

plain flesh tone or lightly pearlized for a moist look, while the blush should be available in a wide range of translucent reds. Changes in fashion and location will determine different requirements in degree of translucence, depth, and type of color, and other appearance characteristics.

4.4.2 *Face powders*

4.4.2.1 *Formulation* Face powders are available in loose and pressed forms—applied with either a brush or puff. A typical *loose formulation* is as follows:

	Wt%
Talc	80.50
Zinc stearate	3.60
Kaolin	8.80
Magnesium carbonate	2.40
Cosmetic oxide	2.50
Octyldodecanol	2.00
Fragrance	0.20
Preservatives	

Talc As in all colored cosmetic formulation bases, face powder is a vehicle to apply color and, in most formulations today, talc is the major ingredient. Talc is chosen because of its excellent slip, its transparency, and its light feel on the skin. Its drawbacks are that it is not very water-repellent, it does not adhere well to the skin, and is somewhat too shiny.

Zinc stearate Zinc stearate adds the water repellancy to the formulation and adheres well to the skin. It improves the wear and has a creamy, almost unctuous feel, which aids the consumer in gaining an even application. Other metallic stearates, particularly magnesium and lithium, are used to impart these same qualities. Generally, zinc stearate is added below 10%, since at higher levels it can cause caking.

Kaolin Kaolin is added to reduce the shine and add a more natural look to the product lay-down (layer left on the skin).

Magnesium carbonate Magnesium carbonate serves two purposes in the standard face powder formulation: (i) it has a tremendous capacity to absorb oil, which helps prevent the formulation from forming lumps; and (ii) it has the reputation for extending the fragrance topnote—for this reason the fragrance is often added to the magnesium carbonate as a separate step before its incorporation into the batch.

Octyldodecanol The octyldodecanol is used to add lubricity to the formulation and to reduce the possibility of dusting. Since this applies to the feeling of the powder on the skin, the addition of such an ingredient depends purely on esthetic considerations. Acetylated lanolin alcohols, mineral oil, lanolin oil, dimethicone, and a multitude of esters are also used for this purpose.

Formulations for *loose pearlescent* face powders can be designed to give a moist look to the skin. This is generally obtained by incorporating a nacreous material (e.g. Mibiron N-50). A typical formulation is as follows:

Loose pearlescent

	Wt%
Talc	to 100
Magnesium carbonate	1.50
Iron oxide pigments	*q.s.*
Zinc stearate	5.00
Mibiron N-50*	25.00
Preservatives	

*Rona Pearl

It is generally not necessary to incorporate an emollient since the pearlescent pigment itself is a lubricant, and usually a heavy material that dampens the potential for dusting. It should be noted that the opacity of pearlescent materials varies considerably and, since most applications require a natural, moist look, the pearls with the most translucency should be chosen.

When formulating *pressed compact powders*, the best results are generally obtained by using low-micron talcs, which compress more easily. In addition, a compact powder needs to be 'wetter', therefore requires increased levels of emollient (mineral oil). The disadvantage of zinc stearate in loose powders (i.e. caking) now becomes an advantage, and its use in pressed powders improves the ability to press. A typical formulation for a pressed face powder is shown below.

	Wt%
Lo-micron talc	79.00
Zinc stearate	5.00
Magnesium carbonate	2.50
Kaolin	6.00
Pigments	2.50
Mineral oil	5.00
Preservatives	

More recently, several speciality items have become available and have gained popularity in these types of formulations. *Sericite* is a special grade of mica that has replaced talc in some loose and pressed formulations. For the most part, however, it is used as an adjunct to talc. Sericite repels water better than talc, and has less slip but a more emollient feel. The use of *spherical silica* imparts tremendous slip to any formulation. Its incorporation in pressed powders also reduces the possibility of 'glazing'—the development of an oily crust on the surface of the pan. The following formulation illustrates a typical example of the use of sericite and spherical silicas.

	Wt%
Lo-micron talc	76.50
Sericite	14.00
Spherical silica	2.00
Zinc stearate	5.00
Hydrogenated polyisobutene	2.50
Pigment	2.50

Pigments Since the desired effect of a face powder is a natural looking flesh tone, the chosen pigments are usually inorganic colors. Titanium dioxide or zinc oxide are used to provide coverage, while iron oxides and ultramarine blue are used to provide the flesh tones.

4.4.2.2 *Manufacture* The preferred method of manufacturing a face powder is by using a PK blender or Gemco-type mixer, although a standard ribbon blender will suffice. The advantage of a PK or Gemco-type mixer is that the equipment is designed to lessen the possibility of dead space.

The first step is to mill the pigments. This is usually done by combining the pigment blend with an equal portion of the talc and pulverizing it through a hammer mill equipped with a low micron screen. All the other ingredients, with the exception of the pearls and the liquid phase, are then filled in the main mixer. When a uniform dispersion has been achieved, the oil phase can be added. With the PK and Gemco mixers this is done through the central mixing bar, which sprays the liquid phase on to the powder and reduces the chances of clumping. With ribbon blenders, the oil phase is often sprayed in with a nozzle and pump arrangement. Uniform blending is followed by the addition of the nacreous materials, if required. If pearlescent pigments are added, the amount of energy required to finish the blending should be as low as possible since there is a danger of breaking the platelets or at least coating them with other ingredients, which will reduce their lustre. If it should be necessary (which can be so when using a ribbon blender) to obtain better dispersion by passing the batch through a pulverizer, this should be done through a jump gap or high micron screen. For the same reason, when using PK and Gemco mixers, the high intensity bars should be avoided as much as possible.

4.4.2.3 *Special considerations* During the filling process most powders, whether loose or pressed, are volumetrically dispensed into their container from cone-shaped hoppers. Considering this, it is easy to see that control of the flow and the density of the product are vital to the ultimate cost effectiveness. This control is not easily quantified, and is usually attained through trial and error by a team of experienced and skilled operators. Nevertheless, there

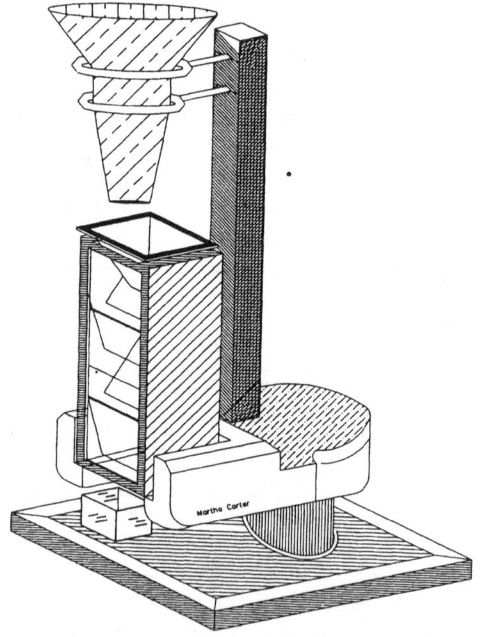

Figure 4.7 Scott volumeter. The powder is sifted through the cone. It then falls over a series of oblique glass plates which serves to randomize the packing into the unit cube receptacle. The weight of the powder is determined by difference. Bulk density is determined by the formula: bulk density = weight/unit volume.

are two evaluations, specifically the bulk and the apparent densities, which are useful aids to predicting the suitability of a powder to be filled.

Bulk density, measures the weight per unit volume of the loose fluffy bulk. In practical terms, this test will predict the powders predisposition, in loose form, to fit into its container. The apparent density is the evaluation of the propensity of a powder to compress under its own weight. It is really a measure of the change in bulk density over time or after undergoing a variable amount of stress. Almost universally, it is an increase in the bulk density that is being measured. The apparent density is important because it will give an indication to the manufacturer of how the volume of the loose powder will change during shipping. Obviously, even though a product may meet its legal weight, an unesthetic fill height is not recommended.

Meanwhile, both the bulk and apparent densities can be used to predict if a specific pressed powder bulk will easily compress into its pan and meet its' legal weight. However, skillful pressing machine manipulation can compensate for some deviation from the standard.

For more detailed information on bulk density the reader is referred to the 'Scott Volumeter Method' ASTM B 329-6 (Figure 4.7) or CTFA Method C-7-1, and for the apparent density, CTFA Method C-8-2 (Figure 4.8).

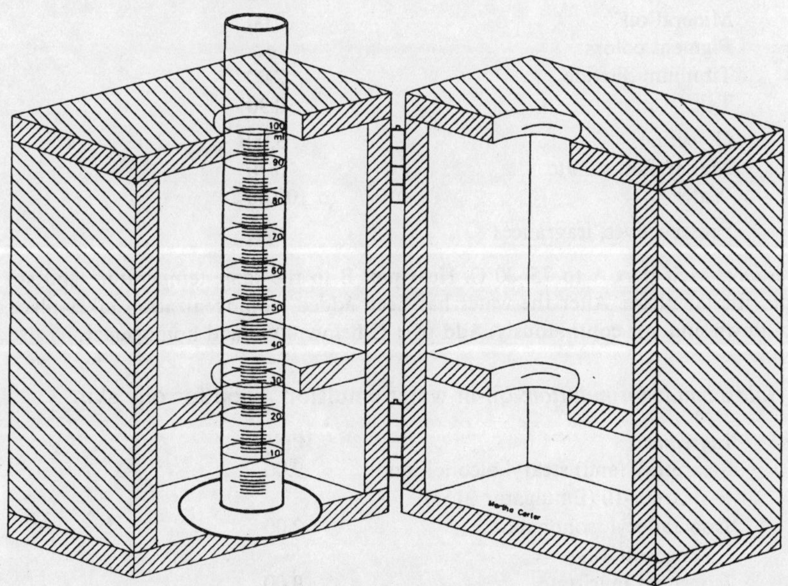

Figure 4.8 Bulk and apparent density. The powder is poured loosely into a volumetric cylinder. The height is noted. The cylinder is then raised manually or mechanically to a fixed height and then allowed to fall freely. The volume of the powder will decrease as it packs. This process is repeated until there is no significant change in the volume. The change in volume is noted. The apparent density is usually written as the percentage of change in the volume.

4.4.3 *Liquid foundations*

To some degree, liquid foundations could be described as powder base suspended in liquid emulsion or gel base. They are the most popular type of face make-up, probably because they are easy to spread, therefore the coverage can be controlled, and any existent skin blemishes hidden more easily. They are often sold to suit particular skin types—gels and oil-in-water emulsions for dry skin, and water-in-oil emulsions for oily skin.

4.4.3.1 *Formulation and manufacture* The formulations given in this section illustrate the variability of liquid foundations, and may be used as introductory formulations that can be worked on to achieve the qualities required.

(i) *Henkel formulation*: water-in-oil cream, recommended for dry skin

Part	*Ingredient*	*Wt%*
A	Glyceryl oleate (Monomuls 90–018)	2.00
	Polyglyceryl-3 diisostearate (Lameform TGI)	4.00
	Dioctylcyclohexane (Cetiol S)	10.00
	Cetearyl alcohol (Lanette O)	1.00
	Microcrystalline wax	7.00
	Mineral oil	10.00
	Pigment colors	*q.s.*
	Titanium dioxide	2.00
	Talc	0.90
B	Glycerine	3.00
	Magnesium sulfate	0.70
	Water	to 100.00
C	Preservatives, fragrances	*q.s.*

Procedure: Melt part A to 75–80°C. Heat part B to the same temperature and stir it into the fat phase. After the water has been added, homogenization takes place. Cool down and stir continuously. Add part C below 40°C and homogenize.

(ii) *Goldschmidt formulation*: oil-in-water emulsion, good for oily skin

Part	*Ingredient*	*Wt%*
A	Steareth-7 (and) stearyl alcohol (and) steareth-10 (Emulgator 2155)	9.00
	Isooctadecyl isononanoate (Tegosoft 189)	2.00
	Isopropyl myristate	9.00
	Mineral oil	8.00
	Stearyl dimethicone (Abil Wax 9800)	2.50
	Dimethicone (Abil 100)	0.50

B	Water	10.00
	Glycerine	2.00
	Talc, USP	4.00
	Titanium dioxide	6.00
	Iron oxides	3.40
	Ultramarine blue	0.15
C	Water	43.45
	Preservatives, fragrances	*q.s.*

Procedure: Heat part A to 70°C with mixing. Heat the water and glycerine of part B to 50°C and grind in talc, titanium dioxide and pigments. Heat the water of part C to 70°C; add to part A. Begin cooling while homogenizing. When homogenized, stir in part B. With a slow sweep agitation, cool to 30–35°C. Add fragrance and preservatives. Slow sweep until cool.

(iii) *Mearl Corporation pearly foundation: helps produce a natural lustre on the skin*

Part	*Ingredient*	*Wt%*
A	Laneth-10 acetate (Solulan 98)	3.00
	Isopropyl lanolate (Amerlate P)	5.50
	Acetylated lanolin alchohol (Acetulan)	5.30
	Glyceryl stearate SE (Tegin)	3.50
	Stearic acid	2.70
B	Propylene glycol	5.00
	Triethanolamine	1.00
	Water	64.00
C	Titanium dioxide coated mica (Flamenco Satina)	10.00

Procedure: Separately heat parts A and B to 80°C. Stir part B into part A until homogeneous. Cool to 40°C with slow stirring. Disperse part C into warm base.

(iv) *National Lead Industries formulation*

Part	*Ingredient*	*Wt%*
A	Talc 1621 (W, C, D)	5.00
B	Titanox 1020 (TGA)-328	5.00
C	Pure oxy yellow 3170 (W, C, D)	0.40
D	Pure oxy umber 3315 (W, C, D)	0.20
E	Lo-micron pink 2511 (W, C, D)	0.40
F	Water (deionized)	79.95
G	Methylparaben (Goldschmidt)	0.25
H	Propylene glycol USP	5.00
J	Cerasynt 840 (PEG-20 stearate)	2.00
K	Bentone LT	1.50
L	Fragrance	0.30

Procedure: Blend the powder phase (parts A to F) and pass through a micro-pulverizer. Heat the water phase (parts G to J) to 65°C in a stainless steel vessel equipped with a homogenizer. Adjust the homogenizer to a medium speed and add the powder phase. Add the bentone and allow to mix for 30 min. Cool the batch before adding the perfume.

This formulation uses Bentone LT (as an organically modified clay) as its gelling agent. PEG-20 stearate is added as an emollient. The thixotropic qualities of bentone allow for excellent pigment-suspending characteristics and smooth, even application.

Modern formulations often contain volatile silicone oils to improve feel and application, and sunscreen to protect the skin against ageing effects of sunlight.

4.4.4 *Blushers*

Most blushers are currently sold in the pressed powder form, although cream versions also constitute a significant segment. Liquid blush is very like its counterpart liquid foundation. The difference is primarily in the choice of pigments. The goal of blusher is to impart a healthy glow to the skin, therefore the major difference compared with regular pressed powder is the choice of pigments.

4.4.4.1 *Formulation and manufacture*

Pressed powder 1: Nikko formulation

	Wt%
Talc	38.90
Sericite (Nikkol Sericite J)	30.00
Mica (Nikkol Mica G)	5.00
Nylon-12	5.00
Mica (Flamenco pearl)	10.00
Magnesium stearate	3.00
Cetyl octanoate (Nikkol C1O)	3.00
Dimethicone	2.00
Mineral oil	2.00
Propylparaben	0.05
Butylparaben	0.05
Fragrance	*q.s.*

Procedure: Mix all ingredients, except liquid oils and fragrance, in a blender. Spray or add liquid oils and perfume. Mix and pulverize. Press into pans. (Any blender suggested for face powder could be satisfactory.)

Pressed powder #2	*Wt%*
Kaolin	1.21
Lithium stearate	1.94

Zinc stearate	3.65
Talc	40.09
Magnesium carbonate	1.70
Iron oxides	12.82
D&C Red #30 lake	4.09
Mineral oil	2.11
Octyl palmitate	1.76
Timica golden bronze (Mearl)	15.00
Gemtone amber (Mearl)	15.00
Preservatives	*q.s.*

4.5 Eye make-up

Eye make-up consists of three major categories: (i) eyeshadow; (ii) mascara; and (iii) liners. Total US and Europe sales for 1991 are estimated to be in the region of 2.5 billion dollars (Figure 4.9).

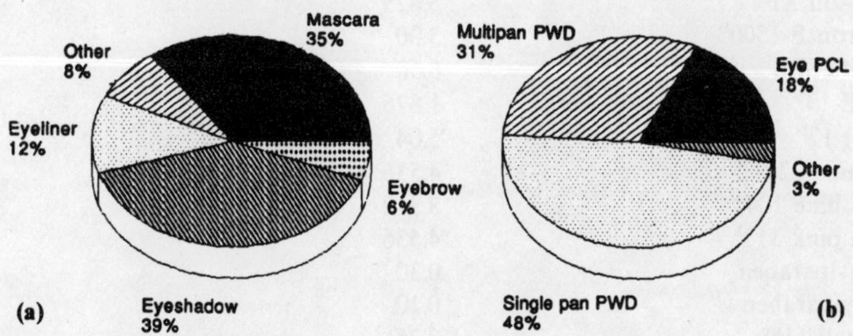

Figure 4.9 (a) Share of eye make-up sales (dollars) in the US and Europe during 1990; (b) share of eyeshadow sales by form.

4.5.1 *Eyeshadow*

4.5.1.1 *Consumer expectations* Eyeshadow should blend easily, give good coverage, yet look natural. It should be available in a wide range of colors, matte and frost (pearl), and should not crease when worn.

4.5.1.2 *Formulation and manufacture* The most popular form of the eyeshadow is the pressed powder cake, but pigmented creams also constitute a significant market share.

Pressed powder cakes The formulation will not differ greatly from a standard compacted pressed face powder, but the range of shades is far more extensive and frost/pearlescent colors are more popular. A standard formulation for a pink pearlescent eyeshadow is as follows:

	Wt%
Talc 141	17.30
Zinc stearate	7.00
Duocrome RY	40.10
Ultramarine pink	4.40
Ultramarine blue	0.60
Flamenco Superpearl 100	3.00
Timica brilliant gold	19.00
Mineral oil	8.00
Preservatives (*q.s.*)	0.60

Due to its superior compressibility, a low-micron talc is chosen but, due to the difficulty in compacting the pearlescent materials, the zinc stearate and mineral oil are present in larger percentages. The following formulation uses spherical silica for slip, and sericite to improve emolliency and waterproofness.

Part A	*Wt%*
141 talc[a]	34.676
Sericite S-100[b]	8.00
Press-aid XF[b]	5.625
Spheron P-1500[b]	3.00
Sericite SI-S[b]	6.50
PTFE-19[b]	1.875
Violet 12[b]	5.04
Carmine 224[b]	4.536
Ultra blue 104[b]	3.312
Ultra pink 113[b]	4.536
Methylparaben	0.30
Propylparaben	0.10
Germall 115[c]	0.25
Part B	
Pearl I[b]	18.00
Part C	
Permethyl 102A/104A (50%/50%)[b]	4.25

Suppliers:
[a]Whittaker, Clark & Daniels
[b]Presperse, Inc.
[c]Sutton

Procedure: Combine ingredients of part A and mix well, then pass through micro-pulverizer. Place back into ribbon-type blender and spray in binder, part C. Pass entire batch through pulverizer, mix well again while adding pearl, part B. This can also be pulverized one final pass to keep a smooth consistency and a matte finish.

Cream eyeshadows Although not the most popular form, these are often chosen when long wear is desired. The following Sutton formulation incorporates a wax-resin base dispersed in a volatile hydrocarbon solvent. When the solvent evaporates, a relatively water-impervious film will be left on the eyeshadow lid.

Part A	*Wt%*
Cetyl lactate (Ceraphyl 28)	2.0
Octyldodecyl stearoyl stearate (Ceraphyl 847)	2.0
Beeswax, white	7.8
PVP, eicosene copolymer (Ganex V-220)	6.5
Trihydroxystearin (Thixcin R)	3.5
Petroleum distillate (Shell Sol 71)	33.0
Part B	
Petroleum distillate (and) Quaternium-18 hectorite (and) propylene carbonate (Bentone gel SS-71)	14.7
Zinc stearate	2.0
Magnesium stearate	1.0
328 titanium dioxide	6.0
Talc 141	14.5
Mica (and) bismuth oxychloride (and) ferric ammonium ferrocyanide (Chroma-lite blue)	6.0
Part C	
Germaben II	1.0

Procedure: Combine and mix part A with propeller stirrer while heating to 70°C or until melted. Add part B ingredients and mix, with homo-mixer at medium speed. Stir with propeller stirrer while cooling and add part C at 60°C. Cool to just above congealing point and pack.

The following highly frosted pearlised formulation (Mearl Corporation) achieves the same effects using a volatile silicone fluid in place of the hydrocarbon.

Part A	*Wt%*
Acetulan[a]	11.05
Syncrowax HRC[b]	4.60
Schercemol ICS[c]	3.00
Part B	
Silicon fluid 344[d]	17.00
Part C	
Flamenco Superpearl[e]	21.00
Cloisonne or Gemtone color[e]	18.00
Part D	
Methylparaben	0.20
Propylparaben	0.10
Butylated hydroxy anisol (BHA)	0.05
Part E	
Bentone gel SS-71[f]	14.00

Part F
Silicon fluid 345[d] 11.00

Suppliers:
[a]Amerchol Div. CPC International, Inc.
[b]Croda, Inc.
[c]Scher Chemicals, Inc.
[d]Dow Corning
[e]Mearl Corporation
[f]NL Industries

Procedure: Heat ingredients of part A to 70°C with gentle stirring until all ingredients are melted. Remove from the heat and add part B with stirring; then mix in the pre-blended part C. When all the pigments are thoroughly dispersed, add preservatives and part D. While stirring, blend in the gel, part E, and, finally part F.

4.5.2 *Mascara*

4.5.2.1 *Consumer expectations* A mascara is designed to make the eyelashes look thicker and longer. Coverage should be good, but the mascara should not clump on the lashes, flake during wear, or feel brittle after drying. In addition, the mascara should be tear-resistant, waterproof or water-resistant, and must not smear or smudge. The most popular color is black but other dark shades (e.g. blue, green and brown) are also sold.

Mascara is available in two basic categories: (i) water-resistant; and (ii) waterproof. Total sales in the US and Europe during 1991 were estimated to be in the range of one billion dollars and over 75% of these sales fall into the water-resistant category.

Mascaras are now almost universally applied from a tube with a specially designed applicator (see Figure 4.10). Cake mascaras are now virtually obsolete.

4.5.2.2 *Formulation and Manufacture*

Waterproof mascaras Waterproof mascara formulations have changed very little for many years. For the most part they consist of a blend of waxes and pigments in a volatile hydrocarbon solvent. The choice of waxes determines the final characteristics of the formulation. A typical formulation is as follows:

	Wt%
Pigments	5.0–10.0
Beeswax	26.00
Ozokerite	4.00
Lanolin	0.50
Preservative	0.25
Aluminum stearate	2.50
Hydrocarbon solvent	to 100.00

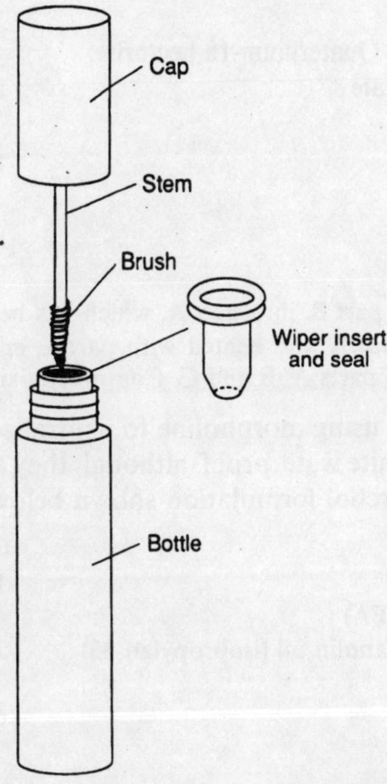

Figure 4.10 Mascara bottle.

Procedure: Add the aluminum stearate to the solvent with stirring while the mixture is heated to approximately 90°C. Maintain at that temperature until solution and gelation are evident. Melt the waxes together and add to the solvent. Grind the pigments in a portion of the solvent–wax mixture and add to the remainder of the batch. To avoid settling of pigments while the mixture is still warm and fluid, continue stirring until the mixture is cool.

Although mascaras of this type are extremely waterproof, they also tend to be very difficult to remove. This is probably why they are less popular than the water-resistant types. The following Tevco formulation attempts to ease removal by combining a mascara solvent in a beeswax borax emulsion.

Part A	*Wt%*
Petroleum distillate	to 100.00
Beeswax	18.00
PEG-6 sorbitan beeswax	6.00
Ozokerite 170-D	4.00
Carnauba wax	6.00
Propylparaben	0.10
Glyceryoleate (and) propylene glycol	1.50
Part B	
Iron oxides	15.00

Part C	
Petroleum distillate (and) Quaternium-18 hectorite (and) propylene carbonate	12.50
Part D	
Deionized water	15.00
Methylparaben	0.30
Sodium borate	0.60
Quaternium-15	0.10

Procedure: Mill pigment, part B, into part A, which has been heated to 90°C. After part C has been added slowly and heated with part A, emulsify by adding part D at 90°C to the mixture of parts A, B and C. Continue mixing until cool.

Emulsion formulations using morpholine to neutralize a fatty acid oil phase are also used and are quite waterproof, although they are generally becoming less popular. The Amerchol formulation shown below is an example.

Part A	*Wt%*
Carnauba wax	1.00
Lanolin acid (Amerlate LFA)	4.00
Isopropyl palmitate and lanolin oil (isopropylan 33)	2.00
Beeswax USP	8.00
Ozokerite	6.00
Part B	
Water	37.20
Magnesium aluminum silicate, 4% aqueous	12.50
Cellulose gum	0.70
Part C	
Iron oxide pigments	10.00
Part D	
Morpholine	1.60
Water	5.00
Part E	
Methacrylol ethyl betaine/methacrylates copolymer (Amersette)	12.00
Preservatives	*q.s.*

Water-resistant mascaras By far the most popular form of mascaras today are formulations based on a triethanolamine stearate, or oleate, soap system. These formulations can be quite water-resistant, feel soft on the lashes, and are relatively easy to remove. Just as importantly, they also have less potential to cause eye irritation. The following Tevco formulation illustrates this approach. The inclusion of hydroxyethyl cellulose in the water phase reduces the potential to smudge.

Part A	*Wt%*
Deionized water	43.00
Hydroxyethyl cellulose	1.00

Methylparaben	0.30
Triethanolamine	1.00
Ammonium hydroxide, 28%	0.50
Preservative	2.00
Part B	
Iron oxides	10.00
Ultramarine blue	2.00
Part C	
Isostearic acid	2.00
Stearic acid	2.00
Glyceryl monostearate	1.00
Beeswax	9.00
Carnauba wax	6.00
Propylparaben	0.10
Part D	
Quaternium-15	0.10
Part E	
30% Acrylic/acrylate copolymer solution in ammonium hydroxide	20.00

Procedure: Mill the pigments of part B in the water phase, part A. Heat to 80°C. Heat the oil phase, part C, to 82°C. Emulsify. Cool to 50°C. Add part D, then part E. Cool to 30°C.

4.5.3 *Eyeliners*

4.5.3.1 *Consumer expectations* On application, eyeliners should 'glide on' without tugging or pulling on the eyelid. The color coverage should be good and should blend easily. The formulations should be such that the depth of color can be controlled during application. The line should last all day without smudging or smearing. The color range can be limited to the basic black, black-blues, greens, and violet.

4.5.3.2 *Formulation and manufacture*

Wax-based crayons The majority of eyeliner formulations are sold in woodcased pencil form, although a significantly smaller, but growing segment of the market is dispensed in the form of a mechanical pencil (Figure 4.11). Most products are extruded like lip pencils, therefore the manufacturing process is the same (see section 4.2.3.3). However, as the eye is a more sensitive area than the lip, the formulations tend to be softer. Typical examples are as follows:

Ingredient	*Wt%*
Myverol 18-50 (Eastman Chemical)	14.35
Stearic acid	13.86

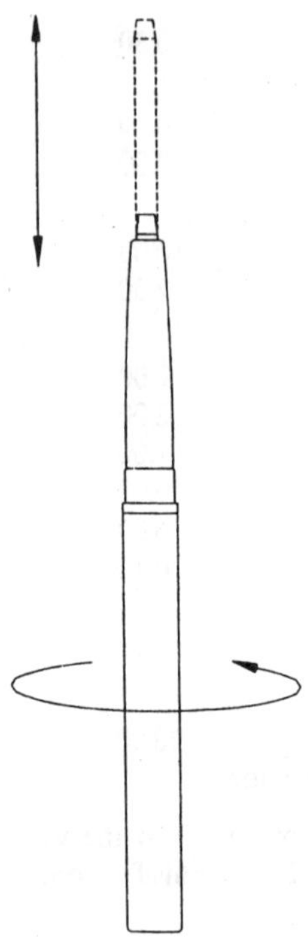

Figure 4.11 Mechanical pencil.

Japan wax	13.50
Paramount XX (Durkee)	7.95
Cutina HR (Henkel)	7.44
Softisan 100 (Kay Fries)	4.51
Cutina CP (Henkel)	4.27
Mineral oil	1.59
Miglyol 812 (Kay Fries)	1.29
Talc 141 (W, C, D)	1.17
Neobee 62 (PVO)	1.05
Crodamol W (Croda)	0.82
Veba wax (Dura Commodities)	0.82
Black iron oxides	16.94
Titanium dioxide 3328 (W, C, D)	7.77
Ferric ammonium ferrocyanide	0.89
Mica	1.05
Preservatives	to 100.00

Ingredient	*Wt%*
Lipocire CM (Gattefosse)	49.69
Eutanol G (Henkel)	3.91
Polyethylene 6 (Allied)	12.56

Carnauba (Strahl & Pitsch)	1.66
Japan wax	5.25
Ferric ammonium ferrocyanide	7.00
Ultramarine blue	8.00
Titanium dioxide 3328	1.05
Black iron oxides	3.44
Chromium hydroxide green	7.00
Preservatives	to 100.00

Liquid eyeliners Liquid eyeliners have recently become more popular due to the introduction of pen liners. In this package, a wick is soaked with a liner formulation containing an ultra-fine pigment. This solution is fed by capilliary action to an applicator nib. The major disadvantage of this system is that, due to the pigment size necessary to feed the tip, a true black is difficult to obtain. The system is, however, very convenient to use. The more traditional liquid formulations remain popular with consumers who like a darker more defined line. These formulations are dispensed from a tube by means of the thin brush. A typical example is shown below.

Part A	*Wt%*
Magnesium aluminum silicate (Veegum)	2.50
Deionized water	75.50
Part B	
PVP (PVP K-30)	2.00
Deionized water	10.00
Part C	
Iron oxides	10.00
Part D	
Preservative	*q.s.*

Procedure: Slowly add magnesium aluminum silicate to part A water while agitating with maximum available shear. Continue mixing until smooth and free of visible particles. Dissolve the PVP in part B water with mild heating. Add part B to part A and mix until uniform. Add parts C and D in order, and mix until fully dispersed and uniform.

Very recently, new pen devices using a cartridge system and conventional pigment technology have been introduced. In these units the formulation is dispensed to the tip using a shaking motion. For further details of pen devices, the manufacturers should be consulted.

4.6 Preservation

It is important that cosmetics are adequately preserved, the degree of preservation being dependent on several variables, which include the nutrient composition of the formulation, the part of the body to which the cosmetic is to be applied and the nature of its intended use. For example, a formulator may wish to be more discriminating in developing a preservative system for mascara, than say, a face powder. The rationale here is that a mascara is probably water based, is used around the eye and by the very method of its use, is subject to repeated insults. By 'repeated insults', it is meant that the applicator, after use, is returned into the product. Face powders, meanwhile, although they share this same bane, are anhydrous and are used on a less sensitive part of the body.

The type of preservative that is to be used is also important. Again, the composition of the formulation is integral to making this decision as well as other factors such as the pH of the product, and last, but not least, where the product is to be distributed and sold. This last point is particularly important because there is a lack of consensus both regionally and around the world as to which preservative agents are preferred.

In summation, preservation is a complex issue and certainly cannot be covered in great detail here. Fortunately, there is aid for the formulator which comes in the form of supplier's literature and several excellent reference books. The reader is also recommended to review the October '93 edition of 'Cosmetics and Toiletries', [4] wherein several distinguished writers cover the topics mentioned here, and several others which are not, in some detail.

4.7 Color coating

One problem that has always been present in the formulation of cosmetic products, particularly items which have high solid levels (e.g. pressed eyeshadow), is the variability of texture from shade to shade. For instance, a shade containing a high level of pearlescent pigment would feel different from a product containing an elevated level of, say, iron blue. Consequently, it is becoming increasingly popular to pre-coat solid materials, particularly pigments, with a common film. Experience has shown this will lead to a more uniform texture across the total shade line. The choice of coating will also have a profound effect on the final attributes of the product and can, when skillfully chosen, help the formulator achieve the pre-set performance characteristics of the finished product more easily. Some of the more common coatings available are silicone, polyethylene, lecithin, teflon, metal soap, and dimethicone copolymer. Several of these processes are patented. Details of available alternatives in this rapidly developing technological area, and ways of using the materials in formulations can be obtained from pigment and color manufacturers.

4.8 General considerations

All cosmetic color products depend on a thorough knowledge of the colors and pigments used. This is true no matter what product form is being worked on. The safety, general stability and physical form of the ingredients is of paramount significance and there is a wealth of detail available from regulatory offices, suppliers, and trade literature.

The handling of colors and pigments, particularly when producing pressed powders, is of fundamental importance, and the setting of machinery to obtain the correct product flow and pressure for a stable pressed cake is a skilled operation. A great deal of attention to trial and error may be necessary to achieve the correct balance of pigment load and binder for the system under study. Pilot work is essential in cost-effective development.

Finally, the essence of good formulation is the precise handling of the colors and color matching. Color matching is a skilled technique only learned with experience, and a good color matching technician can save a great deal of time and money in adjustments to products in manufacture. Close attention must be paid to the consistency of the color shade of the incoming raw materials; observance of quality control is essential to ensure consistency in final product.

References

1. Talbot, G. (1990) Fat migration in biscuits and confectionery systems. *Confectionery Production* (April) **56** (Issue 4), 255.
2. Laustsen, K. (1991) The nature of fat bloom in molded compound coatings. *The Manufacturing Confectionery* (May) **71** (Issue 5), 137.
3. Bennett, H. (1975) *Industrial Waxes*. Chemical Publishing Company, New York.
4. (1993) Preservation. *Cosmet. Toilet.* **108**(10) 10.

Further reading

Belgian Patent 892, 174. Anhydrous nail polish containing styrene or unsaturated amide and acrylic polymers. Assigned to L'Oreal SA (August 17, 1982).

Make-up documentary formulary (1981) *Cosmet. Toilet.* **96**(4).

Make-up documentary (1986) *Cosmet. Toilet.* **101**(4).

Make-up documentary (1989) *Cosmet. Toilet.* **104** (7).

Gels and sticks documentary (1989) *Cosmet. Toilet.* **102**(10).

de Navarre, M.G. (1975) (Ed.) *The Chemistry and Manufacture of Cosmetics*, 2nd edn., vol. IV. Continental Press, Orlando, Florida.

Dweck, A.C. and Burnham C. (1981) Lipstick moulding techniques—comparison and statistical analysis. *Cosmet. Toilet.* **96**(4).

European Patent Application 199,325 A2. Silyl-containing nail enamel. Schnetzinger, R.W., assignor to Revlon, Inc. (October 29, 1986).

Frost and Sullivan (1989) *Marketing Strategies in the US Beauty Products Industry*. Winter 1989.

Frost and Sullivan (1990) *The European Market for Make-up Products*. Spring, 1990.

German Offen 76,076. Fingernail polish. Kabs, U., Ruehle, R. and Petzoid, B. (September 12, 1970).

German Offen 2,830,958. Nail polish. Masters, E.J., assignor to Mallinckrodt, Inc. (July 15, 1977).

German Offen 2,721,456. Nail polish containing hexyl methacrylate–methylacrylate copolymers. Boulogne, J. and Papantoniou, C., assignors to L'Oreal SA (November 24, 1977).

Goldner, T. (1986) Principles of pigment dispersion in color cosmetics. *Cosmet. Toilet.* **101**(4).

Harry, R.J. (1973) *Harry's Cosmeticology*, 6th edn., vol. 1. Chemical Publishing Company, New York.

Japan Patent 3800. Nail enamel. Uekl, K. (September 19, 1952).

Japan Patent 71,43,400. Nail polish containing acryl ester oligomers. Suglyama, I. and Tomozuka, H., assignors to Matsumoto Selyaku Kogyo, Co., Ltd., (December 22, 1971).

Japan Patent 79,129,137. Nail polish composition. Assigned to Kanebo KK (October 6, 1979).

Japan Patent 80,57,512. Stable nail lacquer compositions. Assigned to Kanebo, Ltd (April 28, 1980).

Japan Patent 81,25,107. Linear polyester oligomer resin as nail polish. Assigned to Asunuma Sogyo K.K.; Riken Selyu KK (March 10, 1981).

Japan Patent 86,246,113. Nail lacquer containing nitrocellulose, an alkyd resin, sucrose benzoate, and triethyl acetylcitrate. Yamazkl, K., Soyama, Y., Kotamura, C. and Tanaka, M., assignors to Shiseido Co., Ltd. (November 1, 1986).

Remz, H.M. (1988) Polymers and thickeners in nail-care products. *Cosmet. Toilet.* **103**.

UK Patent 724,041. Nail polish. Assigned to Cosmetic Laboratories, Inc. (February 16, 1955).

US Patent 2,173,755. Nail enamel. Fuller, H.C. (September 19, 1939).

US Patent 2,195,971. Fingernail enamel composition. Peter, R.C., assignor to E.I. Du Pont de Demours & Company (April 2, 1940).

US Patent 2,215,898. Nail polish. Anderson, R.J., assignor to the Vorac Company (September 24, 1940).

US Patent 2,279,439. Lacquer composition. Bowlby, W.B., assignor to Trojan Powder Company (April 14, 1942).

US Patent, 3,483,285. Human nail coating compositions. Michaelson, J.B. and Criswell, A.F. (December 9, 1969).

US Patent 3,786,113. Composition containing an acrylic resin, a polyethylenealmine, and a polyopoxide. Vasslleff, N.I., assignor to National Patent Development Corporation (January 15, 1974).

US Patent 4,097,589. Nail polish. Shansky, A., assignor to Del Laboratories, Inc. (June 27, 1978).

US patent 4,283,324. Nail enamel composition. Duffy, J.A., assignor to Avon Products, Inc. (August 11, 1981).

US Patent 4,545,98. Nail enamel containing polytetrahydroturan as a resin. Jacquet, B., Papantoniou, C and Goetani, Q., assigned to L'Oreal (October 8, 1985).

Wing, H.J. (1975) Nail preparations. In *Chemistry and Manufacture of Cosmetics*. deNavarre, M.G. (Ed.) Continental Press, Orlando, Florida, chapter 49.

5 Baby care

J.L. KNOWLTON

5.1 Introduction

Baby cosmetics and toiletries, intended for use on new-born babies and children of up to five years old, represent a very special category indeed. They are invariably functional, rather than decorative, mainly concerned with keeping the baby or child clean, comfortable and healthy. Since babies and young children are somewhat more vulnerable than their adult counterparts, baby products must be carefully formulated to be extremely mild and gentle to skin and hair. These considerations impose a number of requirements and constraints for the design, evaluation and manufacture of baby-care products. Any person working in the area of baby care must understand these issues and adhere to them accordingly.

Finally, it is important to recognise that baby products have a significant part to play in giving an emotive benefit to the user, that is the parent. It is essential that products used on the young are safe, but the ability to leave the hair and skin feeling soft and clean, and with a pleasant smell is equally important in providing emotional support for protective parental instincts.

5.2 Specific basic requirements for baby products

The basic requirements for baby products are very similar to those for products aimed at the adult sector. However, additional factors, over and above those normally encountered, must be adhered to and, in the opinion of the author, the following list indicates the minimum requirements for any product to be sold in this market sector.

(i) Products should be functional, efficacious and demonstrably capable of delivering the benefits for which they are purchased. Functionality should not sacrifice aesthetics, and pleasantness of use is a mandatory requirement.

(ii) Products should comply with the most stringent standards of safety, such that they can be used with absolute confidence on even the youngest skin.

(iii) Products should contain the purest grades of materials available and the microbiological purity of the finished product should be tightly controlled to eliminate the risk of infection.
(iv) The levels of dyestuffs and fragrances should be kept as low as possible so as to reduce the risk of adverse skin reactions. At the present time an increasing number of fragrance-free products are appearing in the market.
(v) Preservative levels in baby products should be kept to a minimum, whilst ensuring adequate preservation to negate the risk of contamination or infection of the user.
(vi) Product packaging should be carefully designed to minimise the risk of product ingestion or other form of misuse. A number of tamper-proof or childproof enclosures and containers are available, and should be used where possible.
(vii) The platform on which the product is marketed should communicate a high degree of honesty and trust, with emphasis on the benefits of product purity and performance, rather than ephemeral marketing claims.

5.3 Product types and their presentation

5.3.1 *Baby powders*

Powders are amongst the earliest and best known of all baby products, with a very high level of penetration into the target market. The primary function of a baby powder is to absorb residual moisture on the skin and to provide a degree of mild lubrication, thus reducing the possibility of post-cleansing chaffing and soreness.

The main ingredient in baby powder is normally talc, known chemically as hydrated magnesium silicate. It is chosen for its unique ability to provide a high degree of silkiness and lubricity on the skin, due to its unusual morphology. This property is conferred by the flat, hexagonal, platelet structure of the talc crystals, which exhibit a low coefficient of inter-particulate friction, producing a high degree of 'slip'. Raw talc is mined from the earth, therefore must be purified before it can be safely used. A natural contaminant of talc from some sources is asbestos, a fibrous material that can produce a fatal lung disease, known as asbestosis, in humans. Obviously, talc of this type should not be used. Bacteria, natural contaminants in talc, must be controlled. *Clostridium tetani* is the causative organism of tetanus, so the importance of an effective sterilisation process cannot be over-emphasised. Many methods of sterilisation have been tried over the years, but that utilising gamma irradiation is still amongst the most popular [1]. A second method, now becoming far more widespread in use, is steam sterilisation. This involves heating the raw talc, under pressure, with super-heated steam at temperatures exceeding 120°C for periods of 30 min or more. Whilst this method is very effective, it is essential to remove any residual moisture from the raw talc after the sterilisation process, otherwise the likelihood of contamination by adventitious micro-organisms will be significantly increased.

Some powder products are composed simply of talc and a low level of fragrance. Many contain other additives, claimed to give the product additional benefits. Additives that possess a higher degree of moisture absorbency than talc may be included, making the baby powder more effective. Such materials include kaolin, magnesium and calcium silicates, and starch-based products. However, whilst these materials achieve extra absorbency, they compromise the lubricious feel of talc itself, impairing the feel of the finished product. This disadvantage can be minimised by careful selection of the physical form of the absorbent material used. Generally fine particle sizes with uniform spherical shape should be selected. The inclusion of low levels of mild bactericides can improve performance in terms of the prevention of nappy rash. Powdered zinc oxide, with mild bactericidal properties, is often incorporated at levels between 1 and 5% for this purpose. A typical formulation for an absorbent baby powder designed to help prevent the formation of nappy rash is illustrated below.

	Wt%
Talc	77.90
Starch	20.00
Zinc Oxide	2.00
Perfume	0.10

Perhaps the best known baby powder on the market today is that from Johnson & Johnson, which is a high purity talc-based product.

A more recent innovation in this product sector is liquid talc. Typically between 5 and 15% of talc is suspended in an oil-in-water emulsion, yielding a product which provides a convenient method of applying talc to a baby's skin. Apart from the pleasant cooling effect of the product in use, the elimination of dusting and avoidance of unintentional inhalation of airborne particles confers additional advantages. A typical formulation for a baby liquid talc, based on an oil-in-water emulsion, is illustrated below [2].

	Wt%
Cetearyl alcohol	0.60
Cyclomethicone (and) dimethicone	1.00
Polawax NF*	2.50
Carbomer 980	0.20
PEG-7 glyceryl cocoate	0.50
Talc	12.00
Triethanolamine	to pH 7.0
Water	to 100.00
Perfume, preservative, etc.	*q.s.*

* Polawax NF* is a registered trade name of Croda Chemicals Limited

When formulating a product of this type it is important to ensure that good application characteristics are combined with sufficiently high viscosity so that the talc remains homogeneously dispersed within the product, even after storage at elevated temperatures for extended periods of time.

5.3.2 *Lotions and creams*

It is impossible to cover every aspect of baby lotions and creams in this chapter, but the products normally possess one of two primary functions, cleansing and/or protection. Products designed for cleansing are more commonly lower viscosity lotions, while those intended to give a protection benefit are generally higher viscosity creams. Traditional baby-cream products were heavy water-in-oil emulsions. Their modern day counterparts are more commonly oil-in-water. Lotions, on the other hand, are nearly always oil-in-water emulsions, which are more favourable for producing the required aesthetics and cleansing function.

Classic formulations for protective baby creams were based on mineral oil or petrolatum, emulsified with an anionic emulsifier such as triethanolamine stearate. Such products generally possess a relatively high oil:water ratio, making them ideal for use in the nappy area after cleansing during nappy changes. These creams leave a protective film of oil on the surface of the baby's skin, thus reducing the chances of chaffing and soreness, and leaving the skin soft and smooth. In addition, water-in-oil creams leave no sensation of coldness on a baby's skin, as the evaporation of the internal water phase is retarded by the presence of the external oil phase. A traditional protective baby cream, based on mineral oil and incorporating an anionic emulsification system, is illustrated below.

	Wt%
Mineral Oil	30.00
Petrolatum	2.00
Stearic acid	1.20
Stearyl alcohol	1.00
Cetyl alcohol	0.70
Triethanolamine	0.65
Propylene glycol	1.00
Water	63.45
Perfume, preservatives, etc.	*q.s.*

Modern baby creams are frequently based on non-ionic emulsifier systems, and offer protective benefits combined with a high degree of emolliency and much improved aesthetics. A typical formulation is illustrated below.

	Wt%
Mineral oil	15.00
Propylene dicaprate/dicaprylate	5.00

Cyclomethicone	3.00
POE sorbitan monostearate	1.50
Sorbitan monostearate	1.00
Glycerine	2.00
Water	72.50
Perfume, preservatives, etc.	*q.s.*

A special type of protective baby cream, yet one that nevertheless constitutes a very important market sector, is that designed to help the prevention of nappy rash. Nappy rash is a painful inflammatory condition, normally found in the anogenital area of the baby's anatomy. The aetiology of nappy rash is subject to debate but the main causative factor is widely accepted as the microbiological breakdown of urine and faecal constituents into irritant materials, principally ammonia, which are subsequently held in close contact with the baby's skin by the nappy itself. An effective nappy rash cream must therefore be formulated to provide good barrier properties and to inhibit the proliferation of bacteria and fungal micro-organisms in the occlusive nappy area. Products of this type are typically water-in-oil emulsions that are able to provide an effective barrier against moisture and chemical attack, with the inclusion of actives such as zinc oxide, hexamidine and benzalkonium chloride to reduce the levels of micro-organisms present. A typical water-in-oil nappy rash cream, containing zinc oxide, is illustrated below [2].

	Wt%
Petrolatum	13.00
Mineral oil	15.00
Sorbitan isostearate	2.00
Microcrystalline wax	3.50
Glycerine	4.50
Zinc oxide	7.00
α-Bisabolol	0.20
Magnesium sulphate	0.70
Lactic acid/sodium lactate buffer	to pH 5.5
Water	to 100.00
Perfume, preservative, etc.	*q.s.*

Baby lotions are invariably lower viscosity oil-in-water emulsions, with the internal oil phase of the product providing the cleansing property. Mineral oil, with the ability to cleanse by solubilising fats and other lipophilic substances found on the skin, is the most favoured base. Most cleansing lotions also contain emollients to leave the skin feeling soft and smooth after use. Water, present in the product as the external phase of the emulsion, provides a fresh, cooling sensation during use, due to the rapid evaporation of water vapour from the skin's surface. Mixed non-ionic surfactants, frequently used for emulsification of lotion products, offer maximum degree of flexibility in formulation design. A typical formulation for a baby lotion cleanser is illustrated below.

	Wt%
Mineral Oil	12.00
Dimethicone	2.00
Isopropyl palmitate	1.50
POE sorbitan monooleate	1.20
Sorbitan monooleate	0.70
Propylene glycol	1.00
Water	81.60
Perfumes, preservatives, etc.	*q.s.*

5.3.3 *Soaps*

Soap, one of the earliest methods of cleansing the body, is normally a mixture of alkali metal salts of long-chain fatty acids. Triglycerides, for example tallow, palm oil and coconut oil, provide the basic 'fats' from which the fatty acid mixtures used for soap are derived. The finished soap properties are primarily dependent on the mixture and ratio of triglycerides used. Tallow, for example, gives a much harder soap than coconut oil. Potassium soaps are much softer than their sodium-based counterparts, although, in practice, they are rarely used. The finished soap bar can be modified by the addition of other ingredients, such as emollients, opacifiers and chelating agents.

Although there is no evidence to suggest that soap is harmful to a baby's skin, its very alkaline pH (typically 9–10) is a disadvantage, and is implicit in the drying effect that soap is considered to have on skin. Consequently, baby soaps often contain emollients or superfatting ingredients (to help minimise the drying effect), white colour and low fragrance levels.

This disadvantage of soap led to the development of a new type of cleansing bar, with mild properties and superior skin-feel, ideally suited for babies and young children. Such products, commonly known as 'syndet' bars, are made from synthetic detergents, most of which are based on isothionate or sulphosuccinate systems. Various additives, including plasticisers, binders and lather enhancers, can be added to the product to modify the properties of the finished bar.

The mildness of the syndet bar is largely attributable to its 'neutral' pH (5–7). To date, syndet bars have not widely penetrated the market, particularly the baby-care market, for two main reasons: (i) they are much more difficult to manufacture than their soap-based equivalents; and (ii) the production costs are between three and five times that of soap. Syndet bars marketed as baby products should be white in colour, not discolouring with age, and should contain minimum practical levels of fragrance, thus ensuring maximal consumer perception of mildness in use.

5.3.4 *Hair products*

In the context of baby use, hair products can be strictly limited to shampoos and conditioners, the former being far more important than the latter. The most important requirements are mildness and low eye-sting potential. The latter is particularly relevant since the blink response of babies and young children is not fully developed and gives no protection to the cornea against contact with external substances. The low activity of a baby's sebaceous glands means that the potential for soiling of the hair is not particularly high, therefore the cleansing requirements for an effective product are not too severe.

Mildness is normally achieved through the use of non-irritating surfactant systems, all of which have limited detergency properties. One of the earliest, and perhaps still the best known baby shampoo, is that from Johnson & Johnson. This product was the first to make extensive use of amphoteric imidazolines, blended in specific ratios with milder types of alkyl ether sulphate. This combination was used in conjunction with selected non-ionic surfactants to give the now famous 'no more tears' claim.

In more recent years, many new amphoteric surfactants have been developed, allowing much greater flexibility in the formulation of mild baby shampoos. Amphoteric betaines and anionic sulphosuccinates, in combination with alkyl ether sulphates (3 or 4 moles ethylene oxide) produce detergent systems for mild shampoos. Much attention has also been given to effective irritation mitigants which, when added to already mild surfactant systems, further ameliorate the potential for adverse reactions in the eye. Classically, both sorbitan and sucrose fatty acid esters have been used for this purpose. A typical formulation for a mild baby shampoo is illustrated below.

	Wt%
Sodium lauryl ether (3.0) sulphate, 70%	6.00
Cocamidopropyl betaine, 30%	12.00
Polyoxyethylene (80) sorbitan monolaurate	6.00
Polyethylene glycol distearate	1.50
Water	74.50
Perfumes, preservative, colour, etc.	*q.s.*

In recent years there has been increasing concern over the potential toxicity of 1,3-dioxane, an impurity found in most ethoxylated surfactants. Consequently, baby shampoos based on detergent systems containing only amphoteric surfactants have been developed. The perceived need for this change is largely unfounded however, as levels of dioxane found in modern ethoxylated surfactants materials are very low indeed.

With respect to hair conditioners, it must be recognised that the primary function for baby use is detangling, rather than conditioning *per se*. A baby's hair does not require conditioning in the same way as that of an adult as it has not been subjected to regular combing and styling treatments. The detangling attributes are, however, extremely important, as even gentle combing after washing can result in knotting and breaking, causing discomfort and distress for the child. As a result, conditioning agents in baby conditioners are chosen for their ability to impart good wet-combing and slip characteristics, rather than to improve dry hair properties—the main aim of adult products. Relatively low levels of quaternary ammonium salts, such as distearyl dimethyl ammonium chloride are used to achieve this objective. A typical formulation for a baby hair conditioner is illustrated below.

	Wt%
Distearyl dimethyl ammonium chloride	2.00
Ethoxylated stearylamine	0.30
Cetyl alcohol	1.50
Hydroxyethyl cellulose	0.80
Water	95.40
Perfumes, preservative, colour, etc.	*q.s.*

Whilst the potential for eye sting from a conditioner is significantly less than that for a shampoo, mildness should still be paramount in the formulators mind. Typically, quaternary ammonium salts with higher molecular weight hydrophobic chains offer the best potential for mildness on the eye, due to their increased molecular size and distribution of the cationic charge. A material of this type which has more recently become popular is behentrimonium (C_{22}) chloride.

5.3.5 *Bath products*

Baby bath products must provide both gentle cleansing action, and pleasant aesthetics for mother and baby during the bathing process. Like baby shampoos, baby bath products need not possess high activity, as the extent of soiling of a baby's skin will be low by comparison with that of an adult.

The primary requirement of a baby bath is that it should be functional, yet mild on the skin, and be as non-drying as possible. When diluted in bath water, the product is often used to shampoo a baby's hair, therefore low eye irritancy is essential. For this reason, baby bath formulations generally contain low active detergent systems composed of mild alkyl ether sulphates (3 or 4 moles ethylene oxide) combined with a relatively high ratio of imidazoline or betaine foam boosters. In this way, sufficient quantities of light, open foam for the cleansing process, with no significant drying of the skin at the in-use dilution, is obtained. The viscosity of the product should be such that easy and rapid dispersion in the bath water is possible with minimum agitation. A perfume level as low as possible, but sufficient to generate a pleasant odour in use is desirable. Additional ingredients may be used to minimise any drying effect and enhance the softness and feel of the skin after use. Many materials are claimed to be beneficial. Care must be taken when selecting hydrophobic oils for this purpose, so as not to impair foaming characteristics. A popular means of enhancing moisturising properties is to include low levels of polyol fatty acid esters, with which a minimal effect on the foaming capability of detergent systems, and considerable enhancement of pleasant skin-feel may be achieved. The formulation shown below illustrates this.

	Wt%
Sodium lauryl ether (3.0) sulphate, 70%	5.00
Lauryl betaine, 30%	8.00
PEG-7 glyceryl cocoate	2.00
Water	84.50
Electrolyte (viscosity builder)	0.50
Perfumes, preservative, colour, etc.	*q.s.*

A second, although much less important category of bath product, is baby bath oil. Here the emphasis is on mildness rather than cleansing, although the use of such products does provide a pleasant means of removing superficial soil from the baby's skin in a bath environment. A simple baby bath oil formulation is illustrated below [2].

	Wt%
Caprylic/capric triglyceride	80.00
Polysorbate 85	20.00
Perfume, preservative, colour, etc.	*q.s.*

5.3.6 *Moussed products*

The term 'mousse' normally refers to a product that is filled into a sealed container, which is pressurised with a propellant that serves both to 'mousse' the product, and to expel it from its pressurised container. Mousses offer novel ways of improving the convenience of baby products, particularly in view of their suitability for single-handed operation. They also offer the advantage of being 'pre-foamed' at the time of dispensing, providing significant in-use benefits for baby shampoos and body cleansing products. However, at the present time, there are very few moussed baby products on the market. The reason is unclear, but the additional cost, the historical connection with adult hair-care products, and the potential danger of pressurised containers if misused, may be significant factors. An early product was 'Care For Kids' baby shampoo mousse, launched in the US by the Revlon Group in the mid to late 1980s. Despite views that this would lead to an increased demand for similar products, the market remains largely undeveloped.

Products available in the market today are principally skin care mousses, where the functional benefit is moisturisation rather than cleansing. These formulations are normally based on low viscosity oil-in-water emulsions, utilising an oil-soluble propellant system, which provides an instantaneous creamy foam when the product is dispensed.

When formulating baby mousses, the safety considerations normally associated with baby products are paramount. In particular, the propellant system chosen must present no hazard to a baby's skin or respiratory system, even under conditions of predictable misuse.

5.3.7 *Wipes and tissues*

Baby wipes and tissues represent one of the fastest growing sectors of the baby products market. There are essentially two types of baby wipe, categorised by function of either 'cleansing plus freshening' or 'cleansing plus moisturising'. The first type are normally 'aqueous' wipes, while the latter are referred to as 'lotion' wipes, depending on the nature of the wipe impregnant. The wipe itself is normally made from paper or fabric, the latter offering improved strength and aesthetics over the former, at higher cost. Paper substrates may be either 'wet-laid' or 'air-laid' and are manufactured in different weights, depending on the properties required. Fabric substrates come in a variety of different materials, including viscose–rayon, cotton, or various combinations of the two.

Aqueous wipe impregnants are composed largely of water, with the addition of solubilisers to improve cleansing properties and assist in the stabilisation of low levels of fragrance. Lotion wipes, normally much more effective in cleansing than their aqueous-based counterparts, are generally stable, low viscosity oil-in-water emulsions, the internal phase of which provides the cleansing and moisturising properties. Emollient esters and silicones are often added to enhance post-use skin feel. A more recent innovation is the antibacterial or 'total care' lotion wipe. These formulations contain low levels of bactericidal agents which are included to help in the prevention of nappy rash. Additionally, water-repellent materials may be added to help protect the skin in the advent of nappy soilage. The production of a safe and effective wipe product, with adequate preservation throughout the expected shelf-life, is a problem. A wipe with a high water content provides an ideal environment to support microbial growth, if contaminated with adventitious micro-organisms. This is discussed more fully in section 5.8.

5.3.8 *Oils*

Traditional oils, offering gentle and effective cleansing, particularly in the nappy area, are one of the oldest forms of baby product. The oil itself has a softening and moisturising effect on the skin, the residual layer affording some barrier protection against subsequent nappy soilage.

Historically, product design has been simple and straightforward with high purity, low viscosity mineral oil enjoying almost universal use. Additional ingredients are usually restricted to fragrance and a small quantity of solubiliser. A major disadvantage is the inevitable greasiness in use. Overcoming this problem by the use of 'degreasing' additives, for example low levels of fumed silica, have not been successful.

Recently the design approach to baby oils, motivated by the requirement to deliver an effective cleanser with improved aesthetics and lower greasiness, has changed. Many methods have been tried, one of which is the extensive use of vegetable-based oils such as soya-bean oil, jojoba oil or palm kernel oil. Although these are not such effective cleansers, they are often perceived as being less greasy. They are, however, more readily oxidised, leading to rancidity and malodour.

Contemporary baby oils use another approach to improve product aesthetics. High purity mineral oil is mixed with light esters, volatile silicone oils or mixtures of the two. Esters, for example isopropyl myristate, 'dilute' the greasy effects of the mineral oil, while volatile silicones, such as cyclomethicone, evaporate rapidly from the skin's surface after application, leaving behind a thinner oil layer. A light, non-greasy baby oil may be obtained using the following formulation.

	Wt%
Mineral oil	70.00
Octyl palmitate	5.00
Isopropyl myristate	10.00
Cyclomethicone	15.00
Perfumes, solubilisers, antioxidants, etc.	*q.s.*

5.3.9 *Perfumes and colognes*

The popularity of perfumes and colognes for baby use varies in different parts of the world, more so than with any other product. The prime function of a perfume is to leave a pleasant odour on the baby's skin. A cologne achieves a lesser but similar effect, while simultaneously cooling the skin.

Not surprisingly, colognes are particularly popular in hot countries, especially parts of southern Europe around the Mediterranean, where they are often applied several times a day to cool the skin and alleviate discomfort caused by heat. Most baby colognes are water-based products, containing low levels of fragrance and a solubiliser to assist with fragrance dispersion. Other additives include plant extracts, said to impart soothing benefits to the skin. Microbiological contamination from these materials may be a hazard. The cooling effect of the cologne can be greatly enhanced by the addition of relatively low levels of ethyl alcohol. This is rarely done because of the adverse drying effects of alcohol on the skin.

When formulating baby perfumes and colognes, product safety must be the prime concern. Of all cosmetic ingredients, fragrances are most commonly associated with skin allergies and sensitisation reactions. It follows therefore, that any fragrance compound selected for use in this type of baby product should be subjected to the most rigorous safety assessment, before commercialisation is contemplated.

Perfumes for babies are controversial, as the 'image-led' marketing strategy for perfumes and fine fragrances has little relevance for babies, whose social interaction is confined primarily to that of the parent-child relationship. Nevertheless, perfumes and fine fragrances for babies are currently very popular in some parts of Europe, notably in France. Obviously, babies and young children are not sufficiently developed to determine their own social image through the fragrances they wear. Perhaps the perceived need for this type of product is driven largely by the parents' desire to extend their own social values to their offspring.

5.4 Raw materials for baby products

The prime requirement for any baby product is absolute safety. Great care should be taken that raw materials used are not only efficacious but also innocuous, with no incidence of irregularity in their toxicological profile. Only the purest grades of material should be selected, and careful attention paid to the types and levels of any impurities that may be present. This is increasingly important with the current use of pesticides, antioxidants and preservatives, with which even low levels of contamination in the raw material may be potentially harmful to a baby's delicate skin. If any doubt is encountered over raw material selection, significant confidence can be gained by the use of materials which conform to the requirements of the British Pharmacopoeia (BP), European Pharmacopoeia (EP) or the United States Pharmacopoeia (USP). Specifications must encourage tight control over the quality of materials used, such that confidence in the batch-to-batch consistency is high. The concentrations of raw materials must be carefully chosen to eliminate all risks, even of minor skin reactions. The comedogenic nature of any material destined for use in baby products must also be considered. The 'comedogenicity' of a material describes its potential to form comedones, or 'spots', on the skin. Use of comedogenic materials should be avoided.

Particular diligence is required in the selection of raw materials for detergent-based products, which, by their very nature, exhibit a higher propensity for irritation of the skin. Even the so-called 'mild' detergents will extract natural lipids from the skin's surface layers. The formulator's skill in selecting a synergistic blend of detergent materials that exhibits lower irritancy than its component parts is particularly valuable. Materials used should exhibit very low microbial counts, and the absence of any micro-organism that can be classified as pathogenic is mandatory. This is particularly relevant in the case of talc, starches, natural gums and plant extracts, where high levels of contamination by micro-organisms, sometimes pathogenic, are known to occur.

Of particular importance is the quality of water used. The purity of this ingredient, a major component in the vast majority of products, often overlooked, is one of the most common causes of problems in the finished product, particularly those of a microbiological nature.

5.5 Developmental pathways

5.5.1 *Formulation development*

Formulation starts with an evaluation of the marketing concept and accompanying brief, clearly identifying the product design. Early developmental work will focus on establishing a 'base formulation', which can be progressively modified to satisfy the product brief. This base technology can be derived from many sources, including the formulator's personal knowledge, prior technology of a similar type, or the vast quantity of guideline formulations available in raw material suppliers' literature. Product prototypes, satisfying as closely as possible the performance requirements of the original brief are then submitted for initial evaluation of product efficacy, preservative efficacy, product stability and product safety. At this stage, the manufacturing process should be studied (see section 5.10). Optimisation of the product, and subsequent full evaluation, aids the decision as to whether or not to launch the product.

5.5.2 *Practical requirements*

Although developmental pathways can be precisely defined, practical requirements for any new product development must be observed. Aspects of safety have already been highlighted as mandatory. In addition the formulation must support the marketing concept, allow easy and convenient use of the finished product, and be compatible with the chosen packaging presentation. The product should meet the defined cost targets, allowing effective marketing of the finished product, while offering the purchaser a high quality product at a realistic price.

5.6 Product evaluation

Product evaluation is an integral step in any new product development exercise. Three main techniques are used to ensure that the product meets the defined performance criteria.

5.6.1 *Laboratory evaluation*

Laboratory techniques are used to evaluate early development prototypes and, by screening formulations, to assist in the development process before progressing into more sophisticated and expensive evaluation exercises. The many established laboratory techniques available may be modified to suit the formulator, and to extend the range of possibilities. Viscosity, density and pH are readily determined using simple instrumental techniques, and can help to identify correct product design rapidly. In the development of detergent products, the Ross–Miles test is a useful means of determining the quantity and quality of foam generated by products in-use. Skin absorption characteristics for emulsion-based products can be predicted through careful selection of raw materials, combined with rheological evaluation in the laboratory. In baby products containing active ingredients, for example nappy-rash creams, the active material can often be measured spectrophotometrically to determine its availability in the product base.

5.6.2 *Mother and baby panels*

While useful data on product performance may be obtained with laboratory techniques, the most meaningful tests are in-use studies. Classical in-use evaluation techniques involve either panel or consumer testing, but assessment of baby products presents a special problem in that the end-user, the baby or child, is frequently unable to express an opinion. For this reason, a common technique for evaluating the performance of baby products is to use mother and baby panels. These panels may be conducted with the assistance of the local health clinic. The mother uses the test product on her child for a fixed period of time and subsequently acts as the respondent for the research exercise. The mother may also use the product on herself, yielding valuable information regarding the performance of the product on adults, and providing an opinion of the suitability for the baby. This format is one of the most valuable ways of assessing product performance since it is the mother, and not the child, who will ultimately buy the product.

Clinical trials use the same methodology as mother and baby panels but are carried out with a higher degree of medical supervision, monitoring the possibility of adverse skin reactions. Such trials, which normally utilise 50 to 100 respondents, are particularly useful when assessing the performance of 'active' products, such as nappy-rash creams.

Finally, it is essential to note that adequate product safety must be obtained for the product under test before any evaluation using mother and baby panels is undertaken.

5.6.3 *Consumer research*

For the reasons outlined earlier consumer research, the final evaluation step prior to launch of the product, is not always used for baby products. It can, however, provide certain information that the mother and baby panel cannot.

The most valuable data is information about the acceptability of branding for the product. In all human evaluation trials, informed consent must be sought before starting the test, and requirements for product safety must not be compromised in any way.

5.7 Product safety requirement

The need for the highest standards of product safety for baby-care products has been emphasised throughout this chapter. Notwithstanding, there are legal requirements for any toiletry product to be safe, for use as intended, throughout its shelf-life. There is, however, no defined evaluation programme for determining the safety of a product, and this decision is therefore left to the individual. The legal aspects must, of course, be observed but with baby products there is also a high degree of emotive importance in ensuring product safety. This must always be remembered. The perception of vulnerability of babies or young children, and the need to protect them from anything harmful is very strong.

A testing programme, ensuring that the product meets these high standards of safety, is not easy. Historically, such programmes may have included a wide variety of safety tests, culminating with evaluation using animal studies. This approach is expensive and time-consuming, and intense concern has been expressed over the unnecessary use of animals in safety evaluation. Consequently, it has become increasingly feasible to devise a testing programme that avoids animal testing altogether, yet provides a high degree of confidence in product safety. When adopting this approach, careful selection of raw materials and close examination of their toxicological profiles is essential. Fragrances used must conform, as a minimum requirement, to the guidelines laid down by IFRA (International Fragrance Research Association) and RIFM (Research Institute for Fragrance Materials). If sufficient toxicity data is lacking, an increasing number of *in vitro* tests are available to determine product safety. For example, cytotoxicology, using various cell cultures, has already been successfully employed to determine the irritation potential for detergent-based products such as baby shampoos.

When reasonably satisfied with toxicological safety data, studies on human volunteers are normally conducted using adults, not babies. Occlusive patch testing and arm immersion studies are often used to gain more knowledge about the safety profile of a product. Should circumstances demand, a number of more sophisticated methods are available. Human volunteer programmes should reflect the in-use situation, and be carried out over minimum period of two weeks. Testing on babies or young children may then be considered, but must only be carried out under strictly monitored conditions, such as the mother and baby panels referred to in section 5.6.2.

5.8 Product preservation

Product preservation is necessary for two reasons: (i) to prevent the product becoming contaminated during use; and (ii) to protect the user from the possibility of harmful infection. Preservation is particularly relevant in baby-care products often used around the nappy area, where there is normally a high incidence of faecal micro-organisms such as *Escherichia coli*. The potential for contamination of the product and reinfection of the user at a later date is high.

By definition, all preservatives are harmful to living organisms, therefore should be used with extreme caution. A full review of available toxicological data on any preservative used in baby products is mandatory. The preservative level should be kept as low as possible, consistent with adequate activity. Higher preservative efficacy can often be obtained by utilizing the broad spectrum and synergistic activity of preservative blends. In order to minimise total preservative addition levels in a product, it is essential that the raw materials used are microbiologically as clean as possible. A minimum micro-biological specification of less than 100 colony forming units (cfu) per gram is recommended and absence of pathogens is essential. The practice of over-preservation of a product to compensate for poor manufacturing practice must be avoided, particularly in the case of baby-care products.

Preservative efficacy testing is essential. The microbiological quality of toiletry and personal-care products is covered by recommendations and guidelines issued by the CTPA (Cosmetics, Toiletries and Perfumes Association) in 1975 [1]. The reference to baby products specifies less than 100 cfu per gram of product, and absence of pathogenic organisms.

Finally, the media often confounds the issue of product preservation. A particular preservative can become 'unfashionable' overnight, sometimes causing irreparable commercial damage to any products containing it. However, the temptation to produce 'preservative-free' products should be avoided, as the potential for harm to the user, through product contamination, far outweighs the risk of adverse skin reactions to the preservative system.

5.9 Product stability

Before a product can be launched, the producer must ensure that it will remain in good condition, be safe to use, and fulfil its intended function throughout its envisaged shelf-life. In terms of safety, this shelf-life may be defined as three years. Consequently, product stability is an essential consideration when any new product is developed. Clearly, since it is not practical to wait three years before launching a product, it is necessary, as far as is reasonably practical, to predict the product stability. The most common technique is an accelerated ageing study, which is similar for baby and adult products. Product parameters, such as appearance, odour, viscosity, density and active matter, are measured and the product is then stored under a wide variety of conditions, for a period of time. The minimum recommended number of storage conditions are listed below.

(i) Freeze thaw cycling (−30°C to room temperature; six cycles)
(ii) Storage at 5°C
(iii) Storage at room temperature in darkness
(iv) Storage at room temperature in the presence of UV light
(v) Storage at 40°C

Periodically, for example one month intervals, products stored under these conditions are reassessed and changes in measured or sensory attributes checked. During this exercise, microbiological challenge tests are carried out on product stored under various conditions, to see if preservative efficacy has been affected. The decision determining suitability for launch is largely arbitrary; the longer the test duration, the lower the risk in the marketplace. For baby products, maximum levels of caution are advised, and successful completion of ageing studies of not less than three months, and preferably six months, should be considered mandatory.

5.10 Manufacture and quality control

5.10.1 *Manufacture*

The production of baby products, or indeed any cosmetic or toiletry, is normally a simple process. It involves the mixing of the required raw materials, perhaps with heating and cooling, in a variety of ways. Of paramount importance for baby products are the procedures used, general cleanliness in the production area and compliance with GMP (good manufacturing practice) guidelines.

Mixing vessels used for manufacturing baby products should conform to a number of fundamental requirements. Fabrication should be of high quality stainless steel, at least grade 304 and ideally grade 316, and should have the optimum configuration for producing a wide variety of products. Steam-jacketed vessels, with a simple impeller and some form of high-shear mixing device to produce emulsions, are a good starting point. Vessels should be maintained in a clean condition and should be sanitised regularly, according to defined procedures. Associated pipework should not contain any 'deadlegs' and must be continually cleaned and sanitised. Scoops and stirring paddles used for mixing should also be fabricated from stainless steel and should be kept clean and dry at all times. Control of raw materials and manufacturing processes must be specified according to GMP requirements, and all personnel should be adequately trained.

Finally, the quality of water used for the manufacture of baby products should be of the highest order. Water should be chemically purified using a two-stage deionisation plant, and microbiological integrity may be obtained through the use of ultra filtration and UV irradiation treatment.

Water for the manufacture of baby products should be used immediately after purification, ideally on demand and should on no account be stored in static containers prior to use. Finished bulk product should be filled immediately into the primary packaging in order to preserve product integrity. If finished bulk has to be stored for any length of time prior to filling, purpose-designed hygienic storage containers should be used.

5.10.2 *Quality control*

Quality control, the final, extremely important process before a product is released for general sale, must ensure that all product leaving the factory meets the required specification. There is currently an increasing emphasis on quality assurance rather than quality control, the essential difference being to monitor the entire manufacturing process rather than to measure the product it produces. This is the approach of total quality management, and is nowhere more relevant than in the production of baby products (see chapter 10).

References

1. CTPA (1975) Cosmetics, Toiletries and Perfumes Association.
2. Croda Formulary 1 (1994) *Baby Care.* Croda Chemicals Ltd.

Further reading

Harry, R.J. (1982) *Harry's Cosmeticology.* 7th edn. Longmans, London.

Love, P.M. (1989) *What Makes Baby Different*? Unpublished presentation to the Society of Cosmetic Scientists of Great Britain.

6 Afro products

J.F.L. CHESTER

6.1 Introduction

This chapter deals specifically with cosmetic products designed for use by people of African origin. Intensive research and development of products in this field during the 1980s has resulted in the acceptance and recognition of this class of product as a legitimate, and now established part of the general skin, and in particular hair care market. The development of such products is therefore a relatively recent consideration. It must above all take into account fashion and the different characteristics of the skin and hair of this ethnic group.

6.2 Hair structures

The fundamental properties and chemistry of hair are outlined in chapter 2. The purpose of the present chapter is to point out important differences that have a bearing on the approach to formulating Afro products. First impressions suggest that Caucasian and Afro hair are structured very differently, Caucasian being straight and Afro tightly curled. Under the microscope, however, the two hair types appear similar, except that Afro hair is somewhat elliptical in transverse section whereas Caucasian hair is circular. Chemical analysis reveals that the amino acid contents are very much the same. The distribution of sulphur-containing amino acids, however, is different and accounts for the structural variations that give rise to the quite different appearance. In wavy, tightly curled hair there is a non-uniform distribution in that the disulphide bridges are more prevalent on one side of the hair fibre.

The principal reason why specific products are required for treatment of Afro hair lies in the widespread use of chemical treatments to restyle or straighten it. These treatments rupture the bonds that link protein fibres, and thus weaken the structure of the hair. Indeed, the more drastic treatments, which cleave peptide bonds, can cause the destruction of the fibres themselves. Consequently, formulations must be chosen with specific recognition of these facts. Products in this range can be divided into three categories: (i) products for preconditioning, which ensure well-conditioned hair prior to subsequent chemical treatments; (ii) less-aggressive chemical treatments than those normally designed for the straightening process; and (iii) products that will restore lustre, body and moisture content to chemically treated hair.

6.3 Skin characteristics

In contrast to the relatively subtle chemical differences between Caucasian and Afro hair, there are more fundamental differences in the skin types. These must be considered when designing products for the care and decoration of the skin.

Colour is the most obvious difference and is due to the variation of pigments—haemoglobin, carotene and melanin—in the epidermis. The level of melanin is the crucial factor in determining the colour of skin. The greater quantity of melanin in black skin is due to the greater activity of the melanocytes that produce the grains of melanosomes. These 'mature' into melanin in the epidermis, developing from colourless to black and moving closer to the skin surface. In white skin the maturing process occurs only to a limited extent, and smaller, round melanosomes tend to clump together in bundles. In black skin the larger, oval, mature melanosomes are singly dispersed and evenly distributed throughout the epidermis giving it its characteristic colour. There is considerable variation in the quantity of melanin present in 'black' skin, and this gives rise to a much wider range of skin tones. Consequently, in order to represent the full range of skin tones found in dark-skinned people, some 35 shades are required.

Black skin characteristically possesses a tough outer layer of the epidermis. This is prone to scarring, which can result in serious disfigurement (see below). Black skin also contains a greater number of sebaceous glands than white skin (40–60% more). This does not, however, automatically result in more oiliness, although it often produces a shiny surface. All basic skin types (dry, sensitive, normal, oily and combination) are observed for black skin just as they are for white.

Many problems associated with black skin arise from changes in pigmentation. For example, loss of colour (hypopigmentation) is common to all races and occurs in disorders such as the little-understood vitiligo (leucoderma). This effect, unsightly in white skins, can be devastating in black skins. Darkening of the skin or hyperpigmentation, again occurring in all races, can be a particular problem for black women as a result of hormonal changes during pregnancy, or the use of the contraceptive pill. Both darkening and lightening of the skin (due to local increased pigmentation or destruction of melanocytes) can occur as the horny layer of the skin heals after damage caused, for example, by spots and rashes. Common skin conditions such as acne can lead to serious disfigurement for those with black skin. When formulating facial care products for black skin it is therefore especially important to select raw materials shown to possess low comedogenicity so as to avoid irritation or the initiation of comedones 'acne'.

Ashy skin (xerosin) is a particular problem resulting from excessive drying of the skin. This can be caused by harsh detergents or environmental effects such as extreme cold spells or excessive dry heat. The condition manifests itself as dry white patches, commonly on elbows, knees and backs of hands. In areas where conditions are extremely dry, occlusive preparations inhibiting water loss are particularly valuable, although they would be entirely inappropriate for areas of high humidity. For these conditions, light moisturising non-occlusive preparations, allowing passage of moisture to and from the skin, are required.

6.4 Hair products

The formulations described in the following sections are examples designed to offer products suitable for restyling Afro hair using treatments which take account of the need to condition and remoisturise before, during, and after treatment. Where applicable, suppliers of formulation ingredients are given. Full details of these, and other suppliers, can be found in Appendix I.

6.4.1 *Relaxing and restyling products*

Hair-straightening treatment using caustic alkaline chemicals has a severely detrimental effect on the hair shaft. Such treatment should never be used on hair in poor condition. The treatment procedure therefore comprises three stages:

(i) a preconditioning programme;
(ii) a relaxer treatment (sodium, potassium or guanidine hydroxide); and
(iii) a post-conditioning treatment.

6.4.1.1 *Preconditioning programme* A conditioning programme prior to the relaxing treatment is necessary to improve the physical state of the hair. Ideally, this should be started two weeks previously to allow build-up of sebum and protective oils on both hair and scalp. The oily build-up slows the relaxation process and lessens damage to the hair. A three-stage process is recommended:

(i) initial shampoo with protective conditioning shampoo;
(ii) moisturising with hot oil treatment; and
(iii) deep conditioning using rinse-off conditioners containing substantive actives.

Table 6.1 Protective shampoo C1454

	wt%
Empicol AL30T (ammonium lauryl sulphate, 28%)[a]	30.00
Crotein Q or Crodacel QM (steartrimonium hydrolysed animal protein) or (cocodimonium hydroxy ethyl cellulose)[b]	0.50
Crodafos SG or N3N (PPG-5 ceteth-10 phosphate) or (DEA oleth-3 phosphate)[b]	3.00
Incronam 30 (cocoamidopropyl betaine)[b]	10.00
Triethanolamine, 99%	0.30
Deionised water	to 100
Perfume, preservatives, colour	*q.s.*

[a]Albright & Wilson; [b]Croda Chemicals Ltd
Method of manufacture: dissolve Crotein or Crodacel in water along with the Crodafos and triethanolamine; add Empicol followed by Incronam 30. Make final adjustment to pH 5.0–5.5.

Table 6.2 Conditioning shampoo C1219

	wt%
D-Panthenol	0.50
Crodasinic LS30 (sodium lauroyl sarcosinate)[a]	20.00
Incronam 30 (cocoamidopropyl betaine)[a]	10.00
Empilan CDE (cocoamide DEA)[b]	5.00
Incroquat SDQ25 (stearalkonium chloride, 25%)[a]	0.20
Croquat L (lauryldimonium hydroxypropyl hydrolysed animal protein)[a]	1.00
EDTA	0.10
Deionised water	to 100
Lactic acid	*q.s.* to pH 6.7
Perfume, preservatives, colour	*q.s.*

[a]Croda Chemicals Ltd; [b]Albright & Wilson
Method of manufacture: warm together all components to 60–65°C. When homogeneous, stir to cool.

Table 6.3 Conditioning shampoo C1343 (high concentration)

	wt%
Incronam 30 (cocoamidopropyl betaine)[a]	13.00
Empicol ESB3 (sodium laureth sulphate, 28%)[b]	46.00
Crodalan AWS (polysorbate (80) (and) cetyl acetate (and) acetylated lanolin alcohol)[a]	1.00
Crotein Q (steartrimonium hydrolysed animal protein)[a]	1.00
Empilan CDE (cocamide DEA)[b]	1.00
Deionised water	to 100
Lactic acid	to pH 6.00
Perfume, preservatives, colour	*q.s.*

[a]Croda Chemicals Ltd; [b]Albright & Wilson
Method of manufacture: simple blend, with warming and mechanical agitation as necessary. Note this is a high concentrate shampoo; if desired, the detergent content may be reduced to make a cheaper milder product.

Conditioning shampoo Conditioning shampoos (see Tables 6.1 to 6.5) contain a protective protein and are designed to clean, increase the strength of the hair and reduce the porosity of the hair fibre. They may also be used as post-relaxing conditioners.

Table 6.4 Conditioning shampoo C1571M3

	wt%
Empicol AL30T (ammonium lauryl sulphate, 28%)[a]	40.00
Incronam 30 (cocoamidopropyl betaine)[b]	6.30
Incromate SDL (stearamidopropyl dimethylamine lactate)[b]	2.00
Empilan CDE (cocamide DEA)[a]	2.25
Dow Corning 193 surfactant (dimethicone copolyol)[c]	2.25
EGMS PE3127 (glycol stearate)[b]	1.00
Croquat L (lauryldimonium hydroxypropyl hydrolysed animal protein)[b]	0.50
Crodacol CS/BP (cetostearyl alcohol)[b]	0.50
Deionised water	to 100
Lactic acid/sodium lactate buffer	to pH 5.50–6.00 (approx. 1.0%)
Perfume, preservatives, colour	*q.s.*

[a]Albright & Wilson; [b]Croda Chemicals Ltd; [c]Dow Corning Ltd
Method of manufacture: combine all ingredients with half of the quantity of water and heat to approximately 70°C. With stirring, add remaining cold water. Stir to cool.

Table 6.5 Pearlised conditioning shampoo C1571M3

	wt%
Empicol AL30T (ammonium lauryl sulphate, 28%)[a]	40.00
Incronam 30 (cocoamidopropyl betaine)[b]	6.30
Empilan CDE (cocamide DEA)[a]	2.25
Dow Corning 193 surfactant (dimethicone copolyol)[c]	0.50
EGMS PE3127 (glycol stearate)[b]	1.00
Croquat L (lauryldimonium hydroxypropyl hydrolysed animal protein)[b]	0.50
Crodacol CS (cetostearyl alcohol)[b]	0.50
Deionised water	to 100
Lactic acid/sodium lactate buffer	to pH 5.00–6.00
Perfume, preservatives, colour	*q.s.*

[a]Albright & Wilson; [b]Croda Chemicals Ltd; [c]Dow Corning Ltd
Method of manufacture: combine all ingredients (excluding the buffer) with half the quantity of water and heat to approximately 70°C. With stirring, add remaining cold water. Stir to cool. In addition to pretreatment, this shampoo represents an excellent conditioner shampoo for post-treatment cleansing of the hair.

Hot oil treatment This treatment (Table 6.6) can use either a traditional oil-based product, a light mineral oil alone, or an aqueous-based product, which relies on the substantive effects of a cationic material (i.e. cocotrimonium chloride and quaternary proteins or gums) combined with the moisturising effects of a humectant.

Table 6.6 Hot oil treatment C1463M1

	wt%
White mineral oil (25 cSt at 25°C)	to 100
Crodamol ICS (isocetyl stearate)[a]	15.00
Crodamol PMP (PPG-2 myristyl ether propionate)[a]	5.00
Incroquat SDQ-95 (stearalkonium chloride)[a]	0.20
Novol (oleyl alcohol)[a]	2.50
Isopropyl alcohol	1.00
Perfume, preservative, colour	*q.s.*

[a]Croda Chemicals Ltd
Method of manufacture: simple mixing with gentle heat to dissolve the solid material. Cool. Add perfume.

Table 6.7 Deep pre-conditioner C5004

	wt%
Lanolic acid (lanolin acid)[a]	2.00
Polawax NF (non-ionic emulsifying wax)[a]	3.00
Novol (oleyl alcohol)[a]	1.00
White petroleum jelly	4.00
Crodacol S95 (stearyl alcohol)[a]	4.00
Incroquat SDQ-25 (stearalkonium chloride, 25% active)[a]	10.00
Crotein O (hydrolysed animal protein)[a]	1.00
Crotein HKP/SF (hydrolysed keratin)[a]	0.50
Croderol GA7000 (glycerine)[a]	3.50
Deionised water	to 100
Perfume, preservative	*q.s.*

[a]Croda Chemicals Ltd
Method of manufacture: melt oils and waxes together and heat to 70°C. Dissolve the protein and amino acids in water. Heat to 70°C and add oil phase. Cool with stirring.

Table 6.8 High activity deep pre-conditioner C5019

	wt%
Novol (oleyl alcohol)[a]	1.00
Crodacol C90 (cetyl alcohol)[a]	2.50
Incroquat SDQ-25 (stearalkonium chloride)[a]	1.80
Croquat L (lauryldimonium hydroxypropyl hydrolysed animal protein)[a]	1.00
GMS A/S ES0743 (glyceryl stearate (and) PEG-100 stearate)[a]	3.00
White mineral oil (25 cSt at 25°C)	4.00
Deionised water	to 100
Perfume, preservative	*q.s.*

[a]Croda Chemicals Ltd
Method of manufacture: heat oil phase at 75°C. Heat the aqueous phase to 80°C. Add the water to oil phase slowly with stirring. Stir to cool. Fill off at 30°C.

Deep conditioning Deep conditioning treatments use creams containing a high level of quaternary (cationic) product (e.g. stearalkonium chloride), humecant, fats and oily material (Tables 6.7 and 6.8). The conditioner should be applied for 10–20 min and then rinsed off.

6.4.1.2 *Relaxer treatments* These consist of two types: (i) sodium/potassium hydroxide or guanidine hydroxide hair-straightening formulations; and (ii) thioglycollate-based chemical relaxation treatment.

Sodium/potassium hydroxide or guanidine hydroxide hair-straightening formulations These formulations are effective but aggressive. In addition to the ameliorating preconditioning programme outlined in section 6.4.1.1, damage to the hair is lessened by the inclusion of emollient oils in the products.

Sodium/potassium hydroxide based relaxers (see Table 6.9) are formulated as viscous creams. The emulsification systems are often based on non-ionic, self-bodying emulsifying waxes, together with auxiliary emulsifiers

Table 6.9 Hair relaxer creams

	wt%			
	CM5042M1	C1624	C834M1	BW29
Polawax NF (non-ionic emulsifying wax)[a]	7.50	—	—	—
Crodacol C90 (cetyl alcohol)[a]	1.00	—	—	—
White petroleum jelly BP	20.00	20.00	20.00	21.00
White mineral oil (25 cSt at 25°C)	10.00	10.00	10.00	15.00
Crodafos N10N (DEA oleth-10 phosphate)[a]	1.50	—	—	—
Crodafos CES (cetearyl alcohol (and) cetearyl phosphate)[a]	—	—	—	7.00
Crodacol S70 (stearyl alcohol)[a]	—	—	—	2.00
Crodacol C70 (cetyl alcohol)[a]	—	—	—	1.00
Volpo S10 (steareth-10)[a]	2.50	—	—	—
Solan E (PEG-75 lanolin)[a]	—	0.50	0.50	—
Polawax A31 (non-ionic emulsifying wax)[a]	—	11.00	—	—
Volpo CS15 (ceteareth-15)[a]	—	1.00	—	—
Liquid base CB3929 (mineral oil (and) lanolin alcohol)[a]	—	5.00	—	—
Polawax GP200 (non-ionic emulsifying wax)[a]	—	—	15.00	—
Crodacol S95 (stearyl alcohol)[a]			1.00	
Volpo N15 (oleth-15)[a]	—	—	2.00	—
Polychol 5 (laneth-5)[a]			—	1.00
Solan 50 (PEG60 lanolin)[a]				0.50
Propylene glycol	2.00	5.00	2.00	2.00
Deionised water	to 100	to 100	to 100	to 100
Croquat M (cocodimonium hydroxypropyl hydrolysed animal protein)[a]	1.00		—	—
Sodium hydroxide, 25% aqueous solution	6.80	8.00	8.50	8.40
Perfume, preservatives, colour	*q.s.*	*q.s.*	*q.s.*	*q.s.*

[a]Croda Chemicals Ltd

Note: (1) For mild strength product, 7.0–7.5% of a 25% solution of sodium hydroxide is suggested.

(2) For medium strength product, 8.0–8.5% of a 25% solution of sodium hydroxide is suggested.

(3) For high strength product, 9.0–9.5% of a 25% solution of sodium hydroxide is suggested.

Method of manufacture: heat oil and water phases separately to 65°C. Add water to oils with stirring. Add sodium hydroxide solution at 45°C; add perfume. Continue stirring and cool to 30–35°C, homogenise by milling, cool and fill into containers.

such as ethoxylated alcohols. A sodium hydroxide level of 1.9–2.4% is used in the formulation depending on the strength required. C5042M1 contains a protein derivative, for its excellent conditioning effects. High levels (25–30%) of protective emollients, i.e. hydrocarbon oils and waxes, must be included in these formulations to slow down the reaction of the caustic on the hair, and to reduce burning or irritation of the scalp. It is also important that only materials chemically stable at the high pH resulting from the NaOH are employed in this class of product.

Because of the difficulty in formulating products of this class, which are effective, safe and both chemically and physically stable, various raw material suppliers now offer pre-compounded emulsifying bases. To these bases the appropriate (prescribed) levels of protective emollients, active ingredient (Na or KOH), conditioner and water are added. The manufacturing technique involved in the preparation of these products is essentially the same as that described for the formulations illustrated in Table 6.9.

Alternative alkaline relaxer treatments use guanidine hydroxide in place of sodium hydroxide. Guanidine hydroxide is created *in situ* from guanidine carbonate and calcium hydroxide, necessitating a two-component system in which the active chemicals are kept separate until required (Tables 6.10 and 6.11). The guanidine relaxer is prepared by mixing four parts of the cream base (Table 6.10) with one part of the liquid activator (Table 6.11). This causes release of guanidine hydroxide to relax the hair.

Relaxation treatment using either sodium hydroxide or guanidine hydroxide systems requires application of the cream relaxers to the hair for approximately 20 min, after which it is removed by rinsing and shampooing. Extreme caution should always be exercised both in the development of and use of these products. Appropriate trials and tests must be performed in order

Table 6.10 Guanidine hydroxide hair relaxer C5002: cream base

	wt%
Polawax (non-ionic emulsifying wax)[a]	15.00
Crodacol CS/BP (cetostearyl alcohol)[a]	1.00
Volpo N10 (oleth-10)[a]	2.00
White petroleum jelly	8.00
White mineral oil (25 cSt at 25°C)	9.70
Deionised water	to 100
Propylene glycol	5.00
Calcium hydroxide	5.00
Perfume, preservatives	*q.s.*

[a]Croda Chemicals Ltd

Method of manufacture: heat the oil phase and water separately to 60°C. Add water to the oils with efficient agitation. Continue to stir until the cream starts to thicken (around 45°C). Add calcium hydroxide and stir with scraping to disperse.

Table 6.11 Guanidine hydroxide hair relaxer C5002: liquid activator

	wt%
Guanidine carbonate	25.00
Keltrol F (xanthan gum)[a]	0.20
Deionised water	to 100
Preservatives	*q.s.*

[a] Kelco/AIL Ltd
Method of manufacture: dissolve the Keltrol and preservatives in the water with stirring and heat to 70°C. Cool to 45°C and add the guanidine carbonate.

Table 6.12 Neutralising shampoo C5003

	wt%
Empicol AL30T (ammonium lauryl sulphate, 28%)[a]	20.00–40.00
Incronam 30 (cocoamidopropyl betaine)[b]	5.00
Incromine Oxide C (cocoamidopropyl amine oxide)[b]	3.00
Crodafos SG (PPG-5 ceteth-20 phosphate)[b]	3.00
Crotein HKP S/F (hydrolysed keratin)[b]	1.00
Citric acid	1.00
Deionised water	to 100
Perfume, preservatives, colour	*q.s.*

[a] Albright & Wilson; [b] Croda Chemicals Ltd
Method of manufacture: add the Empicol Al30T to water at 65°C. Add the Incronam 30 and the Incromine Oxide C; mix until uniform. Add the Crodafos SG. Cool to 50°C and add the Crotein HKP S/F and citric acid. Cool to room temperature and fill off.

to ensure the product is not over-aggressive, thus being destructive to the hair or irritating to the scalp. Following this, residual alkaline material on the hair must be completely neutralised. This is accomplished by washing the hair with an acid-based shampoo (Table 6.12). In order to ensure stability of the detergent base in the acidic environment, ammonium salts (e.g. ammonium lauryl sulphate) and amphoteric products (e.g. cocamidopropyl betaine) are incorporated into the shampoo. The use of an acid–base indicator will give a change of colour in the lather when all the residual alkali has been neutralised.

Thioglycollate-based chemical relaxation treatment An alternative chemical treatment giving a waved, less curly appearance to the hair, instead of the aggressive straightening achieved with sodium hydroxide relaxer products, is based on sulphur-containing reducing agents, principally thioglycollates. Thioglycollates disrupt sulphur–sulphur bonds in the hair protein by their reducing action; subsequent application of an oxidising agent (neutraliser) results in reformation of the protein sulphur–sulphur linkages. During the reforming of these bonds, the hair fibre is styled into a new, relaxed structure,

Table 6.13 Ammonium thioglycollate hair-relaxer cream C5001

	wt%
Crodacol C90 (cetyl alcohol)[a]	2.00
Crodafos N10N (DEA oleth-10 phosphate)[a]	1.50
Volpo S2 (steareth-2)[a]	0.50
White mineral oil (25 cSt at 25°C)	13.00
White petroleum jelly BP	11.50
Crosterene SA4310 (stearic acid)[a]	8.00
Deionised water	to 100
Propylene glycol	2.00
Volpo S10 (steareth-10)[a]	2.50
EDTA	0.50
Croquat M (cocodimonium hydrolysed animal protein)[a]	1.00
Ammonium thioglycollate, 60%	9.00
Ammonium hydroxide 88	4.63
Perfume, preservatives, colour	*q.s.*

[a]Croda Chemicals Ltd

Method of manufacture: heat oil and water phases separately to 65–70°C. Add water phase to oils under stirrer. Add ammonium hydroxide to ammonium thioglycollate. Cool emulsion to 45°C and add ammonium thioglycollate mix. Final pH must be adjusted to 9.5 ± 0.2 by further addition, if necessary, of ammonium hydroxide.

resulting in a less tightly curled appearance. As with the more drastic caustic-based treatments, thioglycollate relaxing of the hair has an adverse drying effect on the fibres—the hair will feel rough and have a greater porosity. An elaborate, conditioning, aftercare programme is therefore essential to keep the treated hair in optimum condition. Note that in the US hair is straightened with caustic, a process which can result in much hair fall and, if drastic, in baldness at the scalp. Soft curls are provided using thioglycollate treatment to relax the hair and then rolling onto rods (Jerry curls).

Ammonium thioglycollate products are formulated as gels or viscous creams (Table 6.13). They are applied to the hair for up to one hour, prior to wrapping the straightened hair on curling rods to style it. A slightly weaker ammonium thioglycollate solution (often called a curl booster or charger) is also applied. This is similar to thioglycollate permanent-wave treatments for Caucasian hair.

The incorporation of a quaternised protein provides beneficial conditioning effects, but these are merely temporary and will be removed when the hair is next shampooed. A more permanent conditioning treatment (e.g. the Kerasol* process) can also be incorporated into the thioglycollate treatment. This procedure provides a conditioning action, which is effective through numerous shampooings, by covalently bonding the conditioner protein 'Kerasol' to the hair between the reduction and oxidation steps in the waving process. In addition to imparting general condition, the hair is left more

* Croda Chemicals patented process.

Table 6.14 Clear gel permwave C279

	wt%
Deionised water	to 100
Procetyl AWS (PPG-5 ceteth-20) or Solan E (PEG-75 lanolin)[a]	1.00
Carbopol 941 (carbomer 941)[b]	1.86
Thioglycollate acid, 70%	7.24
Ammonium hydroxide, 28%	8.35
Sodium heptanoate dihydrate[a]	0.10
Perfume, preservatives, colour	*q.s.*

[a]Croda Chemicals Ltd; [b]B.F. Goodrich

Method of manufacture: dissolve Procetyl AWS/Solan E in a little water. Dissolve sodium heptanoate into 5% of water. Hydrate Carbopol in rest of water; when hydrated add ammonium hydroxide solution, then Procetyl AWS/Solan E solution; lastly add sodium heptanoate dihydrate solution. pH should be 9.5 ± 0.1.

Note: hydroxyethyl cellulose may be employed as the gellant in place of the Carbopol if desired.

Table 6.15 Kerasol permanent conditioner C1401

	wt%
Kerasol (soluble animal keratin)[a]	5.00–10.00
Crotein Q (steartrimonium hydrolysed animal protein)[a]	0.50
Deionised water	to 100
Perfume, preservatives, colour	*q.s.*

[a]Croda Chemicals Ltd

Method of manufacture: simple blend of components in water.

receptive to moisture, thus enhancing the effect of subsequent treatments such as curl activators, and achieving maximum beneficial effect. The overall process is as follows:

(i) Perm reduction step (Table 6.14)
(ii) Rinse
(iii) Kerasol conditioning treatment (Table 6.15)
(iv) No rinse
(v) Neutraliser oxidation step (Table 6.16)

The neutralising process, during which the sulphur linkages are reformed and the hair is 'locked' into its new shape or style, uses products based on peroxides or bromates. This is exemplified by the neutraliser shown in Table 6.16, which is based on sodium bromate.

The thioglycollate treatment is normally concluded with a conditioner, which must be light to avoid weighting down the curls. The product shown in Table 6.17 combines a fatty quaternary and a humectant to aid in conditioning and moisturising the hair. The use of a moisturising spray, combining humectant (e.g. propylene glycol) and acetamide MEA, together with protein additives, for remoisturising and improving the hair condition,

Table 6.16 Sodium bromate neutraliser C5045/hydrogen peroxide neutraliser no. 5

	wt%	
	C5045	no. 5
Croduret 40 (PEG-40 hydrogenated caster oil)[a]	3.00	—
Cellobond HEC5000A (hydroxy ethyl cellulose)[b]	00–0.30	—
Sodium bromate	10.00	—
Disodium phosphate	0.50	—
Deionised water	85.20	89.50
Crillet 1/perfume[a]	1.00	1.00
Hydrogen peroxide (30%)	—	6.50
Dow Corning Q2-7224[c]	—	4.00
Trimethylsilylamodimethicone (and) dimethicone (and) octoxynol 40 (and) isolaureth-6 (and) glycol		
phosphoric acid	—	to pH 3.4

[a] Croda Chemicals Ltd; [b] BP Chemicals; [c] Dow Corning Ltd.
Method of manufacture C5045: dissolve HEC in water and heat to 80°C, when hydrated, add Croduret and Crillet perfume blend, at 40–45°C add disodium phosphate and sodium bromate, adjust pH to 7.00 + or − 0.30.
Method of manufacture no. 5: mix all ingredients in the order given. Adjust pH to 3.4 with phosphoric acid.

Table 6.17 Hair conditioner C1410

	wt%
Conditioner base CB0967 (proprietary blend)[a]	4.00
Incromectant AMEA-70 (acetamide MEA)[a]	5.00
Deionised water	to 100
Lactic acid	to pH 4.00–4.50
Perfume, preservatives, colour	*q.s.*

[a] Croda Chemicals Ltd
Method of manufacture: heat all components together to 65–70°C, except perfume and lactic acid. Stir to cool, perfuming and adjusting pH to 4.0–4.5 at 45°C. Fill off at 40°C.

is also popular. Further possible treatments include curl activator products, hair-grooming creams and lotions, and moisturising and oil sheen sprays, all of which can be used to impart gloss and sheen, and to enhance the lustre of the hair. Curl activators based on Carbopol (see Table 6.18), contain a high level of humectants to activate the curl. In temperate zones; where a lighter treatment is required, propylene glycol in combination with glycerine is the humectant; in hot climates, a heavier, more durable, higher gloss preparation is recommended, with partial or total replacement of the glycol with glycerine. The preferred ratio of glycerine to glycol is generally 2:1. Extremely hot climates require the inclusion of suitable block copolymers (e.g. Ucon 50 HB660) to give a more lasting effect. The curl activator gel

Table 6.18 Curl activator gels based on Carbopol

	wt%	
	C1391	C1398
Carbopol 940 (carbomer 940)[a]	0.50	0.66
Propylene glycol/glycerine	25.00	12.00
Crillet 1 (polysorbate 20), Procetyl AWS (PPG-5 ceteth-20), or Croduret 40 (PEG-40 hydrogenated castor oil)[b]	6.00	5.00
Deionised water	to 100	to 100
Triethanolamine, 99%	to pH 6.0–7.0	—
Croderol GA7000 (glycerine)[b]	—	10.00
Diisopropylamine	—	to pH 6.5–7.0
Perfume, preservatives, colour	*q.s.*	*q.s.*

[a]B.F. Goodrich; [b]Croda Chemicals Ltd

Method of manufacture: hydrate Carbopol in water (60–65°C). Add propylene glycol and preservatives, neutralised to pH 6.5–7.0 with amine. Combine perfume and Crillet, and add to Carbopol gel. Stir until a clear homogeneous gel is obtained. Fill off. Note: for the preparation of heavier texture/improved durability, the propylene glycol may be wholly or partially replaced with glycerine.

Table 6.19 Curl activator gel C1128

	wt%
Crodafos N3N (DEA oleth-3 phosphate)[a]	7.48
Volpo N5 (oleth-5)[a]	10.88
White mineral oil (25 cSt at 25°C)	16.05
Incromectant AMEA (acetamide MEA)[a]	0.20
Sorbitol, 70% solution	11.00
Croderol GA7000 (glycerine)[a]	7.00
Hexylene glycol	2.00
Deionised water	to 100
Perfume, preservatives, colour	*q.s.*

[a]Croda Chemicals Ltd

Method of manufacture: heat oil phase to 65–70°C. Heat water phase to similar temperature. Slowly add water to oils with stirring, and cool rapidly.

shown in Table 6.19 is based on a microemulsion gel, a formulation which is suitable for all climatic conditions. Again, high levels of propylene glycol/glycerine are used. These systems are, however, sensitive to minor changes in raw materials and, due to the inherently high surfactant level, can be irritating to the eye.

Recommended formulations for emulsion-based hair-grooming products (combining humectants and emollient oils), a moisturising spray, and a comb-out oil are shown in Tables 6.20 to 6.23. These products are used at the end of the thioglycollate treatment to help to maintain the hair in its optimum condition.

Table 6.20 Hair grooming cream/lotion C1450

	wt%
GMS A/S ESO743 (glyceryl stearate (and) PEG-100 stearate)[a]	4.00
Crodamol MM (myristyl myristate)[a]	4.00
Liquid base CB3929 (mineral oil (and) lanolin alcohols)[a]	3.00
Crodalan LA4028 (cetyl acetate (and) acetylated lanolin alcohol)[a]	1.00
GMS GEO803 (glyceryl stearate)[a]	2.50
Volpo CS20 (ceteareth-20)[a]	1.00
Squalene	0.60
Deionised water	to 100
Croderol GA7000 (glycerine)[a]	10.00
Incromectant AMEA-70 (acetamide MEA)[a]	1.00
D-Panthenol	0.40
Incroquat SDQ25 (stearalkonium chloride)[a]	1.00
Perfume, preservatives	*q.s.*

[a]Croda Chemicals Ltd

Method of manufacture: combine oil phase and water phase and heat to 65–70°C. Add water to oils with stirring. Stir to 45°C and perfume. Stir to cool.

Table 6.21 Hair-softening spray lotion C1129

	wt%
Croderol GA7000 (glycerine)[a]	10.00
Propylene glycol	20.00
Perfume	0.20
Crillet 1 (polysorbate 20)[a]	2.00
Deionised water	to 100
Preservatives, colour	*q.s.*

[a]Croda Chemicals Ltd

Method of manufacture: blend Crillet and perfume, add Croderol, glycol, water and preservative. Stir till clear.

6.4.1.3 *Post-conditioning treatment* Post-relaxation treatment to combat the defatting nature of the relaxer includes use of a conditioner as in the preconditioning programme (Tables 6.1 to 6.5), together with a scalp grease (Table 6.24) to lubricate the scalp and prevent dryness and flaking between the roots.

To complete the relaxation treatment the hair can be styled in the normal way. Examples of a setting lotion and styling gels are given in Tables 6.25 and 6.26. A hair-groom cream, which is applied to provide a moisturising, glossy effect, is shown in Table 6.27.

Table 6.22 Moisturising spray mist C5016

	wt%
Propylene glycol	7.00
Croderol GA7000 (glycerine)[a]	1.00
Incromectant AMEA-70 (acetamide MEA)[a]	1.50
Crotein HKP/SF (hydrolysed keratin)[a]	0.50
Crodacel QM (cocodimonium hydroxy ethylcellulose)[a]	0.50
Deionised water	to 100
Perfume, preservatives, colour	*q.s.*

[a]Croda Chemicals Ltd
Method of manufacture: dissolve Crodacel QM in water. Add the protein and amino acids. Blend in the remainder of the ingredients. Fill off. This product is designed to be dispensed from a mechanical pump applicator.

Table 6.23 Comb-out oil C5008

	wt%
Lanexol AWS (PPG-12 PEG-50 lanolin)[a]	4.00
Crotein Q (steartrimonium hydrolysed animal protein)[a]	1.00
Croderol GA7000 (glycerine)[a]	10.00
Crodacel QM (cocodimonium hydroxy ethyl cellulose)[a]	0.10
Dow Corning 193 fluid (dimethicone copolyol)[b]	0.50
Deionised water	to 100
Perfume, preservatives, colour	*q.s.*

[a]Croda Chemicals Ltd; [b]Dow Corning Ltd
Method of manufacture: combine ingredients under good agitation.

Table 6.24 Scalp grease C5048

	wt%
Paraffin wax (140/145°C)	5.00
Crodamol IPM (isopropyl myristate)[a]	6.00
White mineral oil (25 cSt at 25°C)	30.00
Lanolin[a]	10.00
Super Hartolan (lanolin alcohol)[a]	3.00
White petroleum jelly	46.00
Perfume, preservatives, colour	*q.s.*

[a]Croda Chemicals Ltd
Method of manufacture: heat all ingredients together with the exception of the perfume. Cool, stir in fragrance and fill.

Table 6.25 Setting lotion C1444

	wt%
PVP/VA S630 (PVP/VA copolymer)[a]	1.50–3.00
Gafquat 755N (polyquaternium 11)[a]	1.50–3.00
Deionised water	to 100
Crillet 1 (polysorbate 20) or Crovol 70 series (ethoxylated vegetable oils)[b]	*q.s.*
Preservatives	*q.s.*

[a]GAF Europe; [b]Croda Chemicals Ltd
Method of manufacture: dissolve PVP/VA in water, followed by Gafquat. Pre-dissolve perfume in Crillet or Crovol (ratio 3:1) and add to water phase with stirring. Add preservative and colour and fill off. Note: The ratio of perfume to Crillet or Crovol may have to be altered to achieve clarity depending on perfume used. One part perfume to four parts solubiliser is given as nominal starting point.

Table 6.26 Hair styling gels

	wt%	
	C1123	C1465
PVP K30 (polyvinylpyrrolidone)[a]	2.00	3.00
Carbopol 940 (carbomer 940)[b]	0.50	0.50
Volpo N20 (oleth-20)[c]	1.00	—
Triethanolamine, 99%	—	0.75
Procetyl AWS (PPG-5 ceteth 20)[c]	0.50	—
Ethanol 74.OP	—	15.00
Crotein O (hydrolysed animal protein)[c]	0.50	—
Uvinol D50 (benzophenone-2)[d]	0.10	—
Crillet 1 (polysorbate 20)[c]	—	0.50
EDTA	0.10	—
Diisopropanolamine	0.30	—
Deionised water	to 100	to 100
Perfume, preservatives, colour	*q.s.*	*q.s.*

[a]GAF Europe; [b]B.F. Goodrich; [c]Croda Chemicals Ltd; [d]BASF (UK) Ltd
Method of manufacture:
(1) C1123: hydrate the Carbopol in one third of the water (60–65°C). Dissolve the remaining components in the rest of the water. Mix together—the product will gel. Adjust to pH 6–7.
(2) C1465: hydrate the Carbopol in one third of the water (60–65°C). When fully hydrated, neutralise with amine to form the gel. Dissolve PVP K30 in ethanol and add to gel. Pre-dissolve perfume in Crillet, add to gel and stir to clear. Fill off.

6.4.2 *Hair pomades and grooming aids*

Hair pomades—the original hair products—are examples of mild products offering modern formulations for use with a very traditional African hairstyle.

Table 6.27 Hair groom cream W/O C953

	wt%
Polawax GP200 (non-ionic emulsifying wax)[a]	5.00
White mineral oil (25 cSt at 25°C)	40.00
White petroleum jelly BP	4.00
GMS N/E GEO803 (glyceryl stearate)[a]	0.50
Deionised water	to 100
Carbopol 934 (carbomer 934)[b]	0.50
Propylene glycol	5.00
Triethanolamine, 99%	to pH 6.8–7.0
Perfume, preservatives	*q.s.*

[a]Croda Chemicals Ltd; [b]B.F. Goodrich

Method of manufacture: combine oil phase and heat to 70°C. Hydrate Carbopol in water phase, and heat to 70°C. Add oil to water with efficient agitation. Add triethanolamine at 60°C followed by perfume at 50°C. Fill off at 40°C.

'Canerow' is the art of plaiting the hair close to the scalp in a variety of patterns, a style which has passed from generation to generation amongst African women. Traditionally, natural oils such as shea butter and coconut oil are used to keep the scalp supple while the hair is being teased and stretched into tight plaits. The modern formulation for a hair pomade, C1411 (Table 6.28), consists of a blend of oil, petroleum jelly and waxes. This pomade is also suitable for use during the 'hot-comb' procedure for temporarily straightening the hair, a treatment which runs a considerable risk of burning the scalp and also has a damaging effect on the hair shaft. The temporary straightening achieved in this way only remains until the hair comes into contact with moisture, when the crinkly appearance will re-occur.

Table 6.28 Hair pomade C1411

	wt%
Lanolin[a]	4.00
White mineral oil (25 cSt at 25°C)	to 100
Microcrystalline wax (78–82°C)	28.00
Aluminium stearate G[b]	1.20
Crodamol IPM (isopropyl myristate)[a]	2.50
BHT	0.01
Perfume, preservatives, colour	*q.s.*

[a]Croda Chemicals Ltd; [b]Durham Chemicals Ltd

Method of manufacture: add aluminium stearate to cold mineral oil whilst stirring. Stir well until thoroughly dispersed. Maintain mixing and heat oil to 105–110°C, adding microcrystalline wax. When the aluminium stearate has completely dissolved, a slightly gelatinous solution will form. Begin cooling, adding the rest of the ingredients, except perfume, at 90°C. Add perfume at 75°C. Fill off at 70°C.

6.5 Skin products

Dermatological problems can often arise due to the use of cheap semi-refined materials; the use of reliable suppliers, coupled with an understanding of the significance of raw materials specifications, is an important step in producing satisfactory results. Careful raw materials selection plays an important part in designing formulations for any product, but is especially significant for products designed for black skin. The cosmetic qualities of all materials must be ensured.

6.5.1 *Raw material selection—factors for consideration*

6.5.1.1 *Emollient comedogenicity* The skin condition of acne (or comedone formation) causes particular problems to black skin. Work on the comedogenicity of common cosmetic emollients to assess their potential to cause (or aggravate) comedones, has resulted in the grading of emollients according to low, medium or high comedogenic categories (Tables 6.29 to 6.31). This provides a useful guide when formulating. Low comedogenic rated products (Table 6.29) are recommended for use in facial care products, materials with

Table 6.29 Emollient materials with a low comedogenic effect

Avocado oil	Beeswax (25% in mineral oil)	Candelilla wax
Castor oil	Glycerine	Cetyl alcohol
Cholesterol	Lanolin oil	Isocetyl stearate
Lanolin	Pentaerythritol tetraisostearate	White mineral oil
PEG-200	Stearyl alcohol	Penterythritol tetracaprylate
Spermaceti wax	Diisopropyl adipate	Propylene glycol
Stearic acid	PPG-2 myristyl propionate	Dimethicone
Octyl palmitate (5%)		
Cyclomethicone		

Table 6.30 Emollient materials with a medium comedogenic effect

Glyceryl monostearate	Myristyl myristate	Octyl palmitate (50%)
Ethoxylated lanolin	White petroleum jelly	Capric/caprylic triglyceride
Olive oil	Arachis oil	
Sesame oil	Lanolin alcohols	

Table 6.31 Emollient materials with a high comedogenic effect

Cocoa butter	Isopropyl myristate	Isopropyl palmitate
Acetylated lanolin	Oleyl alcohol	Crude petroleum jelly
Alcohols	Isopropyl lanolate	Sweet almond oil

a medium rating (Table 6.30) can be used with caution, while a high comedogenic rating (Table 6.31) identifies materials that are preferably avoided or kept to a minimum.

6.5.1.2 *Occlusivity* Another important aspect of cosmetic products is the degree to which passage of water (or gaseous vapour) to and from the skin is controlled. Cosmetics that impede insensible perspiration are deemed 'occlusive', a phenomenon attributed to the emollient phase having a linear structure, which acts as a barrier on the skin. Occlusive preparations can lead to uncomfortable sensations and give rise to localised heating of the skin and prickly heat. However, if correctly manipulated, such a characteristic can be used to advantage in creams designed for treatment of ultra-dry skin caused by adverse physical and environmental conditions. By preventing moisture loss, skin can be kept hydrated and moisturised from within. Materials suitable for such applications (e.g. treatment of ashy skin at elbows and knees) are high quality white mineral oils and petroleum jelly. Lanolin and certain derivatives are similarly valuable since they are capable of retaining moisture in an emulsified form at the skin's surface. Materials such as lanolin, liquid lanolin, lanolin alcohols and lanolin absorption bases are all suitable for use in preparations requiring occlusive properties. The traditional use of natural vegetable oils such as shea butter and palm oil, which applied neat to the skin, has generally been replaced by petroleum jelly and mineral oil. These achieve a similar but more comfortable effect.

The formulation of products for hot humid climatic conditions, where water loss from the skin is important, requires the use of ingredients that will operate to reduce the occlusivity of the cosmetic product. In particular, inclusion of a porositone (a branched chain ester) such as 2-ethyl hexyl palmitate (deemed non-occlusive) serves to provide low occlusive, light-textured creams and lotions. Other suitable branched chain esters are di-2-ethyl hexyl adipate, cetostearyl octanoate, 2-ethyl hexyl cocoate, 2-ethyl hexyl succinate and acetylated lanolin alcohol.

6.5.2 *Emulsification systems*

The traditional emulsification system involves the use of soaps, the properties of which naturally confer an alkaline pH to the cosmetic product. This can theoretically damage the acid mantle of the skin—part of the skin's natural protective system—leaving it susceptible to bacterial attack and skin disorders. Despite this, many market leader products with good safety records have been based on amine soap systems, particularly the triethanolamine stearate system with a pH of 7.5–8.5. The aforementioned use of highly refined materials for cosmetic products is particularly important in the case of amines; low quality triethanolamine can contain secondary amines as impurities. These materials are recognised for their potential for forming nitrosamines—

carcinogenic chemicals resulting from the interaction of secondary amines and nitrites.

Preferred emulsification systems for more sophisticated products are non-ionic materials. These give the formulator greater flexibility in terms of the type of emulsion (oil-in-water or water-in-oil), pH and inclusion of additional materials (e.g. cationic germicides). Favoured non-ionic emulsifiers (and emulsifier combinations) involve sorbitan esters and their ethoxylated derivatives, often in combination with glyceryl monostearate, self-emulsifying waxes, acid-stable grades of glyceryl monostearate, ethoxylated castor oil derivatives, and surface-active lanolin derivatives.

6.5.2.1 *Humectant selection* Most cosmetic creams and lotions contain humectants to prevent the product drying out at the surface; the humectants also act as moisturisers on the skin. In dry climates, the occlusivity approach described in section 6.5.1.2 is a practical method for retaining skin moisture.

Table 6.32 O/W hand and body lotions

	wt%		
	C1076M1	C1282B	C1437M1
Crodamol IPM (isopropyl myristate)[a]	5.00	—	3.00
Polawax GP200 (non-ionic self-emulsifying wax)[a]	—	2.00	2.70
Crosterene SA4310 (stearic acid)[a]	—	1.00	0.18
Crodamol CAP (cetearyl octanoate)[a]	2.00	—	—
Silicone fluid 200/100 cSt (dimethicone)[b]	1.00	—	—
Crodafos CDP (DEA cetyl phosphate)[a]	—	0.50	—
GMS S/E GE0802 (glyceryl stearate S/E)[a]	—	—	1.50
Lanolin[a]	—	0.50	—
White mineral oil (25 cSt at 25°C)	5.00	2.00	2.10
Crodamol W (stearyl heptanoate)[a]	—	2.00	—
Crodamol OHS (octyl hydroxystearate)[a]	—	—	2.70
GMS A/S ES0743 (glyceryl stearate (and) PEG-100 stearate)[a]	6.00	—	—
Refined cocoa butter	1.50	—	—
Deionised water	to 100	to 100	to 100
Propylene glycol	2.00	—	2.20
Croderol GA7000 (glycerine)[a]	—	3.00	—
Sodium lactate/lactic acid buffer solution (pH 5.5)[a]	1.00	—	—
Carbopol 941 (carbomer 941)[c]	—	0.10	0.10
Triethanolamine, refined	—	0.60	0.10
Perfume, preservatives, colour	*q.s.*	*q.s.*	*q.s.*

[a] Croda Chemicals Ltd; [b] Dow Corning Ltd; [c] B.F. Goodrich

Method of manufacture: heat oil and water phases separately to 65–70°C. Where included, Carbopol should be hydrated in the water phase first, omitting the amine. Add water to oil phase under stirrer and stir to cool. The Carbopol should be neutralised with amine at 60°C. Perfume at 40–45°C and fill off at 30°C.

In the lighter preparations, the cream itself can moisturise due to the inclusion of suitable moisturising components in the formulation. Typical humectants used to achieve this are glycerine, sorbitol, polyethylene glycol, sodium pyrrolidone carboxylate, sodium lactate/lactic acid buffers and urea. Acyl alkanolamides have also been shown to be effective moisturisers.

Other highly successful materials for use in moisturising cosmetic products are proteins and amino acids. These have the ability to retain high levels of moisture while imparting a pleasant feel to the skin. Classically, collagen has been employed in this role, but other useful protein derivatives include hydrolysed animal- and plant-derived proteins, together with the somewhat more evocative proteins derived from silk, egg and milk. In more up-market products, special complexes of proteinaceous materials, such as hyaluronic

Table 6.33 All-over body creams

	wt%		
	C1433	C1409	C1416
Liquid base CB3929 (mineral oil (and) lanolin alcohol)[a]	6.50	—	—
Crodamol OP (octyl palmitate)[a]	—	4.00	2.00
Crodamol LA (cetyl acetate (and) acetylated lanolin alcohol)[a]	3.50	—	—
Crodamol IPM (isopropyl myristate)[a]	—	6.00	—
White mineral oil (25 cSt at 25°C)	—	—	2.00
GMS A/S ES0743 (glyceryl stearate (and) PEG-100 stearate)[a]	4.30	—	—
GMS N/E ES0803 (glyceryl stearate)[a]	—	3.00	5.00
Volpo CS10 (ceteareth-10)[a]	—	1.00	—
Refined cocoa butter	—	—	2.00
Fluilan (lanolin oil)[a]	—	—	1.00
Crosterene SA4310 (stearic acid)[a]	4.30	3.00	2.00
Crodamol CAP (cetearyl octanoate)[a]	3.00	—	—
Polawax GP200 (non-ionic self-emulsifying wax)[a]	—	3.00	3.00
DL-Alpha tocopherol[b]	—	0.50	—
Silicone fluid 200/100 cSt (dimethicone)[c]	—	—	1.00
Almond oil	—	1.00	—
Cithrol 6MS (PEG-12 stearate)[a]	—	—	1.00
Deionised water	to 100	to 100	to 100
Propylene glycol	1.50	—	—
Croderol GA7000 (glycerine)[a]	—	2.50	4.00
Carbopol 934 (carbomer 934)[d]	—	0.25	0.25
Triethanolamine, refined	—	0.60	0.60
Perfume, preservatives, colour	*q.s.*	*q.s.*	*q.s.*

[a]Croda Chemicals Ltd; [b]BASF (UK) Ltd; [c]Dow Corning Ltd; [d]B.F. Goodrich

Where included, Carbopol should be hydrated in the water phase first, omitting the amine. Add water to oil phase under stirrer and stir to cool. The Carbopol should be neutralised with amine at 60°C. Perfume at 40–45°C and fill off at 30°C.

acid/protein complex and chondroitin sulphate/protein complex, have been shown to improve extensibility and elasticity of the skin. Lanolin (see section 6.5.1.2) is also a particularly valuable additive in this class of moisturiser.

Table 6.34 W/O general-purpose creams

	wt%		
	C741	C436	C489M
Crill 6 (sorbitan isostearate)[a]	2.50	—	—
Liquid base CB3929 (mineral oil (and) lanolin alcohol)[a]	5.00	12.00	—
White microcrystalline wax (78–82°C)	9.00	5.80	3.40
Hartolite (modified lanolin alcohols)	—	3.16	—
Aluminium stearate G[b]	0.10	—	—
Satulan (hydrogenated lanolin)[a]	—	—	2.50
White mineral oil (25 cSt at 25°C)	16.50	8.70	16.00
White petroleum jelly BP	—	—	2.50
Ozokerite wax 65–70°C	—	2.20	—
Lanolin[a]	—	—	9.50
BHT	0.01	0.01	0.01
Crodacol CS/BP (cetostearyl alcohol)[a]	—	0.30	—
Croderol GA7000 (glycerine)[a]	1.50	2.50	3.50
Magnesium sulphate	0.70	0.70	0.70
Deionised water	to 100	to 100	to 100
Perfume, preservatives, colour	*q.s.*	*q.s.*	*q.s.*

[a]Croda Chemicals Ltd; [b]Durham Chemicals Ltd

Method of manufacture: heat oil and water phases separately to 65–70°C. Add water to oils with stirring. Stir to cool, adding perfume at 45°C. When cool homogenise with triple roll mill, colloid mill or other suitable equipment.

Table 6.35 O/W glycerine silicone hand creams

	wt%	
	C1432M1	C1502
Silicone fluid F111/200 cSt (dimethicone)[a]	3.00	3.00
Crodacol C90 (cetyl alcohol)[b]	4.00	9.00
GMS N/E GE0803 (glyceryl stearate)[b]	2.00	—
Crodex A (anionic emulsifying wax BP)[b]	7.00	10.00
Crodamol IPM (isopropyl myristate)[b]	5.00	—
Crodamol OHS (octyl hydroxystearate)[b]	—	3.00
Deionised water	to 100	to 100
Croderol GA7000 (glycerine)[b]	10.00	35.00
Perfume, preservatives, colour	*q.s.*	*q.s.*

[a]Dow Corning Ltd; [b]Croda Chemicals Ltd

Method of manufacture: heat oil and water phases separately to 65–70°C. Add water to oils with mechanical stirring. Stir to 45°C and perfume. Stir to cool and fill off.

6.5.2.2 *Skin creams and lotions* Skin creams and lotions fall into two categories: (i) those for use in warm humid climates; and (ii) those for use in dry climates or during the dry season. The first category comprises relatively light-textured products with low occlusivity, exemplified by all-over skin-care lotions (Table 6.32); and all-over body creams (Table 6.33). (Note that for these, and all other formulations given in this section, full details of suppliers can be found in Appendix I.)

For the second category, water-in-oil emulsions are favoured. The skin-cream formulations shown in Tables 6.34 include a super-emollient moisturising cream (C489M) designed for treating 'ashy skins'. The moisturising hand creams shown in Table 6.35 are characterised by high levels of glycerine and silicone incorporated into the emulsion system. These hand creams deposit a wax-like film, which gives a dry emollience and avoids excessive oiliness.

An additional product that assumes importance in the black populace is generally termed complexion or fade cream. This contains hydroquinone, a chemical product that is used to lighten the skin, and is commonly misunderstood as making black people whiter. In fact, hydroquinone creams are more often used to lighten specific darker areas of skin (the result of hyperpigmentation), thus 'evening-out' the skin tone. Formulations for hydroquinone creams are shown in Table 6.36.

Table 6.36 Hydroquinone creams

	wt%	
	C1420	C1564
Liquid base CB3929 (mineral oil (and) lanolin alcohol)[a]	7.00	7.00
Crodacol CS/BP (cetostearyl alcohol)[a]	5.00	5.00
GMS A/S ES0743 (glyceryl stearate (and) PEG-100 stearate)[a]	10.00	10.00
Polychol 15 (laneth-15)[a]	3.00	—
Deionised water	to 100	to 100
Croderol GA7000 (glycerine)[a]	2.00	—
PEG-1500	—	4.00
Ascorbic acid	0.20	0.20
Citric acid	to pH 4.00	—
Hydroquinone	2.00	2.00
Sodium metabisulphite	0.10	0.20
Perfume, preservatives, colour	*q.s.*	*q.s.*
EDTA	0.30	0.35

[a]Croda Chemicals Ltd

Method of manufacture: heat the water and oil phases to 65–70°C. Add the water phase to the oil phase with stirring. Stir until 40–45°C. Add, in order, sodium metabisulphite, ascorbic acid, hydroquinone and citric acid. Perfume at 40°C (the perfume must not react with hydroquinone or bisulphite). Adjust the pH value, if necessary, to 4.00–4.50 with citric acid.

6.6 General practical considerations

Many of the precautions outlined in this chapter apply to the general manufacture of make-up (colour cosmetics). The manufacture of ethnic products, however, necessitates a complete range of different colour shades that are complementary with black skin (see chapter 4). All the normal rules for testing and evaluating product stability and safety apply, but particular attention to the special marketplace conditions are essential. Severe conditions of humidity, heat and UV radiation occurring in parts of Africa should be carefully considered when determining test conditions and microbiological preservation.

Further reading

Balsam, M.S. and Sagarin, E. (1974) (Eds) *Cosmetic Science and Technology*, 2nd edn, Vol. 3, p. 178. John Wiley and Son, New York.
Chester, J. and Dixon, M. (1987) *Manufacturing Chemist* **58**(5) 30.
Chester, J. and Mawby, G. (1987) *Manufacturing Chemist* **58**(4) 53.
Coupland, K. and Smith, G. (1986) *Speciality Chem.* **6**(3).
Fulton, J.E. *et al.* (1976) *Cutis* **17** 344.
Fulton, J.E. *et al.* (1984) *J. Am. Acad. Dermatol.* **10**(96).
IFSCC (1986) An approach to permanent hair conditioning. *Proc. 14th IFSCC Congress*, September 1986. Preprints Vol II.
Kligman, H. and Mills, O. (1972) *Arch. Dermatol.* **106** 843.
Morris, W.E. *et al.* (1983) *J. Soc. Cosmet. Chem.* **34** 215.
Munroe, L. (1986) *Cosmet. Toilet.* **101** 63.

7 Dental products

M. PADER

7.1 Introduction

As the world's economy expands, so does the demand for improved preventive health services and products. One of the consequences is that substantial progress has been made in the development of over-the-counter (OTC) cosmetics and toiletries, with an increasing trend to endow them with properties which are biologically active, e.g. anti-dandruff shampoos. Oral-care products have taken a similar trend, that is towards therapeutic and/or biological activity. The new trend has had a strong influence on the formulation and other aspects of preventive oral care products worldwide. These present both a challenge and an opportunity to marketers of dentifrices and related products, particularly those who must depend largely on local resources and must accommodate local customs and regulations. There are a tremendous and growing number of oral-care products on the market. It is obviously impossible to discuss them individually in this chapter alone. Hence, this chapter will define only important principles governing the more prominent oral-care products. The reader must seek other references for details of the chemistry of the raw materials and finished products, manufacture, etc. (Two reference works which are complete enough for most of the detail needed to understand successfully, formulate and market dentifrice are by Pader [1, 2]).

The role of oral hygiene products has traditionally been to keep the teeth free of stain and to freshen the breath. To this must now be added the prevention of diseases which can result in the loss of teeth, e.g. dental caries. New ingredients and procedures have been introduced. Now, the emphasis is on maintenance of a healthy dentition; stain removal and breath freshening are features of oral hygiene which have generally been relegated to the category of problems which have been solved. New claims for improved tooth cleaning occasionally appear, but the principles of formulation of products which clean teeth and breath are well established. Maintenance of a healthy dentition, however, has achieved only limited success with current dental science, and is the driving force behind most of today's dental research on OTC oral-care products.

The two major arenas of preventive self-applied dental formulations have been reduction of dental caries and prevention of gum diseases. Dental caries affected as much as 95% of people in industrialized populations. The incidence is now down to a small hard core of susceptible people; some individuals have gone through childhood without experiencing the disease. Periodontal diseases are still rampant in all populations, but this must be put into perspective. There are many types and stages of periodontal disease. If the disease is categorized in terms of disease which threatens tooth loss or the need for major surgical measures, the incidence is relatively low in industrialized societies, it has been suggested as even less than 2%. If *all* conditions which cause mild inflammation or irritation are considered to be periodontal disease, the incidence is considerably higher.

Good self-applied oral hygiene measures can go far in reducing substantially the incidence and severity of caries, periodontal diseases and other dental disorders among those not afflicted with special dental disease problems. This chapter will discuss the OTC products and practice of oral hygiene which will help achieve the goal of a healthy mouth. They need not be overly sophisticated, depending on the goals specified. Certainly, a clean dentition is a first, necessary step towards a healthy dentition [2].

7.2 The human dentition and its environment

7.2.1 *Teeth and associated oral structures*

Figure 7.1 presents the anatomy of a tooth, as classically represented by Schour [3]. Oral hygiene practices are designed to maintain the crown enamel, dentin (if exposed) and gingiva in healthy states. Maintenance of these structures is expected to result in the general health of the dentition, including other hard and soft tissues and the supporting bony structure.

During childhood, humans have a complement of 20 teeth, which erupt from 6 months of age (lower central incisors) to 20–24 months (second molars). Adults normally have a complement of 32 teeth, erupting from 6–7 years of age (first molars, central incisors) to 17–21 years (third molars). The teeth are supported by a bony structure. The principle diseases causing loss of teeth are dental caries and periodontal diseases. Good oral hygiene and dietary habits are effective preventive measures. If left undisturbed, the carious lesion will penetrate to the pulp and expand to destroy the crown. Similarly unless tended to, periodontal disease will result in inflammation and destruction of the soft tissue surrounding the tooth, and eventual loss of the tooth.

7.2.2 *Saliva and crevicular fluid*

Saliva is a mixture of the secretions of the parotid, submaxillary, sublingual and accessory salivary glands, occasionally along with the crevicular fluid.

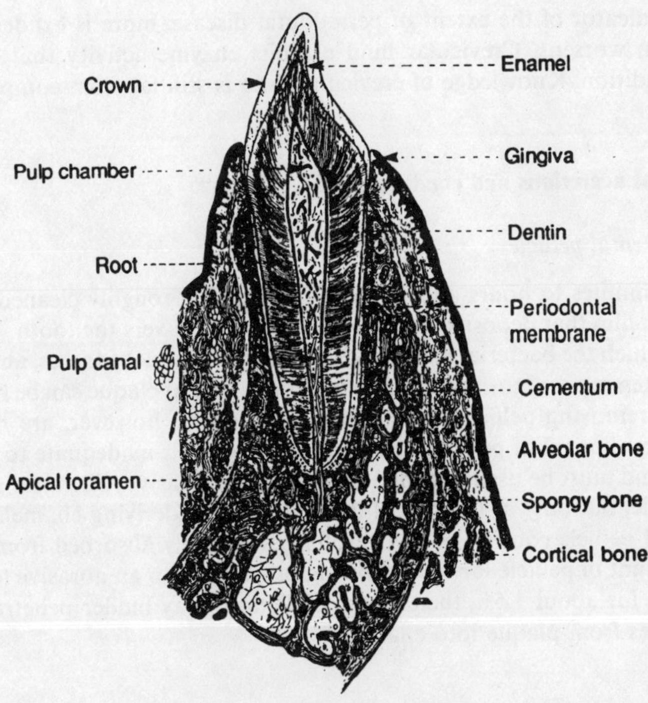

Figure 7.1 Diagrammatic representation of the tooth and periodontal tissues. From [3].

Saliva performs two functions: (i) it is involved in protection of the oral cavity through several factors related to the bacterial population of the oral cavity; and (ii) it is involved in the initial processes of food digestion. Dawes [4] suggested that saliva could be anti-cariogenic by: (i) increasing the rate of carbohydrate clearance; (ii) reducing acid production by fermentative pathways; (iii) buffering the drop in pH caused by acid production in plaque; (iv) increasing the rate of glycolysis in plaque, thereby reducing the time the plaque is below a critical pH for attack on enamel; (v) increasing enamel resistance to demineralization by acid; (vi) increasing the degree of saturation of plaque fluid with respect to hydroxyapatite or fluorapatite; (vii) promoting remineralization of initial subsurface carious lesions; (viii) increasing bacterial clearance from the oral cavity; and (ix) increasing the protection possibly afforded by the dental pellicle that selectively deposits from saliva immediately following cleaning of the teeth. Calcium ions present in saliva appear to play an important role in microbiological processes involved in both protection and destruction of the tooth by oral bacteria.

Crevicular fluid is introduced into the salvia through the gingival crevice. It is not noticeable in a healthy dentition or the absence of teeth, and can be an indicator of the extent of periodontal disease; more is exuded as the condition worsens. Crevicular fluid exhibits enzyme activity that changes with condition. Knowledge of crevicular fluid is still far from complete.

7.3 Oral accretions and conditions

7.3.1 *Dental pellicle*

Within minutes to hours after a tooth has been thoroughly cleaned, dental pellicle, a film that deposits selectively from saliva, covers the tooth. It is this film to which the bacterial mass (known as dental plaque) adheres, and which stains when exposed to chromogenic materials. Dental plaque can be removed without removing pellicle. Substantive tooth stains, however, are removed only with removal of pellicle. Toothbrushing alone is inadequate to remove pellicle and must be used in conjunction with an abrasive. Chemical removal is possible, but only at the risk of damaging the underlying enamel.

Dental pellicle consists of glycoproteins selectively absorbed from saliva. The amount of pellicle formed following brushing with an abrasive (pumice) increases for about 1.5 h, then levels off. Pellicle may hinder penetration of substances from plaque into enamel.

7.3.2 *Dental plaque*

Dental plaque is primarily a bacterial accumulation. It occurs surpragingivally (above the gum line) and subgingivally (below the gum line). Figure 7.2 shows a scanning electron microscope photograph of supragingival plaque, magnified 300 ×. The actions of self-administered oral hygiene procedures are usually limited to the supragingival plaque. Subgingival plaque is best cared for by the dental professional.

Existence of a relation between dental plaque, periodontal disease and caries is firmly established. Freedom from dental disease is usually a consequence of the absence of plaque. Some investigators may not be very precise in their definition of plaque, and different investigators measure plaque in different ways with different results. For example, Bhaskar [5] reported improvement in gum condition with use of an oral irrigator, even though there was no major reduction in plaque as measured by a commonly used technique.

Dawes *et al.* [6] defined dental plaque accurately, but inadequately, as a soft and tenacious material found on surfaces of the teeth, readily removed by mechanical means such as brushing or flossing, but not by rinsing with water and other solutions. A more comprehensive definition is that of a mass of oral bacteria, which initially accumulates around the teeth at the cervical margin and then grows apically. In addition to a fairly characteristic microbial

Figure 7.2 Dental plaque: scanning electron microscope, 300 × . Courtesy of Unilever Research, Port Sunlight Laboratory.

population, dental plaque contains leucocytes, desquamated cells, and other oral debris. It undergoes maturation, which involves changes in the microbial flora, possibly including calcification and loss of viability of some of the bacteria. It resists removal by water and, when mature, can be removed only by mechanical means. This definition acknowledges that plaque is a dynamic system and that quantitative measurements must be reported only in the context of that dynamism. A corollary would be that only a totally plaque-free tooth can confidently be expected to be disease-free.

De la Rosa *et al.* [7] proposed a model for plaque development over time (see Figure 7.3). Starting with a clean tooth, plaque accumulates to a plateau level following a pattern of alternate build-up and removal by toothbrushing. Each oral site may follow a different pattern, depending on efficacy of brushing at that site. Plaque measurements at specific sites may be of greater value than average values to elucidate clinical plaque status. Clearly the status of the plaque at the start of a clinical experiment can influence the apparent anti-plaque efficacy of an agent.

Inadequate control of plaque can lead to either or both of two serious disease conditions, periodontal disease and dental caries [8]. A cause and effect relationship between plaque and gum inflammation (gingivitis) was convincingly demonstrated by Löe *et al.* [9]. These workers showed that refraining from toothbrushing for several weeks, with subsequent build-up of plaque, resulted in gingivitis. This was rapidly alleviated when the plaque was removed by oral hygiene measures. Despite about three decades of further study, however, there are many questions still to be answered about the development and activity of dental plaque. Pader [10] has reviewed the

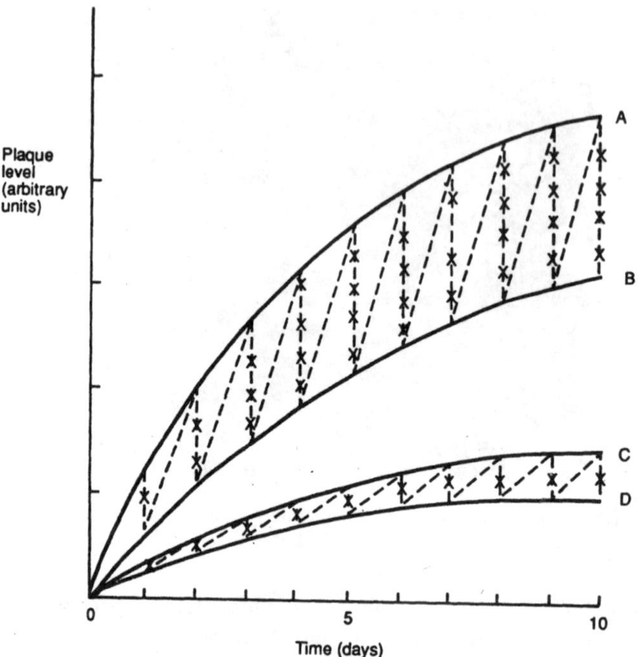

Figure 7.3 Plaque growth and removal assuming only one brushing per day. **A** represents plaque accumulation by a 'normal' individual before toothbrushing; **B** represents plaque remaining after toothbrushing by a 'normal' individual. Curves **C** and **D** have the same meanings for an unusually good brusher. (– – –) daily plaque accumulation; (—×—×—×—) daily plaque removal by toothbrushing. After [7].

problem and concluded that there is no simple proven *quantitative* relationship between plaque and gum disease. Many questions remain unanswered. Gingivitis does not develop in the total absence of plaque. But it has been shown that some sites covered with plaque do not develop gingivitis, that a clinical reduction in plaque need not be accompanied by an improvement in total gingival health [11, 12] and that in some instances removal of no more than a minor amount of plaque reduces gingivitis significantly. But gingivitis, itself, is not necessarily tooth threatening. Formulators of OTC dental-care products must keep in mind the fact that gingivitis and tooth-threatening periodontal diseases represent different levels of concern. The latter is still understood only incompletely, including the role of anti-plaque agents in OTC products in their control. Overall oral hygiene status, rather than plaque status, has been postulated by Attström to be the better predictor of total periodontal health [13].

The relationship between dental plaque and caries is clear. Dental plaque generates acid from carbohydrate, and this acid attacks the tooth enamel and initiates formation of the carious lesion. The carious process has been discussed at some length by Nikiforuk [8], Pader [2] and others.

7.3.3 *Dental calculus (tartar)*

Schroeder [14] examined dental calculus as understood up to about the year 1968. In 1986, Mandel and Gaffar [15] re-examined the subject. Schroeder defined two types of dental calculus, supragingival and subgingival. The former was explained as a hard, mineralized dental plaque and/or materia alba permeated with crystals of various calcium phosphates covered with a layer of 'vital, non-mineralized plaque'. Subgingival calculus was defined as an organic structure of micro-organisms and intermicrobial matrix, containing major amounts of crystalline calcium phosphates, different in amount and distribution from those of supragingival calculus. A major impetus for the study of dental calculus was the belief that this hard, rough accumulation, (or the plaque concentrated on it), in direct contact with the gum tissue, was an irritant leading to periodontal disease.

Supragingival calculus arises from the nucleation of calcium phosphate, particularly in areas where the large salivary gland ducts secrete their saliva. The calcium phosphate crystals in both supra- and subgingival calculus consist of octacalcium phosphate, brushite, whitlockite and hydroxyapatite. Subgingival calculus has a major amount of whitlockite and a reduced amount of brushite compared to supragingival calculus. This suggests that they have different origins. The contents of the periodontal pocket are apparently able to fulfil the conditions necessary for nucleation and subsequent crystal growth.

Dental calculus is formed fairly quickly. Schroeder [14] reported that a four day old deposit (on a Mylar strip mounted on anterior teeth) was mineralized about 40% (assessed by ash content), a twelve day old deposit over 60%, and a two year old deposit about 80%. Mandel and Gaffar [15] subsequently made a number of observations.

(i) Gingivitis can develop in the absence of supragingival calculus, although it is not known to what extent the presence of the calculus can increase gingival inflammation.
(ii) While calculus formation can be inhibited by chemical agents, the reduction need not be accompanied by a reduction in gingivitis.
(iii) Subgingival calculus results from the interaction of subgingival plaque with the influx of minerals that form part of the serum transudate and inflammatory exudate.
(iv) Calculus is porous and can retain both bacterial antigens and easily available toxic stimulators of bone resorption. This, together with

the increased accumulation of plaque on calculus, can produce more severe destructive effects than the presence of plaque alone.

(v) Subgingival calculus contributes to the chronic character and to the progression of periodontal disease, even though it may not be the cause of the disease.

7.3.4 *Periodontal diseases*

Periodontal disease merits special attention in this chapter because it is of major research interest to marketers of oral hygiene products. On the one hand, dental caries has to all intents and purposes been conquered with OTC fluoride toothpastes and rinses used in conjunction with water fluoridation and other fluoride treatments. On the other hand, OTC products offering the same degree of protection against tooth-threatening periodontal diseases as fluoride does against caries are not yet available.

Van Dyke and Zinney [16] characterized periodontal diseases as "... a family of chronic inflammatory infections affecting the supporting tissues of the dentition." They stated that: "Modern periodontal therapy is based on the tenet that supragingival plaque causes gingivitis, which is the precursor of more advanced periodontal breakdown arising from subgingival growth and apical extension of bacterial plaque." Pader [2] delineated several stages in periodontitis. First, bacterial plaque forms at the gingival margin. If not removed, it spreads. The bacteria generate toxins that inflame the soft gum tissue, which becomes soft, puffy and red, and bleeds easily. Following this, a periodontal pocket (a pathologically deepened gingival sulcus) may form in susceptible individuals, without destruction of tissue. In the next stage, the inflammatory process is aggravated: the pockets become enlarged and form the receptacles for subgingival plaque bacteria, debris and exudates. The surrounding connective tissue degenerates. With increasing challenge by the bacterial mass, the pockets continue to deepen, and the teeth start to loosen. Subsequently, the gums recede from the tooth crowns. Finally, microbial, plaque toxins and other materials in the pockets intensify the inflammatory process, supporting bone tissue is destroyed, and the teeth loosen and eventually fall out or are extracted.

The relation between dental plaque and periodontal disease has been discussed in detail in the literature [2, 10]. Some of the more important findings relating to product development have been outlined by Attström [13]:

(i) Removal of supragingival plaque once every 24–48 hours is adequate to preserve gingival health.

(ii) There is a close relationship between the subgingival location of the advancing bacterial front and the level of connective tissue attachment. The bacteria produce an inflammatory reaction apically/laterally to the junctional epithelium.

(iii) Periodontal disease may proceed by bursts of increased inflammatory activity. Dental plaque, gingival redness and bleeding on probing, and initial pocket depth probing do not relate to the occurrence of probing attachment loss. The incidence of periodontal disease sites in untreated patients is low. The progression of periodontal disease does not correlate with the presence of supra- and subgingival bacterial colonization, and is probably related to the presence of specific bacteria.

(iv) The level of oral hygiene correlates with the severity of periodontal disease. Periodontal pathology and attachment loss are most severe interproximally and least severe in the buccal sites. Supragingival plaque removal by self-administered oral hygiene procedures slows the progression of periodontal disease.

(v) The composition of the subgingival bacterial flora, and the progress of periodontal disease varies from individual to individual, tooth to tooth, and site to site on the same tooth. The pathological process can be arrested by removal of subgingival bacteria along with supragingival bacteria. Subgingival plaque is generally not affected by supragingival cleaning once the subgingival plaque has established itself in the more apical regions of the dentogingival area. However, in untreated patients the composition of the subgingival flora is affected by the presence of supragingival plaque.

(vi) Regular supragingival plaque removal can prevent infection of the subgingival sites and maintain periodontal health, but not as effectively in persons with a compromised defense system.

Burt has noted that epidemiological and clinical evidence indicates that most gingivitis does not proceed to periodontitis, but periodontitis has not been reported without a preceding gingivitis [17]. (Elucidation of this effect would be useful in putting the value of gingival health products into better perspective.)

7.3.5 *Dental stain*

Vogel [18] proposed that the tooth stains accumulated by a significant number of individuals are extrinsic in nature and result from discoloration of pellicle and/or plaque. Eriksen and Nordbø [19] proposed that extrinsic staining may involve at least three mechanisms: (i) the production of colored substances in plaque by chromogenic bacteria; (ii) the retention of colored substances either from dietary factors passing through the oral cavity (berries, tea, coffee) or from smoking; and (iii) the formation of colored products resulting from the chemical transformation of pellicle components. Besides tea, wine, tobacco smoke and other pellicle-reactive colored materials, oral use of antimicrobial agents (including agents which have clinically exhibited plaque control) has been implicated in stain formation.

7.3.6 *Dental hypersensitivity*

Dentinal or cervical hypersensitivity is experienced by some individuals after their roots are exposed by gingival recession, or surgical or other periodontal treatments. The condition, called 'dentinalgia', is characterized by pain when the exposed dentin is subjected to thermal, osmotic, electrical or dehydrating stimuli. Berman [20] and others have proposed theories to explain the phenomenon. These are beyond the scope of this chapter, other than to note that movement of fluid within the dentinal tubule can explain several phenomena associated with hypersensitivity.

Desensitizing agents have successfully been incorporated into dentifrices. Diverse agents are claimed to have desensitizing efficacy in a toothpaste vehicle, indicating that hypersensitivity can possibly have multiple origins. Certainly, different responses were obtained to different agents.

7.3.7 *Oral malodor*

Pader [2] has discussed oral malodor in terms of its bacterial origins. Bacteria from virtually all sources in the oral cavity have been implicated. 'Morning breath' has been attributed to the overnight putrefaction of food deposits, salivary deposits, desquamated epithelial cells, and other types of oral accumulation and debris. Malodor of other than bacterial origin (for example due to the intake of odoriferous foods or to diseases which cause the exhalation of odoriferous metabolic products) cannot be eliminated by oral hygiene measures.

The main cause of oral malodor is probably the putrefaction of sulfur-containing protein substrates, predominantly by gram-negative bacteria. The process is not limited to specific bacterial populations. Micro-organisms indiginous to dental plaque, saliva, gingival crevice and tongue can participate, and mixed bacterial populations are most capable of odor production. The fermentation process releases hydrogen sulfide and methyl mercaptan (which together account for 90% of the volatile compounds generated), dimethyl sulfide and dimethyl disulfide. The mouth air also contains aliphatic and aromatic alcohols, and indole. Mouth air and putrefying saliva from persons with periodontal disease show higher than normal levels of sulfur compounds.

Mouth malodor can be controlled by vigorous oral hygiene practice. Most of the hydrogen sulfide and methyl mercaptan in periodontitis-free individuals derive from the dorsoposterior surface of the tongue and can be considerably reduced by tongue brushing, toothbrushing, food ingestion and use of an oral rinse that reduces oral bacterial populations.

7.4 Oral-care products

7.4.1 *Product categories*

Oral-care products can be divided broadly into mechanical devices and chemical formulations. Additionally, one can include products designed for use with prostheses, such as dentures, since some of the conditions affecting the natural dentition can also affect those products as they are worn in the mouth (for example, plaque and stain can accumulate on dentures).

The toothbrush is the most important part of the oral hygiene regimen. Toothbrushing alone, without supplements such as toothpaste, can maintain a reasonable state of oral health. Its value in this regard, however, is greatly enhanced by supplementation with toothpaste, which contains materials to remove stain, freshen the breath and accomplish specific functions. The toothbrush is a mechanical device. Other mechanical devices include dental floss, water irrigators and pics of various designs.

Increasingly, toothpaste is becoming a preferred vehicle to apply special active substances to the dentition. This enhances the role of toothbrushing in oral care. Toothpastes are now being marketed with special active agents to combat dental caries, reduce the formation of dental calculus, combat gum disease and reduce tooth hypersensitivity. Wide acceptance of toothpaste has made it the vehicle of choice for applying therapeutic or special cosmetic additives to the dentition. Toothpaste brands are proliferating as new health benefit additives are discovered, and older concepts of toothpaste formulation are no longer applicable.

The selected additive(s) must, of course, be compatible with and active from the total formulation.

Oral-care products have been categorized as cosmetic or therapeutic, depending on whether or not they contain a health benefit agent. The definition of a health benefit agent is a very complicated matter and varies from regulatory agency to regulatory agency around the world.

7.4.2 *Toothpaste*

Toothpaste fulfills two primary functions; removal of stain from the teeth and freshening of the breath and mouth. Mild abrasives and appropriate flavors, respectively, accomplish these objectives. These basic and essential functions are being supplemented, but not reduced in importance, as it has become recognized that additional functions can be beneficially built into toothpastes.

Toothpastes on today's market present a wide range of rheological properties and appearances. The most common rheology is that of a paste that can be extruded from a container onto a toothbrush as a ribbon which can maintain its shape for the necessary time prior to brushing. Toothpastes are marketed in a wide variety of physical characteristics and dispensers. These include viscous liquids, pastes extrudable from a tube, pastes extrudable from pumps, clear products (frequently referred to as 'gels'), products with visual signals (such as sparkles or soft plastic particles with colors different from the main toothpaste base), and products which are dispensed simultaneously as multiple pastes with different colors. The rheological properties of the paste components are key to the obtention of the various toothpastes (toothpaste rheology will be discussed below).

Conventional toothpaste can remove accretions which dull the tooth surface, that is stained pellicle, plaque. The consumer and regulatory agencies recognize that toothpaste can not whiten teeth in the sense of improving tooth color, which is determined mainly by the color of the dentin, enamel playing only a minor role [21]. The dental professional has successfully made teeth whiter by the use of bleaching agents, primarily peroxides. Although questions about the safety of prolonged, repeated application of peroxide are still heard, the American Dental Association (ADA) has recently approved as safe and effective two peroxide-containing products, one for dental surgery use and one for daily use. Both products have been extensively safety-tested and full reviews provided for the ADA. Nonetheless the use of peroxide in OTC toothpaste, for as yet unspecified claims, is the subject of major research and new dentifrice brands.

7.4.3 *Tooth powder*

Tooth powder is a forerunner of toothpaste. Once popular, it now has a very low level of popularity. Toothpowder is a physical mix of dental abrasive, flavor and foaming aid, packed in a rigid container equipped with an orifice from which the powder can be shaken. Some users prefer to pour the powder onto the hand and scoop it up onto the toothbrush, others to dispense it directly onto the toothbrush. The disadvantages of tooth powder compared with toothpaste and the difficulty of obtaining predictable doses of health-benefit agents at each use, has led to the virtual demise of tooth powder as a major oral-care product.

7.4.4 *The toothbrush*

The toothbrush is the real workhorse in oral hygiene. (It has been discussed in detail in the first edition of this book). Toothbrushes are available as hand operated devices and as mechanically driven devices. The major functions of the toothbrush are to remove dental plaque and to work with toothpaste to remove dental stain. The toothbrush also provides a mechanism with which to apply toothpaste for its secondary benefits, such as anti-calculus activity. The toothbrush, used alone, can remove soft dental plaque; dentifrice is not necessary. The consumer, however, has been taught to use toothbrush and toothpaste together and adheres to that practice. (In a study reported by de la Rosa [7], use of dentifrice with the toothbrush for a period of 14 days resulted in an increment of only 5% in plaque removal compared with use of a toothbrush without dentifrice.)

Claims are frequently made for superiority of one type of toothbrush or design over another. Such claims are difficult to support at the consumer level because of the critical importance of non-structural factors in toothbrush performance.

7.4.5 *Oral rinses*

Oral rinses supplement, but cannot substitute for the toothbrush. Most on the market today are formulated to freshen the breath directly and perform at least one secondary function, usually related to a health benefit, e.g. destruction of plaque and other oral bacteria. The sale of oral rinses around the world is variable. In the United States, oral rinse sales are about one-half of toothpaste sales on a monetary level.

Some 'active' agents have defied incorporation into toothpaste because they interact with toothpaste excipients. This problem has been overcome by formulating the agents into oral rinses. In some instances, where the agents are compatible with both paste and rinse, the rinse has shown greater efficacy.

7.4.6 *Mechanical devices*

Mechanical devices to cleanse the teeth are becoming increasingly sophisticated. Mechanical devices are available which not only cleanse the teeth, but also are able to deliver agents to the dentition, e.g. anti-plaque agents from water irrigators. The purpose of special mechanical devices (other than the toothbrush) is to compensate for the inability of most consumers to realize the full potential of the toothbrush (mechanical cleansing aids have limited popularity and will not be discussed in this chapter).

7.5 Consumer practices

The average consumer, including those in more industrialized countries, is non-compliant in following dental care regimens advocated by dental professionals [2]. A major objective of providers of oral-care products is to make them easy and pleasant to use and more effective at the same given level of compliance. Dental diseases and conditions caused by non-compliance are not generally life threatening, and this is reflected in poor adherence to oral hygiene measures.

Plaque removal has come to be recognized as a most important objective of self-applied oral care; breath freshening and stain removal are expected functions of toothbrushing with paste. Studies have shown that the average consumer does far less than is necessary for maximum plaque control [2], this despite the large number of highly effective devices and chemical treatments available. Reasons for non-compliance in disease treatment have been studied [2]. In oral hygiene practice, major factors are lack of motivation to follow the regimen properly, not being able or willing to follow label instructions, not understanding the regimen and cost of adhering to a regimen. In toothbrushing, for example, individuals frequently do not brush long enough, do not attempt to reach the more inaccessible parts of the dentition (e.g. interproximal spaces), generally are not adept at manipulating the toothbrush, do not fully understand the reasons for toothbrushing, use old, worn-out toothbrushes which provide only reduced efficacy and do not use enough toothpaste.

Individuals frequently disregard use directions provided by oral rinse marketers. Sometimes the reason lies with refusal to endure the discomfort of maintaining the rinse in the mouth for a long time (e.g. 30 seconds) especially if it has a high alcohol content and burns. But not infrequently, the consumer does not differentiate between oral rinses, and even will use one specified for pre-toothbrushing in the same way as most rinses are used, viz. after brushing. The benefits of oral rinses, in particular, are usually established via clinical trials wherein use conditions are specified, especially frequency and time of use and dosage. It is obvious that non-compliance with use directions can seriously compromise the benefits studied in the clinical trial.

Supplements to toothbrushing fare no better than toothbrushes themselves. Dental floss, for example, is recommended as a routine procedure, but in fact not more than a small percentage of a population in an industrialized society even attempts to use it. Flossing requires a degree of patience and skill which severely decreases motivation. Years of research have failed to produce a dental floss which overcomes the basic failures of common flosses.

Water irrigators and mechanically driven toothbrushes have been introduced to overcome the deficiencies of hand-operated tooth brushes. These products have developed a substantial market. They are effective, but very expensive, and it has yet to be demonstrated convincingly that they can perform substantially better than a conventional toothbrush when the latter is operated by a knowledgeably motivated brusher. Nonetheless, their efficacy in the hands of the average brusher can be expected on the theoretical grounds to be superior to that of hand brushing.

Many studies have been conducted to determine how long people brush (see Rugg-Gunn and MacGregor [22] and Kleber *et al.* [23]). The results of such studies have been conflicting. This is not unexpected, because brushing time is influenced by the brushing environment (laboratory or at home), observed versus hidden, gender of study participants, non-conformity among investigators in measuring the end-point (which frequently is residual plaque level) and other factors. Times reported have been from a few seconds to several minutes. Many dental investigators have, despite this wide variation, elected to have subjects brush for 60 seconds in studies in which brushing time can influence the results, e.g. comparisons of performance of different toothbrushes. An average of 30 seconds is a good guess, but only a guess and of little significance considering the interaction of many factors in toothbrush effect.

7.6 Oral-care product marketing

7.6.1 *Targeting consumer groups*

The oral-care market is much more competitive today than it was only a few decades ago because new innovations, in products and packages, has vastly increased the number of product niches. At one time, the major concern of the consumer was whether to use a product for its therapeutic benefits (fluoride) or its purely cosmetic benefits (flavor, etc.) [2]. Almost all toothpastes now contain fluoride, so realistically fluoride can be eliminated as an influence on the toothpaste purchase decision. New 'therapeutic' functionalities have been introduced into toothpastes and oral rinses, however, and the consumer still is swayed to purchase on the basis of interest in a product's 'therapeutic' benefits as opposed to satisfaction with cosmetic properties, such as flavor.

Products are available which fall into the following categories:

Toothpastes

- With low abrasion
- For tooth 'whitening'
- To inhibit calculus (tartar) formation
- To relieve hypersensitive teeth
- To reduce plaque
- To freshen breath (mouthwash effect)
- To provide a special 'clean feeling' (as claimed for baking soda)
- To provide special beneficial ingredients for confidence in using products used historically by the dental profession, e.g. peroxide
- For children (special appearance and flavor)
- With combinations of one or more of the above
- With 'natural' ingredients

Oral rinses

- To freshen the breath
- To destroy bacteria that can cause bad breath
- To treat the dentition with fluoride

	To help inhibit calculus
	To help remove plaque, especially when used before toothbrushing
	To reduce plaque and plaque-related gingivitis
Toothbrushes	Which are manipulated by hand
	Which are driven mechanically
	With special configurations of bristles and arrangements thereof
	With special head designs
	With inflexible or flexible shafts
	For children
Dental floss	Thread-like or ribbon-like
	With or without fluoride
	Waxed or unwaxed
	Flavored or unflavored
Denture cleansers	With toothpaste characteristics
	To remove plaque and debris by soaking
	To remove plaque and debris in a foaming medium

The huge array of oral-care products now available makes targeting towards specific but large target groups very difficult, especially when consumer preference is frequently influenced by his or her concern for 'therapeutic' versus cosmetic attributes. Fortuitously, 'therapeutic' agents used in toothpaste are for the most part compatible with toothpaste excipients. As a result, toothpastes are being marketed which claim to be effective simultaneously in one or more of cleaning teeth, protecting against caries, combating plaque and gingivitis and inhibiting calculus. In some circumstances, advertising may concentrate on only one facet, perhaps calculus inhibition, in the hope that the consumer recognizes other attributes without prompting.

There is considerable disarray overall in the oral care area. Promoters of oral-care products in the US (mainly new to the market) are introducing new products which can provide the ultimate in what the consumer seeks in oral care, namely, teeth which are truly whiter optically, actually bleached. This is the ultimate toothpaste function. A toothpaste which achieves this objective conveniently and at reasonable cost undoubtedly will be the OTC dentifrice product with greatest consumer demand.

7.6.2 *Regulation of the industry*

The oral-care product industry is regulated to greater or lesser extents depending on the country involved. Usually, the ultimate authority is the national government. Indirect regulation, however, not infrequently is exerted by non-governmental bodies and by marketing pressures. In some instances, however, there is no substantial regulation at all and ludicrous claims may be made.

Elements of the working of the regulatory process can be illustrated by reference to features of the United States system. The authoritative body is the Food and Drug Administration (FDA), an arm of the federal government. The FDA categorizes the OTC oral-care products as either drugs or cosmetics. Toothbrushes and other mechanically operated products are considered to be drug devices. Chemical agents, such as toothpastes which combat dental caries, are also categorized as drugs. The two types of drugs are regulated by different groups within the FDA. The FDA is responsible for both efficacy and safety. Products which do not affect disease status of the oral cavity are treated as cosmetics, and regulation thereof by the FDA is minimal, depending on individual situations.

A next level of regulation is the dental profession, usually acting through a central association or society. Examples are the American Dental Association (ADA) and the British Dental Association. These groups frequently consult with the government regulatory body. Some formally endorse oral products. The ADA gives its 'seal of acceptance' to products which have demonstrated safety and efficacy to the ADA's satisfaction. This level of regulation is especially powerful when the endorsement can be used in advertising. The ADA's seal is valuable because it can be used in commercial advertisements. The ADA has a long history of product approval via the seal, and from time to time has issued guidelines for proving safety and efficacy to the ADA's satisfaction. Among OTC drug products, there are available guidelines, for example, for proving out fluoride products and making changes in them [24], and for demonstrating the safety and efficacy of oral rinses which reduce gingivitis associated with supragingival plaque.

Less direct regulation of oral-care products may be exerted via advertising agencies or advertising review groups. Reputable agencies usually require substantiation of advertising claims. The depth of substantiation varies from agency to agency. Another related regulatory process is challenge of a company's claims by a competitive company. This may be resolved by the judicial system.

The restraints that are placed on advertising OTC oral-care products are substantial, especially in advanced industrial markets. Significantly, quantitative comparisons of product performance, such as product A is twice as effective as product B in performance, are rare. There are several reasons for this, one unfortunately, being lack of consensus on evaluation methodology.

Space prohibits detailed descriptions of guidelines recommended by authoritative bodies for proving the safety and efficacy of new OTC oral-care products. New guidelines are constantly being developed. It behoves promoters of OTC oral-care products to assure themselves that such products can be supported by appropriate guidelines, if they exist. Guidelines prepared by authoritative groups exist for the determination of toothpaste abrasivity, fluoride product efficacy, efficacy of pastes for hypersensitive teeth and products to reduce plaque and gingivitis.

7.6.3 *Product evaluation; support of marketing claims*

Evaluation of product performance is unusually difficult in the arena of oral hygiene products, with few exceptions. Major areas where product performance must be substantiated to support marketing claims are toothbrush performance, tooth cleansing (stain removal), anti-plaque efficacy, anti-caries activity and other areas where the activity concerns specific biological factors, such as anti-calculus and reduction in tooth hypersensitvity. The depth of scientific support required to market a product varies from country to country, and, as discussed above, with the nature of the regulatory agency, e.g. governmental or advertising agency, and whether an authoritative guide to testing has been established.

7.6.3.1 *Hierarchy of claim support* Depending on circumstances, four interrelated levels of claim support can be considered. The lowest is laboratory data (including animal data). Such data may suffice in some regulatory climates, but are weak as support unless backed by a strong correlation between them and more advanced levels of evaluation. A next step above laboratory data is short term use in the human under controlled conditions. The results may be quite misleading in predicting long term efficacy. However, the studies can be structured to incorporate conditions which take them a step closer to real life. The next rung of the testing ladder is long term clinical testing under carefully controlled conditions, wherein compliance with the use regimen is enforced and product distribution, etc. is carefully monitored. Finally, long term clinical trials in which the test product is used under true home-use conditions, i.e. without supervision. There are innumerable wrinkles in the conduct and interpretation of the results of each of these trials to support claims for efficacy. It is beyond the scope of this chapter to discuss these trials fully, e.g. statistical requirements, logistics, etc., but a few examples are provided here for clarification.

7.6.3.2 *Laboratory data* Laboratory data can be very convincing in supporting *in vivo* performance claims, but can be misleading, and even counterproductive, unless they are grounded in an extensive data base generated over the course of time by reliable investigators, and have been clearly correlated with the results of more advanced evaluation procedures. Several examples can be cited. The relationship of abrasivity of toothpaste to removal of tooth stain has been studied for at least a century. The *in vivo* work preceded the *in vitro* and over the course of experimentation the latter was developed to a fine art and it was established that a minimum radioactive dentine abrasion (RDA) for a toothpaste of about 15 was prerequisite for removal of stained pellicle [2]. Today, the RDA value (or equivalent) can usually be relied on to assure that a toothpaste can clean away stained pellicle. The addition of fluoride to toothpaste is now common. The possible anti-caries efficacy of fluoride was recognized as the result primarily of epidemiological observations. The mechanisms of fluoride action were eventually established by laboratory studies (including animal studies) of high investigative quality. Reputable dental investigators now accept specific *in vitro* studies of fluoride toothpaste as being reliable predictors of the anti-caries activity of a toothpaste in the human, and forego the costly, lengthy clinical trials once required to prove efficacy.

The discouraging side of *in vitro* testing of oral hygiene products is that most relevant *in vivo* aspects are difficult, if not impossible, to mimic *in vitro*. Quantitative comparisons of product performance based on *in vitro* data are generally invalid because of the uncertainties associated with duplicating real life conditions. Measurement of toothbrush performance is an outstanding example. It is simple enough to design a brushing machine, an arrangement of teeth or another substrate, and an artificial soil, but the almost infinite possibilities of variation in consumer brushing habits and tooth accumulations make it implausible to generalize quantitatively on the results obtained with only one variable usually in play in an *in vitro* study.

Laboratory data, alone, have occasionally been used as claim support. The marketer, however, is obliged to point out to the consumer that the basis for a claim is laboratory data.

7.6.3.3 *Short term human studies* Short term human clinical studies can yield valuable predictive data, and are preferred by far to *in vitro* tests. A classical example is the Löe model for gingivitis [9]. Here, the subject refrains from all oral hygiene practice for a period of about 2–3 weeks. This simulates the worst case of poor oral hygiene. Plaque and gingivitis develop profusely. The effect of treatment on these conditions can then be determined. They can be reversed, for example, by returning to a good toothbrushing regimen, or by an effective anti-plaque agent, such as chlorhexidine. This type of test and variations thereof have in many circumstances been able to predict whether oral hygiene agents and devices have the *potential* to be useful in oral care.

A variety of other short term clinical studies has been used productively. One is the half-mouth technique. Here, the dentition on one side of the mouth is treated with the test material and the other side acts as control. The plaque removal ability of a product or process can be estimated semi-quantitatively by testing for plaque removal effects using subjects who already had accumulated plaque during normal living. (In the case of plaque studies, the results can be severely confused by inadequacy of the plaque measurement procedure [10]).

7.6.3.4 *Major scale clinical trial* A full blown trial is the ultimate test for efficacy. Individual investigators have tended to use their own designs, with the result that consensus on the test agent has not always been possible. However, guidelines for conducting clinical trials for specific purposes are being developed, e.g. to evaluate the efficacy of agents which reduce plaque-induced gingivitis. Clinical trials are very expensive and time consuming. Early trials of fluoride toothpastes, for example, ran for at least two years and involved hundreds, even thousands, of subjects. Protocols are still controversial. Many investigators prefer to include in the trial design at least one supervised use of product per day when the subjects are available, thereby ensuring a degree of compliance with the use regimen. Others aver that a more realistic situation in the test of an OTC agent is to omit supervised use, making the test conditions closer to real life. There are several ways to conduct clinical trials, depending on the objectives, logistics and ethical considerations. In the long run, it is preferable before starting clinical trials to consult with regulatory agencies which may have a voice in the interpretation of the results.

7.7 Principles of product formulation

7.7.1 *The toothbrush and other mechanical aids*

7.7.1.1 *Toothbrush function* The toothbrush is a device, and those involved in the development of oral hygiene products must understand its functioning if they are to be able to design suitable supplemental products. The toothbrush is the key to oral cleanliness. Alone (or with water) it can maintain the mouth in a satisfactory state of hygiene. However, as mentioned previously, the vast majority of the population do not employ the toothbrush properly, or with adequate frequency. As a result, supplemental products have been developed. These include (i) dentifrice, to be used with the toothbrush; (ii) oral rinses, to be used before or after the toothbrush; and (iii) devices to supplement the mechanical action of the toothbrush, e.g. dental floss.

7.7.1.2 *Plaque removal* The toothbrush alone, used with care and diligence, can remove enough plaque for the maintenance of a satisfactory state of oral hygiene. Conventional toothpaste may improve plaque removal by only about 5%. Toothpaste, however, depending on its composition, can offer a route to substantially increased plaque reduction or other therapeutic function, e.g. inhibition of calculus formation. Figure 7.3 depicts the typical route of plaque removal by toothbrushing. The efficacy of the toothbrushing operation is determined, however, by factors over which only the toothbrusher (not the toothbrush manufacturer) has control:

(i) motivation to brush properly;
(ii) brushing for a long enough time;
(iii) ability to manipulate the toothbrush;
(iv) knowledge of how to brush; and
(v) use of a toothbrush in good condition.

7.7.1.3 *Stain removal* A second major role of the toothbrush is to act in concert with the abrasive to remove stained pellicle. Here, the toothbrush applies the force needed to provide abrasive cleaning.

7.7.1.4 *Toothbrush construction and design* The toothbrush consists of a handle, a neck and a head. The functionality is in the head, which contains rows of bristles, usually nylon filament. Hog bristles, popular at one time, have now been replaced by nylon for reasons of sanitation, cost, safety and reproducibility. Nylon bristles can be made to uniform, strict standards.

Bass [25], an early investigator of toothbrushes, recommended a straight handle, an overall length of about 6″, width about 7/16″, three rows of nylon bristles, six tufts to the row, about 80 bristles per tuft, 0.007″ diameter bristles, straight trim, finished to 13/22″ length, bristle ends rounded. Very few toothbrushes with those dimensions are being marketed today. Rather, toothbrush marketers have let their creativity run amuck, and the range of toothbrush innovation has run from the Bass-type brush to brushes with diamond-shaped head, flexible handle and bristles which vary in length on the same head. Support for superiority of the new designs is meager, usually based on *in vitro* experiments, uniqueness and the investigator's creativity.

Mechanically driven toothbrushes have been developed. These should theoretically perform better than manual brushes, by making the brushing experience less dependent on the individual. The question of whether or not they are more effective in the home-use situation is still to be answered.

7.7.1.5 *Other mechanical aids* Toothbrushing alone, as practiced by most individuals, is inadequate to remove plaque. The interproximal and subgingival sites are the loci of a major proportion of plaque-caused diseases, and clearing those areas of plaque with the toothbrush requires an extraordinary degree of motivation and dexterity. Consequently, a large proportion of the population needs additional help to control plaque. This help is available from a variety of mechanical devices. The available mechanical supplements cannot all be reviewed here, but some of the more useful devices are:

- Oral irrigator—highly effective at reducing gingivitis, despite apparently limited plaque removal.
- Dental floss—can remove plaque, improve gingival condition, but requires high manual dexterity for maximum effect.

- Wooden tips—made from soft wood to prevent injury to the gingiva. Not recommended where interdental space is fully occupied by the papillae.
- Rubber tips—pointed, pliable rubber tips mounted on the toothbrush handle distally to the brush head.
- Interdental brushes—marketed either as a single tufted brush or a miniature bottle brush. Not popular, but highly effective.

7.7.2 *Toothpaste formulation practice*

7.7.2.1 *General concerns* As recently as thirty to forty years ago, dentifrices were designed to fulfil two major functions: (i) to act as vehicles for abrasives to remove stained pellicle; and (ii) to flavor and freshen the breath. Today, dentifrices also act as vehicles to introduce active agents into the oral cavity. Although dentifrice is not a requirement to remove or control dental plaque [7], few people would brush enthusiastically without one. The accumulation of new plaque may be slightly retarded due to the surfactant's effect. Although some investigators claim dentifrice can increase the amount of plaque removed by toothbrushing, most evidence indicates that the same amount of plaque is removed, whether a highly abrasive, a non-abrasive, or no toothpaste is used [2]. The need for abrasive to remove stained pellicle, however, was firmly established by Kitchin and Robinson [26], who asked 'How abrasive need a dentifrice be?' They reported a trial with 113 dental students, in which stains ranging from minor to heavy were readily removed by brushing 1–2 times with an abrasive dentifrice, whereas toothbrush and water alone had little effect.

A breath-freshening quality in a dentifrice is now expected by the consumer. Although the cost of flavoring a dentifrice can be higher than that of all of the other ingredients together, most people would agree that dentifrice usage would be significantly lower today, were the easily perceptible stain removal and breath-freshening performance of dentifrice less than satisfactory. Flavor can determine consumer purchase decision, all other factors being equal. Toothpastes are now being marketed with flavors that appeal only to limited segments of the market.

Dentifrices are marketed as pastes, liquids, and powders. Paste is the predominant form, both because of its inherent consumer benefits, and because it is the preferred dosage vehicle for special therapeutic and cosmetic agents. Powders, once very popular, have fallen considerably in market appeal. Concentrated liquids have maintained a small share of market.

The major ingredients of dentifrices and their functions are outlined in Table 7.1. (For further details, the reader is referred to Pader [2]). The economics of many dentifrice raw materials differ in different parts of the world; the formulator may find one particular type of formulation less

Table 7.1 Dentifrice ingredient functions

Ingredient	Function	Example
Abrasive	Helps remove stained pellicle, acting in concert with the toothbrush.	Precipitated chalk Hydrated silica
Humectant	To provide a bacteriologically-stable, fluid, creamy base, which resists 'drying out' and in which other ingredients can be suspended or dissolved.	70% sorbitol
Binder	To develop a cohesive structure with the desired rheological properties.	Precipitated silica Calcium carrageenate Carboxymethyl cellulose
Surfactant	To provide foam and detersive action during use; possibly may offer anti-microbial and other benefits.	Sodium lauryl sulfate
Flavor	To make the toothbrushing experience pleasant, and to leave a residual freshness and cooling.	Peppermint oil Saccharin
'Active agents'	To provide therapeutic and/or other special benefits.	Fluoride
Water	To assist formulation.	

expensive in an Asian country than in, say, Europe, and vice versa for another type of formulation. The cost difference may be very significant. Consequently, the formulator must carefully balance the type of formula against the cost factor; a dentifrice that has optimal characteristics of, for example, performance and consumer appeal, in one country may not be economically feasible in another.

7.7.2.2 *New toothpaste functions* The sudden proliferation of new toothpaste functions, packaging and visual signals has made formulation much more complicated than heretofore. It is in the best interests of the marketer capable of handling a range of toothpastes to simplify the formulations in the interest of ease of manufacture. It is worthwhile to develop a base formulation which can be modified easily. For example, most toothpastes offered today contain fluoride. It is advantageous to restrict all the toothpaste formulations in the line to one which is compatible with fluoride. That approach is espoused here; multiple types of formulation in the same manufacturing facility can be disastrous economically. Today's technology allows major variations in toothpaste appearance, flavor, etc., by simple changes in the same base formula.

Toothpastes must be stable during shipping and storage (6 months at 125°F). They should be esthetically appealing. They should function as promised by the marketer. A deficiency in any of these parameters can seriously impair acceptance by the consumer, and each brand hopefully will show a unique difference from competitive brands.

Toothpastes are now available with various 'therapeutic' functions. The main functions are:

- to help prevent caries
- to promote gum health (anti-gingivitis)
- to reduce plaque
- to reduce sensitivity of hypersensitive teeth

The same 'active' sometimes is offered in both toothpaste and oral rinse vehicles, e.g. fluoride toothpaste and fluoride rinse. Incompatibility between toothpaste excipients and the agent will prevent this in some instances. A host of agents has been proposed to function in specific oral therapeutic modes. In the usual course of development of an agent, initial studies are conducted with simple liquid systems. The toothpaste formulator must later ascertain that the toothpaste vehicle developed does not negate the positive activity already demonstrated with the agent dispensed as a simple oral rinse. Unfortunately, the selection of a therapeutic active for a toothpaste has been clouded by such factors as the regulatory climate, inappropriate research to support marketing claims, lack of consensus among investigators on the interpretation of laboratory and/or clinical trials, use of conflicting methodologies by different researchers, etc. The agents of greatest current interest in OTC oral-care products are discussed briefly, below (section 7.9).

7.7.2.3 *Toothpaste ingredients and their functions* Toothpaste is generally formulated along either of two lines: (i) low abrasive level/high humectant level (Type I) or (ii) high abrasive level/low humectant level (Type II).

Table 7.2 Example of a Type I dentifrice[a]

		wt%
Active agent	Sodium monofluorophosphate	0.78
Abrasive	Silica xerogel[b]	14.00
Binder/thickener	Carboxymethyl cellulose[c]	0.30
	Hydrated silica[d]	8.00
	Polyethylene glycol 1450	5.00
Humectant	Sorbitol, 70%	46.72
	Glycerin, 96%	20.90
Surfactant	Sodium lauryl sulfate (dentifrice grade)	1.50
Flavor	Mint-type	2.00
	Saccharin	0.20
Color	Aqueous solution of permissible dye	0.50
Preservative	Sodium benzoate	0.10

[a] After Pader [34]
[b] Sylodent 750; W.R. Grace & Co., Davison Chemical Division
[c] Grade 9MX; Hercules, Inc.
[d] Sylodent 15; W.R. Grace & Co.

Table 7.3 Example of a Type II dentifrice[a]

		wt%
Abrasive	Calcium carbonate	48.00
Binder/thickener	Carrageenan[b]	0.94
Humectant	Glycerin	22.00
Surfactant	Sodium lauryl sulfate	2.00
Flavor	Flavor oil	0.80
	Sodium saccharin	0.20
Preservative	Sodium benzoate	0.50
Other	Water	to 100.00

[a] From FMC commercial bulletin (1988)
[b] Viscarin[R] TP 389

Table 7.2 presents an example of a Type I toothpaste, Table 7.3 that of a Type II dentifrice [1].

7.7.2.4 *Toothpaste rheology* It is necessary to understand relevant features of paste rheology to formulate and manufacture toothpaste variants economically and expeditiously. The rheological properties affect many esthetic and functional attributes. The ideal rheological profile is a rigid gelatinous structure which becomes deformed and flowable with the application of minimal stress and which re-establishes its original structure almost instantaneously. The preferred rheological properties of a paste include [27]:

(i) A homogenous ribbon is formed upon extrusion from the container.
(ii) The paste does not flow from the open container (uncapped toothpaste tube) in the absence of an extrusion pressure.
(iii) Only minimal effort is required to extrude the paste.
(iv) The extruded ribbon of paste breaks sharply and cleanly onto the toothbrush.
(v) The extruded ribbon 'stands up' on the brush; it does not sink between the bristles or roll off the sides.
(vi) The paste disperses quickly in the mouth and does not feel gummy.
(vii) The flavor releases quickly.
(viii) The foam quality is pleasant and the amount appropriate.
(ix) The foam rinses from the mouth easily, leaving no residual gummy feeling in the mouth.
(x) Spittle is easily rinsed from the sink.
(xi) The paste is stable during shipping and storage.
(xii) Dispensing of the paste is compatible with the dispensing mechanism.

The best rheological properties are obtained with formulations with relatively high viscosity along with a relatively high yield point [27].

Preferred rheological properties are most readily achieved with Type I toothpaste. Acceptable pastes, however, can be obtained with Type II. The formulator may select either depending on quality standards required or marketing concerns in a local market and/or availability and cost of local supplies. Type I formulations are more versatile with respect to developing a range of products from a single base composition, whether therapeutic or cosmetic.

Type I formulations frequently rely on amorphous silica for cleaning/abrasion as well as rheological structure. The Type II formulation rheology, however, relies largely on the sheer bulk of the abrasive.

Rheological additives for the Type I and Type II toothpastes have been discussed in detail previously [1]. The principles of their selection will be discussed here.

Type I toothpastes are more restricted in the choice of a rheological additive system than Type II toothpastes. In both instances, the toothpaste structure derives from abrasive particles suspended in a three-dimensional network of a polymer, or, more usually, a mixture of polymers. The demands of Type II toothpastes are few, because structure and stability thereof are the result of what may be only a loose agglomeration of abrasive particles in a viscous liquid medium, the concentrations of which limit mobility of the abrasive, as though it were a pile of sand. A polymer is added to improve flow characteristics and prevent separation of aqueous and particulate phases, i.e. to retard settling of the particles by increasing the viscosity of the liquid phase. Any number of polymers have been used for the purpose—sodium carboxymethylcellulose, guar gum, gum arabic, xanthan, locust bean gum, etc.

A different situation prevails in Type I toothpastes. Here, the abrasive or other particles are suspended and immobilized within a three-dimensional network of polymer chains which is essentially rigid when at rest but readily takes on the flow properties of a liquid when it is subject to a stress. These polymeric structures exhibit yield values; they are immobile until a certain stress is exceeded, as by squeezing the tube of toothpaste. They are restored to their original structure when the stress is removed, almost instantaneously.

7.7.3 *Basic dentifrice ingredients*

7.7.3.1 *The abrasive* Pader [2] defined an abrasive as "... a solid, particulate material which, when applied under pressure, can dislodge foreign material from a surface." He also noted that for an abrasive to be effective, it must remove a finite amount of dental enamel along with stained pellicle, and also some dentin if the dentin is exposed to the abrasive's action. Abrasives are preferably chemically inert. They must:

(i) resist interaction with the other dentifrice ingredients and be essentially tasteless;
(ii) maintain their integrity in the dentifrice product—salt and sodium bicarbonate have frequently been proposed and used as dentifrice abrasives, although the latter is used in marketed dentifrice to a very limited extent;
(iii) be safe for their intended use;
(iv) be capable of being formulated into a stable, smooth, extrudable paste that tastes good, is quickly dispersed in the mouth, and has satisfactory rheological properties.

Four physico-chemical characteristics are important in selection of an effective, safe abrasive for dentifrice use: (i) hardness; (ii) toughness; (iii) degree of chemical inertness; and (iv) particle size and shape.

Hardness The ideal abrasive will be no harder than necessary to attack its dental substrate (i.e. stained pellicle). The vulnerable underlying enamel and any dentin that may be exposed must be considered. Hardness is generally stated in terms of the Mohs scale of hardness. In this scale, talc is rated 1 and diamond 10. The Mohs hardness of dental enamel is 4–5, and of dentin 2–2.5. There is no hard and fast rule for selection of abrasive on Mohs hardness, because the safety and efficacy of an abrasive is critically dependent on its performance in the presence of the total dentifrice formulation. As a general rule, however, the prudent formulator will avoid excessively hard crystalline materials for abrasion. The relative abrasivity of an abrasive material in a series of abrasives is not determined solely by its Mohs hardness, but also depends on the substrate used. The prudent formulator will seek an abrasive that yields a formulation, the abrasivity of which is such that it will remove a minimal amount of dentin, compatible with pellicle removal using dentin as a test substrate.

Toughness This is defined as the ability of an abrasive to withstand the forces brought to bear on it when in use. Although not previously given great attention, toughness may become increasingly important. This is because tailor-made dentifrice abrasives are formed by controlled agglomeration of small particles. Crystalline silicon dioxide is extremely abrasive and, as an amorphous, friable particle, is relatively soft and therefore an ideal dentifrice abrasive.

Degree of chemical inertness Dental abrasives should be chemically inert. The importance of this becomes increasingly evident for formulations with therapeutic action; containing potentially chemically reactive materials, such as fluoride.

Particle size and shape These strongly influence performance, safety and esthetics. Particles with diameters of about 30 μm or more will be detected as objectionably gritty in the mouth. An average particle size of about 3–12 nm will generally be acceptable to the consumer. Tight specifications on particle size are critical to the manufacture of a safe, pleasant dentifrice—particular attention must be paid to the particle size range achieved during manufacture by careful milling and sizing.

The ability of an abrasive to remove stained pellicle will generally increase with its concentration in the dentifrice, with particle size, and with hardness. The abrasivity of a given formulation will vary according to external forces (including the toothbrush), the vigor with which it is used, and the characteristics of the individual dentition [28].

Abrasion for dentifrice application can be measured by many methods. The two most popular are the radioactive dentin abrasion (RDA) method and the surface profile method. Both have been described by Pader [2]. In the RDA method (most frequently used), extracted teeth are subjected to a neutron flux under standardized conditions. They are then mounted in a machine equipped to allow controlled brushing with a toothbrush and a dentifrice slurry. After brushing for a given time under a specific toothbrush pressure, radioactivity (^{32}P) in the slurry is measured. A calcium pyrophosphate abrasive system is the reference abrasive standard, set at a RDA value of 100. Generally speaking, a dentifrice with an RDA value of 60–100 is adequate to maintain teeth essentially free of stain. A dentifrice with an RDA value over 250 is probably unnecessarily abrasive.

Cleaning (removal of stained pellicle) is related to abrasivity: the higher the abrasivity (up to a limit), the better the cleaning. Although RDA is considered to be an adequate method for assessing cleaning, it has been difficult to develop an *in vitro* cleaning test due to the lack of a suitable artificial stain. Stookey *et al.* [29] devised a stain which reportedly was adequately resistant. These workers correlated their *in vitro* method with a laboratory 'cleaning ratio'. The American Dental Association [30] conducted a study wherein three dentifrices with abrasivity indices of 85, 170 and 255, respectively, were compared with respect to ability to remove or to prevent the accumulation of stained pellicle *in vivo*. The results validated the premise that abrasivity of a dentifrice to dentin can be used to assess cleaning power.

Several dental abrasives are available. Some of the more popular ones will be discussed in the following sections.

Hydrated silica Intensive investigation of dental silica technology has demonstrated the potential of tailor-made silica products to meet the needs of today's dentifrice market. Figure 7.4, for example, shows the relationship between RDA and silica gel loading for a range of silica xerogel abrasives developed by a single manufacturer. Hydrated silica precipitates, with abrasive properties, are also available, developed as alternatives to the silica xerogel. Xerogels and precipitates can be formulated into dentifrices; both can give clear, abrasive dentifrice gels. One marketed hydrated silica

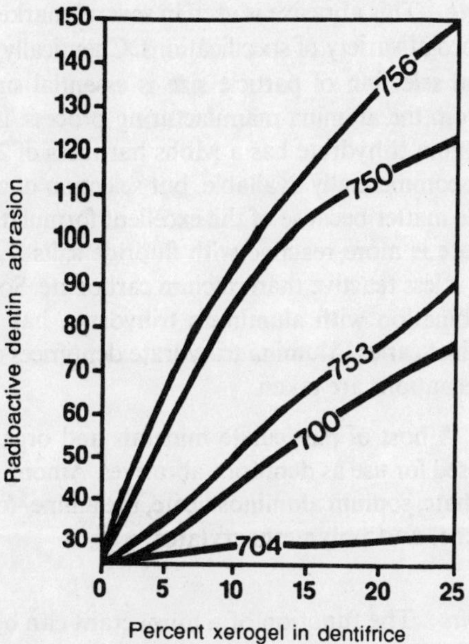

Figure 7.4 Effect of silica xerogel structure and level in dentifrice abrasivity. Courtesy of W.R. Grace & Co., Davison Chemical Division.

precipitate is a mixture of large and small particles; the former provide abrasion, the latter, gel structure.

Calcium phosphates Di- and tricalcium phosphates have been widely used as dentifrice abrasives. They belong to the family of calcium phosphates represented by the $CaO-P_2O_5-H_2O$ system. They are not compatible with sodium fluoride, but are used with sodium monofluorophosphate. Dicalcium phosphate dihydrate (DCPD), $CaHPO_4 \cdot 2H_2O$, is not stable in dentifrice systems. On heating, or even storage, the water is lost and the anyhydrous salt (DCP) forms. DCP is considerably harder than DCPD, therefore more abrasive. In dentifrice, the conversion of DCPD to DCP may be accompanied by firming of the dentifrice, but DCPD can be stabilized by the addition, during manufacture, of tetrasodium pyrophosphate or trimagnesium phosphate. DCPD for dentifrice use has been manufactured with RDA values of 200–1350 (reference calcium pyrophosphate, set at 500). Consequently, it is relatively low in abrasion, but quite adequate for most situations. DCP, being very hard, can be admixed with DCPD to raise DCPD dentifrice abrasivity.

Calcium carbonate Precipitated calcium carbonate is an inexpensive, but effective dentifrice abrasive. Its high reactivity with fluoride limits its use to sodium monofluorophosphate dentifrices. Even then, fluoride bioavailability may be poor. Several commercial grades of precipitated calcium carbonate are available, ranging in particle size.

Alumina trihydrate This abrasive is used in several marketed dentifrices and is available with a broad variety of specifications. Chemically, it is $Al_2O_3 \cdot 3H_2O$ or $Al(OH)_3$. Careful selection of particle size is essential since the material is generally derived from the alumina manufacturing process. In its gibbsite form, used primarily, alumina trihydrate has a Mohs hardness of 2.5–3.5. Excessively abrasive grades are commercially available, but selection of a grade suitable for dentifrice is a simple matter because of the excellent formulation abilities.

Alumina trihydrate is more reactive with fluoride salts than is, for example, hydrated silica, but is less reactive than calcium carbonate. Sodium monofluorophosphate, in combination with aluminum trihydrate, has been shown to be clinically active against caries. Alumina trihydrate dentifrices can corrode aluminum tubes if no precautions are taken.

Other abrasives A host of particulate minerals and organic polymers have been used or proposed for use as dentifrice abrasives. Among them are insoluble sodium metaphosphate, sodium aluminosilicate, melamine–formaldehyde, polystyrene, polyethylene, and polymethacrylates.

7.7.3.2 *Humectants* The function of a humectant can be divided into two categories: (i) bulk dentifrice properties; and (ii) in-use properties. The humectant contributes to the bulk properties of the dentifrice by:

(i) providing a vehicle that carries abrasive, flavor, drug and other components, to give a smooth, homogeneous, functional dentifrice;
(ii) maintaining an extrudable paste that resists 'drying out' when exposed to the atmosphere for prolonged periods of time;
(iii) providing microbiological stability;
(iv) developing a paste structure in conjunction with the thickener;
(v) assisting maintenance of phase stability, i.e. preventing separation of the aqueous and non-aqueous components; and
(vi) providing the means to formulate a translucent/transparent 'gel' dentifrice.

Features of the dentifrice that are humectant-dependent are:

(i) flavor release and impact, and sweetness of the flavor by virtue of the inherent sweetness of the humectant;
(ii) amount and character of the foam; and
(iii) dispersibility and spreadability of the dentifrice in the mouth.

Sorbitol solution and glycerin are the predominant humectants for dentifrices, alone or in combination. The 'sorbitol' solution is usually a 70% aqueous solution of sorbitol, or a sorbitol modification that resists crystallization due to its content of higher hydrogenated polysaccharides. The amount of polysaccharides in the final sorbitol syrup can be determined during the manufacturing process. Pader [31] described a sorbitol with relatively high amounts of hydrogenated polysaccharides, which was used advantageously in toothpastes.

Two microbiological aspects of the humectants are important: (i) ability of the humectant to resist microbial growth in the toothpaste; and (ii) resistance to metabolism and subsequent production of acids by oral organisms. The water activity (a_w), defined as the vapour pressure of the solution divided by the vapour pressure of water at the same temperature, determines whether or not a solution will support microbial growth. Normally, values of a_w above 0.9 are required for growth. The addition of glycerin or sorbitol lower the a_w of the dentifrice to a point where microbial growth is suppressed. In addition, the osmotic pressure is increased, and cannot be tolerated by most organisms. The humectant, however, cannot be relied on entirely. Even properly formulated dentifrices have been known to deteriorate microbiologically (sometimes with gassing) during storage. Contamination must be minimized, from the raw materials through to the final, manufactured dentifrice.

In pure solution, a 30% solution of glycerin or a 40% solution of sorbitol will stabilize against bacterial attack. Moulds and yeasts, being more tolerant to environmental conditions, are kept under control with chemical preservatives (e.g. benzoate and derivatives). The refractive indices of glycerin and 70% aqueous sorbitol are both about 1.47, therefore the preparations can be used interchangeably in so far as a particular refractive index is needed to formulate clear dentifrices (where humectant and abrasive refractive indices must be similar).

The humectant must not be fermentable by micro-organisms resident in the oral cavity, such as *Streptococcus mutans.* As discussed previously, metabolism of carbohydrate by such bacteria, to produce acids, is a route to caries. Hydrogenation of glucose and glucose polymers derived from starch hydrolysis makes the sugars essentially non-fermentable by oral micro-organisms.

7.7.3.3 *Binder/thickener* The importance of rheological control of toothpastes was discussed above (see section 7.7.2.4). The binder/thickener is the key to the achievement of the desired rheological properties.

Besides the functional attributes of the polymer, attention must be given to its ease of incorporation during the manufacture of the toothpaste, its resistance to microbial, enzymatic, chemical and thermal challenges, and its cost-effectiveness. Dozens of organic polymers and inorganics (both natural and synthetic) capable of forming three-dimensional structures have been proposed for toothpaste use [1]. Few (if any) have all of the rheological attributes required, and frequently it is advantageous to mix polymers, for example a polymer with high yield value with one that has no yield value.

Every polymer that is useful in toothpaste formulation is able to form a three-dimensional network in systems containing low water concentrations. Both organic and inorganic materials can be used in such systems. A common feature of organic polymers is the presence, in the dry state, of tightly aggregated molecular chains, which, on solvation, open to form strong gel-like networks of polymeric chains. This phenomenon is pH- and/or temperature-dependent for some polymers. Maximum strength is obtained with certain polymers only in the presence of a sufficient amount of appropriate cations. The strength of the polymeric structure depends on the polymer concentration, the presence of surfactant, flavor and other additives, and the composition of the humectant system. The rheological properties of the binder will have a strong influence on these features of the polymeric structure, and will carry through to the final toothpaste.

Polymers used in major dentifrice brands are: alginates, carboxymethyl cellulose, carrageenan, gum tragacanth, locust bean gum, xanthan gum, polyacrylates and hydrated silicas. Other possible thickening polymers currently available are gum arabic and gum ghatti. As mentioned previously, combinations of polymers are frequently used.

A 'viscosity bonus effect', whereby glycerin addition to aqueous solution increases the viscosity, can be observed with many polymer–lean-solvent systems. For example, sodium carboxymethyl cellulose dispersed in a mixed solvent system (glycerin:water, 60:40) is about ten times as viscous as in water alone. A brief outline of the use of polymers in dentifrice applications will be given in the following sections. For further details, reference [32] should be consulted.

The number of polymers used extensively in major toothpaste brands has decreased in recent years. Some manufacturers, generally in order to cut ingredient costs or because they do not have a need for best toothpaste rheology, have used less than optimal toothpaste polymers. Only those which are most valuable in obtaining the rheology discussed in section 7.7.2.4 are discussed here.

Xanthan gum This is a polysaccharide produced by the micro-organism *Xanthomonas campestris* in pure culture fermentation. The structure of the main chain of xanthan gum is that of cellulose, and contains sodium, potassium and calcium ions. The rheological properties of xanthan gum are very favorable for the manufacture of toothpastes, particularly low abrasive/high humectant products. Aqueous systems exhibit high pseudo-plasticity along with a definite yield value. The viscosity decreases when shear stress is applied, and the gum recovers almost instantly when the stress is removed.

Xanthan gum can exhibit very reproducible and predictable suspending power over a wide range of conditions. It has high viscosity at low shear rates. At very low shear rates, suspended particles remain essentially stationary because of the very high apparent viscosity of gum solutions below the yield value. Shear rates of pouring or squeezing are large enough to reduce apparent viscosity. Under high shear rates, low viscosities are observed, and little apparent viscosity is exhibited during operations such as pumping. Xanthan gum solutions are resistant to even the high shear imparted by a colloid mill or homogenizer. The viscosity of xanthan gum solutions is almost independent of pH or temperature. Xanthan gum will tolerate high levels of water-miscible solvents. It is even soluble in glycerin at 65°C, but the solution is stringy. It is compatible with most salts, although it may gel under prolonged exposure to some salts at sufficiently high concentrations. A small amount of salt helps to develop optimum rheology in xanthan systems. Xanthan gum is compatible with many gums, and the rheology of its solutions can be modified by the addition of other gums. Xanthan gum systems generally require preservation.

Acrylic acid polymers Acrylic acid polymers have excellent rheological properties for high humectant dentifrices. The Carbomer® resins are carboxyvinyl polymers, ranging in molecular weight from about 450 000 to 4 000 000, and containing $-(CH_2CHCO_2H)-$groupings, which may be cross-linked, as with a polyalkenyl polyether. The dry powders in aqueous dispersion show a pH of 2.8 to 3.2, depending on concentration. In this state, prior to solvation, the molecule is tightly coiled and provides only limited thickening. On hydration in water, some uncoiling occurs, but full development of viscosity potential requires uncoiling of the molecule—best achieved by elevation of the pH.

The resins form highly pseudoplastic solutions. Their ability to stabilize suspensions and emulsions is related to high yield value rather than viscosity. Consequently, stabilization of suspensions is achievable at low resin concentration.

The high molecular weight polymers are most effective at pH 6 to 8; viscosity decreases sharply on either side of this pH range. A 1%, neutralized aqueous system is a firm gel-like mass. Soluble salts, including sodium chloride and di- and trivalent cations, considerably reduce the viscosity of the resin solution. The salt effect varies with the system's composition. For example, it is reduced in oil-in-water emulsions compared to in water alone. The resins are generally incompatible with cationic compounds but cationic compounds of a moderate-to-high molecular weight, exhibiting steric hindrance, will not interfere with the polymer's thickening action.

Acrylic acid polymer resins are sensitive to high shear, and permanent loss in viscosity may occur with excessive shear. Temperature has only a small effect on the viscosity of neutralized polymer solutions, which resist freezing and thawing, and solutions of the 4 000 000 molecular weight polymer are stable to prolonged exposure at 70°C. Acrylic acid resins are compatible with other polymeric materials, such as xanthan gum and hydrated silicas and do not support microbial growth.

Silica powders The major silica powders used for dentifrice thickening are the silica aerogels, silica precipitates and pyrogenic silicas. These powders are chemically dissimilar from the organic gums mentioned previously, but have in common with them the ability to form a three-dimensional structure capable of maintaining abrasive particles in suspension. The silica powders have found greatest use in low-abrasive/high-humectant formulations, although they have been used in high abrasive/low humectant dentifrices.

The silica thickeners most widely used in toothpaste in the US are the aerogels and precipitates. All are amorphous and have a refractive index about 1.46, making them suitable for clear-gel toothpastes. They are variable with respect to patricle size, surface area and pore structure, and function primarily by formation of a solvated gel-type structure, which provides a ready route to the establishment of thixotropy in toothpastes. Silica thickeners exhibit a yield point, and are excellent agents for the suspension of abrasive particles. The silica aerogels and precipitates are used in low-abrasive/high-humectant pastes at loadings of about 5 to 15%. The rheological properties of the gelatinous structures formed can be modified by the addition of organic polymeric materials.

Not all silica preparations will develop the desired structures. Thickening effects are observed only when particles of colloidal size form a network through the liquid phase. Maximum network formation and thixotropy occur when small, three-dimensional aggregates are linked together by forces such as hydrogen bonding. These linkages are broken easily by shearing, and are re-established when the shear is removed.

Carboxymethyl cellulose Carboxymethyl cellulose is offered mainly as the sodium salt (SCMC). It does not have the rheology of, say, polyacrylates. It yields non-Newtonian solutions. The degree of thixotropy of lean-solvent systems containing SCMC is a function of many factors, such as degree of substitution (DS) of the polymer backbone, uniformity of substitution, water content, heat, long hold-times and shear. Thixotropy can vary at a given DS, depending on processing conditions and raw materials.

SCMC alone will not provide the rheological properties of a Type I toothpaste, but may find use in such as a supplemental polymer for its water-binding and thickening properties. SCMC (as well as gums such as locust bean gum, gum acacia, carrageenan gum and other vegetable gums) finds extensive use in Type II toothpastes, where the structural nature of the gum is less important.

7.7.3.4 *Surfactant* The surfactant is most important because the amount and character of the foam generated by a dentifrice affects its consumer acceptability. Slight improvement in plaque removal can be attributed to the detergent, which possibly inhibits recolonization of the tooth surfaces [7]. Virtually every detergent developed has been recommended, at one time or another, for use in dentifrice. In the case of anionic detergents, the limitation has frequently been the flavor—even trace amounts of impurities can be detected as a soapy, bitter flavor note. Cationic detergents exhibit inherent bitter, astringent notes, relatively poor foam generation and doubtful safety. Non-ionic detergents have been developed for dental specialities, but not for general use.

Sometimes, addition of a surfactant may result in inexplicable effects on toothpaste rheology, e.g. increasing fluidity. These effects have not all been explained. Reaction of the binder/thickener polymer with the detergent probably is a factor. The detergent may influence solvation of the polymer. It may, thereby, interfere with function of the three-dimensional solvated polymer. It is useful to solvate the polymer(s) completely before using it in toothpaste manufacture, to promote product uniformity and maximum utilization of the polymer.

Three anionic detergents have been widely used in toothpastes. These are sodium lauryl sulfate, sodium dodecylbenzene sulfonate and sodium lauroyl sarcosinate.

Sodium lauryl sulfate Sodium lauryl sulfate (SLS) is the long-chain fatty acid sulfate, RSO_3Na, where R is a mixture of long-chain saturated alkyl chains derived from coconut or similar oils and ranging mostly between C_{10} and C_{16}, principally C_{12}. Some grades are available with almost all C_{12} cut. SLS is synthesized through a fatty alcohol route, and the residual fatty alcohol content must be very low to ensure good flavor. Dentifrice grade SLS is freely available, in a range of particle size and alcohol chain length (R) distributions. Larger particle size is preferred for handling in the plant, to cut down dusting.

SLS is a primary irritant but, used at appropriate levels, is safe in toothpaste. The low toxicity level of SLS when judiciously used in toothpaste is supported by decades of use in dentifrices around the world, as well as by in-depth studies of toxicity (cf. [2]). No studies have shown that SLS reduces the efficacy of fluoride, indeed it has been reported that SLS can enhance *in vitro* fluoride uptake by dental enamel [33].

Sodium dodecylbenzene sulfonate Sodium dodecylbenzene sulfonate (DOBS) is represented by $RC_6H_4SO_3H$, where R is a long-chain fatty acid radical, mostly dodecyl. Commercial preparations comprise both branched- and straight-chain radicals. DOBS is an excellent foaming agent and provided it is specially 'cleaned up', has a low order of toxicity. The preference for SLS can probably be attributed to the development of pure grades of SLS.

Sodium lauroyl sarcosinate Sodium lauroyl sarcosinate is represented by $CH_3(CH_2)_{10}CON(CH_3)CH_2CO_2Na$. It was used extensively in dentifrices, and was claimed to possess anti-caries activity, but clinical testing failed to support this. In 1980, the US Advisory Review Panel on OTC Dentifrices and Dental Care Products observed that sodium lauroyl sarcosinate was a cause of oral mucosal irritation. Following this, the sarcosinate was replaced by SLS in offending dentifrices and, although still employed, its use is limited.

Other anionic detergents A large number of detergents have been proposed for dentifrice use, including α-olefin sulfonates, sulfocolaurate, sodium monoglyceride sulfonate, and others.

7.7.3.5 *Flavor and other cosmetic attributes* Dentifrice flavor plays two roles: (i) provision of a refreshing taste and feel in the mouth and (ii) masking of the natural flavors of the dentifrice raw materials. Tastes vary according to the target market but the most popular flavors are mints (peppermint, spearmint), cinnamon and mixtures (modified as deemed necessary to develop individuality and appeal). The flavor can be a source of irritation or sensitization and should be carefully tested for these and other toxicological properties before being released to the consumer. In some countries there are lists of permissible flavor oils. The creator of the flavor is responsible for compounding the flavor oils according to regulations, but a safety check must still be performed on the final flavored dentifrice.

Visual effects are becoming popular and include colors other than white, sparkles, stripes, speckles, etc. Clear hydrated silica formulations are used advantageously to obtain these effects.

7.7.4 *Type I versus Type II toothpastes*

As stated previously (section 7.7.2.3) two types of dentifrice formulations predominate today: (i) relatively low abrasive/high humectant (Type 1); and (ii) relatively high abrasive/low humectant (Type II). Both types can be formulated over a similar range of abrasivity to dentin. Examples of Type I and Type II formulations are given in Tables 7.2 and 7.3, respectively. The type most suited to a particular market can vary according to local preference, raw material availability, cost, and performance requirements. Products that are intermediate between Type I and Type II in terms of properties may be desired. Advantages of Type I are:

- (i) with silica abrasive (today, a virtual requirement for Type I), bioavailability of fluoride in the product is very high and remains so during storage, regardless of whether the fluoride is derived from sodium fluoride, sodium monofluorophosphate or stannous fluoride;
- (ii) it is possible to prepare transparent or translucent dentifrice;
- (iii) the range of cosmetic or therapeutic 'actives' that can be used is extensive;
- (iv) there is greater flexibility in tailor-making the dentifrice with respect to abrasion level, flavor, and visual effects;
- (v) the specific gravity of the Type I dentifrice is considerably lower than that of the Type II;
- (vi) despite a lower level of abrasive material, high abrasivity for cleaning the tooth surface can be maintained; and
- (vii) there are numerous manufacturers of silicas, many capable of providing technical assistance in dentifrice applications.

Advantages of Type II are:

- (i) there is a wide selection of abrasive materials from which one or more can be chosen with economic advantage, depending on the comparative costs of the abrasive and the humectant;

(ii) the Type II paste has a higher specific gravity than the Type I; and
(iii) less sophisticated, less expensive packaging materials can be employed depending on the Type II formulation, ranging from simple collapsible aluminum tubes to laminates of paper/aluminum foil/polyethylene.

7.7.5 *Dentifrice manufacture*

Toothpaste can be manufactured either automatically, semi-automatically, or batchwise. It is usually manufactured with the aid of techniques and equipment devised by the manufacturer in response to required production schedules. Toothpaste manufacturing processes are outlined in Figure 7.5. Most formu-

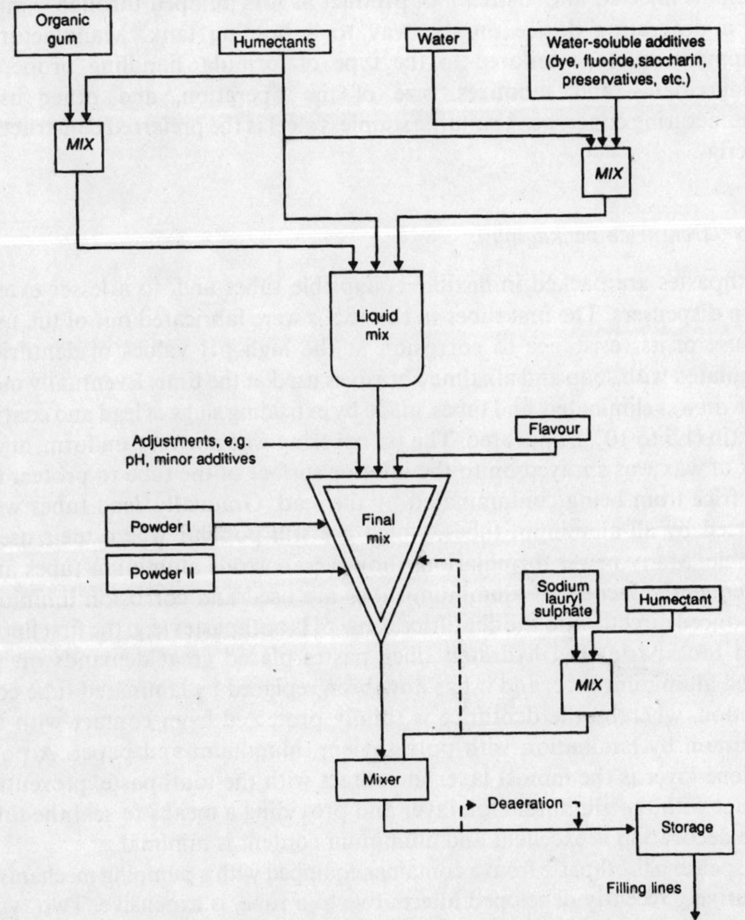

Figure 7.5 Schematic diagram of toothpaste manufacture.

lations can be prepared by simply mixing all the ingredients in one vessel and deaerating the final product. This is wasteful of time, energy and equipment, however, because of the inordinately long mixing time that would be required to achieve an equilibrium state.

Toothpaste can be made at any convenient temperature. It can be made by a cold process, such as outlined in Figure 7.5. Subsolutions are made of the various ingredients, with the exception of the powders, sodium lauryl sulfate and flavor. The subsolutions are pumped into a suitable vacuum mixer, followed by sucking-in the powders. After all the powders are 'wetted' out, the mixer is run until a uniform paste has developed. Following this, either the sodium lauryl sulfate submix and flavor are added and cautious mixing continued to prevent excessive foaming, or only the flavor is added and the submix is injected into a stream of product as it is pumped through a mixer and a deaerating device on the way to a holding tank. Manufacturing equipment must be tailored to the type of formula, handling properties of ingredients and submixes, size of the operation, and other usual manufacturing concerns. A suitable stainless steel is the preferred construction material.

7.7.6 *Dentifrice packaging*

Toothpastes are packed in flexible collapsible tubes and, to a lesser extent, pump dispensers. The first tubes in the 1850s were fabricated out of tin, used because of its resistance to corrosion at the high pH values of dentifrices formulated with soap and alkaline abrasives used at the time. Eventually most of the tin was eliminated, and tubes, made by extruding slugs of lead and coating with tin (1.5 to 10%), appeared. The tin covering was not very uniform, and a layer of wax was sprayed on to the interior surface of the tube to protect the dentifrice from being contaminated by the lead. Gradually, lead tubes were replaced by all-aluminum tubes, which are still popular where their use is possible. Many newer formulations, however, corrode aluminum tubes and, consequently, lacquered aluminum tubes are used and corrosion inhibitors introduced directly into the dentifrice. Low pH toothpastes (e.g. the first fluoridated toothpaste) and hydrated silica pastes placed great demands on the coated aluminum tube, and it has now been replaced by laminated tube construction, whereby the dentifrice is totally protected from contact with the aluminum by lamination with polyethylene, aluminum and paper. A polyethylene layer is the inmost layer in contact with the toothpaste, preventing contact with the aluminum foil layer and providing a means to seal the tube. Tube decoration is excellent and aluminum content is minimal.

Dispensing toothpaste from a container equipped with a pumping mechanism, a relatively recently developed alternative to a tube, is expensive. Two types of pump mechanism are in use, one based on a vacuum principle, the other on mechanical forces. For further details of construction and operation of the devices, the reader is referred to the packaging manufacturers' literature.

Several brands of toothpaste are dispensed from their containers as striped ribbons. There are several ways to achieve the effect. One is to insert a striping device in the nozzle of the tube. A small amount of colored paste is placed at the nozzle and the tube is filled with regular paste. The paste is expressed through a tube which has several orifices in the colored paste. The colored paste is drawn through the orifices as stripes onto the surface of the major ribbon of paste as it is extruded. Another technique involves coextrusion of different colored pastes into the tube as it is being filled [35]. The filled tube consists of separate strips of the different colored pastes aligned side-by-side. On applying pressure to the tube the different ribbons are extruded side-by-side. The pastes have similar rheologies. In yet another procedure, different pastes are packed into separate containers which are side-by-side. The pastes are extruded via a common pump. Each paste is extruded as a separate entity. A multi-colored paste ribbon is obtained by placing the two extrusion orifices adjacent to each other.

7.8 Oral rinses

7.8.1 *Function*

The present discussion of oral rinses will be limited to those that are claimed to have oral hygiene benefits. These oral rinses are characterized by two common features: (i) antimicrobial action; and (ii) ability to freshen the breath. Some also contain fluoride for its anti-caries activity. The antimicrobial activity is credited with several benefits: reduction of dental plaque, maintenance of periodontal health, and reduction of oral malodors of microbiological origin. Breath freshening is attributable to both the flavor, and the cleansing action of the rinse.

Recently, major emphasis has been placed on therapeutic benefits of oral rinses, such as inhibition of gingivitis by the control of plaque. As stated previously, the toothbrush is the key to plaque control, but is inadequate. Mandel and Gaffar [36] observed: 'Despite the proven efficiacy of adequate mechanical removal of plaque (Frandsen 1985), the level of patient involvement is so demanding that only about 30% of the population in the developed countries, and a small fraction of that in the undeveloped countries can be expected to practice adequate plaque removal. Clearly, the availability of chemical agents that supplement or even supplant the purely patient-dependent mechanical regimen are essential (Mandel 1972) if the plaque diseases are to be dealt with on a population basis.'

7.8.2 *Dosage forms and formulations*

Oral rinse products can be categorized broadly as: (i) aqueous solutions, to be used 'neat' or diluted with water; (ii) aerosol sprays, to be directed into the oral cavity; (iii) concentrates, to be diluted with a relatively large amount of water; and (iv) specialty forms, such as powders, to be 'reconstituted' with

Table 7.4 Mouthwash ingredients and functions

Ingredient	Functions	Examples
Alcohol	Adds bite and freshness Enhances flavor impact Helps solubilize some flavor components Contributes to cleansing action and antibacterial activity	
Flavor	Makes mouthwash pleasant to use Adds a refreshing cool quality to oral cavity immediately, and for some time after use Makes breath temporarily pleasant by imposing a pleasant note over breath aroma Some flavors exert significant antibacterial effect	Mint, cinnamon
Humectant	Adds 'body' to product Inhibits crystallization around closure	Glycerin
Surfactant	Solubilizes flavor Provides foaming action Assists removal of oral debris by lowering surface tension Can be antibacterial	Poloxamer 407, sodium lauryl sulfate
Water	Major vehicle to carry other ingredients	
Special ingredients		
Antibacterial agent	Enhances antibacterial efficacy	
Astringent salts	Can interact with proteins of saliva and oral mucosa	Zinc chloride
Fluoride	Anti-caries agent	Sodium fluoride

Table 7.5 Formulation of a mint-type mouthwash

	wt%
Ethyl alcohol	10.0
Flavor (menthol, 30%; methyl salicylate, 30%; peppermint oil, 30%; eucalyptol, 10%)	0.25
Glycerin	10.0
Polyoxyethylene/polyoxypropylene block polymer (Poloxamer 407)	2.0
Saccharin sodium	0.05
Water	to 100.00

From [2]

water by the user. All but the first form have minor markets and will not be considered here.

The composition and function of an oral rinse is described in Table 7.4. A typical formulation for a breath-freshening oral rinse is given in Table 7.5 Alcohol is a key ingredient, but an excessive amount can cause a strong burning sensation, not tolerated by some users.

Oral rinse directions generally call for 'swishing' for approximately 20–30 seconds. The flavor must not be so strong and biting as to reduce compliance with the usage directions. Some marketers claim that strong flavor is appealing because it implies high efficacy. The formulator must strike a balance between flavor strength, marketing image, and compliance with usage instructions.

7.8.3 *Pre-brushing dental rinse*

The pre-brushing, anti-plaque dental rinse (PBDR), introduced to the market in the mid-1980s is not a mouthwash in the classical sense. It is used just before toothbrushing to increase the amount of plaque removed at each toothbrushing event. Its mode of action is radically different from that of conventional mouthwashes based on antimicrobial agents.

The PBDR is based on the premise that plaque removal by toothbrushing can be enhanced if the plaque is 'softened' and 'loosened' prior to toothbrushing, and that at least some plaque can be removed by purely detergent action in the classical sense of separation by detergent of a soil from its substrate. Sodium lauryl sulfate was selected as the detergent. The rinse marketed originally was clinically tested extensively for anti-plaque efficacy. Positive clinical results supported the postulated mechanism of action, i.e. detergency. This suggested that enhancement of efficacy could be obtained by addition of a detergent builder. Consequently, pyrophosphate was added. The new product was quite effective in plaque reduction as shown by clinical trials [11, 12]. Surprisingly, plaque reduction was not accompanied by improvement in gingivitis level. (The results confirmed that simple reduction in plaque does not necessarily lead to a reduction in gingivitis. Obviously, a quantitative change in plaque level cannot be relied upon to predict the effect of a product on gingival status, i.e. gingivitis.)

7.8.4 *Manufacture and packaging*

The manufacture of oral rinses is straightforward. Ingredients are dissolved in the main body of water, with or without the use of submixes as required. For elegance in appearance, the final product may be filtered, the final filtration through sub-micron filter. Where a gum has been added, filtration is not indicated. Despite the presence of antibacterial substances, the manufacture of oral rinses must be carried out with appropriate microbiological precautions.

For packaging of oral rinses, clear glass or plastics are preferred for esthetic value. Two of the more commonly used plastic constructions are polyvinyl chloride and polyethylene terephthalate. Flavor quality is a very important feature of many oral rinses, and should not be compromised by packaging in containers from which off-flavor can be leached into the product, or flavor can be lost through the wall.

7.9 Active agents

Therapy is now an integral part of OTC oral-care products. Agents able to prevent or alleviate the severity of certain problems associated with the dentition are now marketed. Pertinent features will be discussed here briefly.

7.9.1 *Anti-caries agents*

Fluoride has been proven in numerous clinical and epidemiological studies favorably to influence the progress of dental caries. Fluoride is active delivered from a toothpaste or an oral rinse. Three species have been used extensively in marketed products—sodium fluoride, sodium monofluorophosphate and stannous fluoride.

Fluoride acts by making tooth enamel more resistant to attack by plaque-generated acids, promoting remineralization of early carious lesions and attacking bacteria involved in the formation of caries. The active species is the fluoride ion. Sodium monofluorophosphate appears to be hydrolysed to provide free fluoride ion before it exerts anti-caries activity. Free fluoride ion may react with the enamel surface to form a coating of calcium fluoride, which can act as a fluoride reservoir.

Achievement of maximum fluoride effect with toothpaste requires that the fluoride ion in the toothpaste does not react with toothpaste excipients, such as the abrasive. Numerous clinical trials have shown that unsupervised use of a fluoride toothpaste with 1000 ppm fluoride can be expected to reduce caries incidence by about 25%, and a fluoride oral rinse with about 0.05% sodium fluoride by about 40%. Fluoride toothpastes undoubtedly have contributed in a major way to the decline in caries in industrialized societies.

7.9.2 *Reduction in tooth hypersensitivity*

Potassium nitrate (5%) in a properly formulated toothpaste has been shown effectively to reduce tooth hypersensitivity in those subjects afflicted with the condition. Strontium chloride is also effective, but for the efficacy, formulation and use characteristics potassium nitrate is preferred [37].

7.9.3 *Reduction of plaque and calculus and improvement in gingival health* [38]

7.9.3.1 *Chlorhexidine* Chlorhexidine is perhaps the most effective agent available to reduce plaque and gingivitis. It is a bis-biguanide antiseptic. Inability to formulate the compound into a conventional toothpaste has limited its use mainly to oral rinses. It can cause staining of the teeth, taste modification and increased calculus formation.

7.9.3.2 *Essential oils* These chemicals can have strong antiseptic properties. An oral rinse containing a mixture of eucalyptol, menthol, thymol and methyl salicylate in an aqueous–alcoholic base has been 'accepted' by the American Dental Association (ADA) for anti-plaque/anti-gingivitis activity. The essential oils have characteristic flavors which are difficult to mask. Nonetheless, their use as oral antiseptics is widespread.

7.9.3.3 *Quaternary ammonium compounds* Those used in oral-care products include cetylpyridinium chloride, benzethonium chloride and domiphen bromide. They have not been used in toothpastes because they are inactivated by the anionic detergent used for foaming. A cetylpyridinium chloride oral rinse has shown anti-plaque activity. Effects on gingivitis, however, have been equivocal. Cetylpyridinium chloride is effective in reducing oral bacteria in general, including those involved in the generation of 'bad breath'. It is used in a popular US oral rinse, supplemented with domiphen bromide.

7.9.3.4 *Triclosan* Triclosan can reduce both plaque and gingivitis. It has the advantage over, for example the quaternary ammonium compounds, of not having a strong astringent flavor and of being compatible with anionic detergents. Thus, it can be incorporated into toothpaste as well as into oral rinse. It is claimed that the efficacy of triclosan can be enhanced by supplementing it with either zinc citrate or Gantrez, a copolymer of polyvinylmethyl ether and maleic acid [1, 39]. The combined formulations also are reported to reduce dental calculus.

7.9.3.5 *Herbal compounds* Sanguinarine is an extract from a blood root plant. It is available both in oral rinse and toothpaste delivery systems. Claims have been made that it has anti-plaque and anti-gingivitis efficacy. However, these claims have been supported only by mixed clinical results, and the ADA has not granted its seal of acceptance to products containing sanguinarine for gingival health.

7.9.3.6 *Stannous salts* Stannous fluoride is the basis for a recently introduced gum-care toothpaste, with claims of reduced gingival bleeding and gingival inflammation. Stannous fluoride has not been formulated into OTC oral rinses because of stability problems, although preparations in glycerin to be diluted with water just prior to use have been approved in the United States for anti-caries effect.

7.9.3.7 *Zinc salts* Zinc citrate trihydrate (ZCT) can be formulated into commercial toothpaste and a soluble form thereof into oral rinses [1, 2]. Zinc salts have many oral care benefits. ZCT has anti-plaque activity and inhibits the formation of calculus. It can be delivered from toothpaste vehicles. Zinc salts combine with organic sulfides (e.g. dimethyl sulfide) which are major contributors to 'bad breath'.

7.9.3.8 *Complex phosphates* Phosphate salts such as pyrophosphates are the basis for toothpastes which can inhibit the development of calculus. They can be delivered from toothpaste and oral rinse. Pyrophosphate is preferred. It is stable and essentially non-reactive with conventional Type I toothpaste excipients with silica abrasive.

7.9.3.9 *Sodium lauryl sulphate (SLS)* SLS delivered from an oral rinse can reduce plaque. It also can remove oral bacteria from various sites, and helps to keep the bacterial count down for several hours [33]. Clinical evaluation of a pre-brushing rinse containing SLS confirmed its ability to reduce plaque accumulation, but this reduction was not accompanied by improvement in gingival health [11].

7.9.3.10 *Hydrogen peroxide* Hydrogen peroxide has shown some promise in controlling plaque and gingivitis. Long term safety has been demonstrated. It is a relatively new addition to the list of OTC oral care agents. OTC products are formulated with hydrogen peroxide in one base and a second base comprising sodium bicarbonate which is codispersed with the first from a dual pump system. Dentifrice containing hydrogen peroxide and baking soda has in preliminary studies been shown to be safe and on completion of further studies has been approved by the ADA as safe and effective [40]. A toothpaste containing hydrogen peroxide (0.67%), sodium bicarbonate (5.48%) and fluoride (1100 ppm) as sodium fluoride was significantly superior to a fluoride control at reducing the concentration of oral precursors found in saliva [41].

7.10 Specialty products

7.10.1 *Tooth whiteners*

Tooth whitening preparations, once exclusive to the dental profession, are now being marketed with increasing enthusiasm directly to the general public. The consumer market is still very small, but can be expected to grow at an accelerating rate. The potential size of the market in the future is huge if product performance can be made to equal product expectations.

The active ingredient in most of the whitening agents is 10% carbamide peroxide, which reacts with water to release hydrogen peroxide. Also used is calcium peroxide. The unsupervised use of peroxide to bleach the teeth has been of concern to some professional groups. The ADA has questioned the safety of unsupervised use of peroxide for the purpose, reasons including potential carcinogenicity, effect on wound healing and other potential problems. Recently, however, the ADA granted its seal to a tooth bleaching product to be used at home with the proviso that it be dispensed by a dentist.

7.10.2 *Products for the edentulous*

Denture cleansers and adhesives represent two major categories of products for the edentulous. The former are marketed mainly in two forms: (i) an effervescent tablet or powder with a dye marker to signal when the cleansing process is complete; and (ii) a fairly conventional toothpaste. Denture adhesives are generally formulated around a gum (karaya is popular) that swells and becomes sticky, and acts as an adherent between the denture and the oral tissue.

Many of the proprietary powders and tableted denture cleansers comprise a bleaching agent, which is usually a compound that releases the peroxide ion (e.g. sodium perborate, sodium percarbonate, or an alkali metal salt of monopersulfuric acid). Alkali may be added to perborate or percarbonate compositions to enhance release of the peroxide ion. Builders, such as sodium bicarbonate, sodium carbonate, sodium sulfate and sodium chloride, are added to the formulation. Chelating agents may also be added to remove calculus and to prevent precipitation in hard water. In tablets, effèrvescence is needed to facilitate rapid disintegration. This may be produced by anhydrous sodium perborate or a mixture of citric acid and sodium bicarbonate. (In powders, breakdown of the peroxide can yield mild effèrvescence.) Other ingredients include flavor, detergent, tableting aids and binders. A typical tablet composition might contain the following: potassium monopersulfate, 30%; sodium bicarbonate, 65%; sodium perborate, 2%; sodium lauryl sulfate, 0.5%; blue dye, 0.1%; flavor, 0.3%; other builders, to 100%.

Paste denture cleansers are formulated much like toothpastes, taking precautions to avoid a degree of abrasivity which can harm the denture.

Denture adhesives are based on gum karaya or another gum, plus fillers, flavors, and colors. A formulation may contain the following: petrolatum, 50%; gum karaya, 45%; magnesium oxide, 3%; fillers.

References

1. Pader, M. (1992) Dentifrices. In *Chemistry and Technology of the Cosmetic and Toiletries Industry*, Williams, D.F and Schmitt, W.H. (Eds). Blackie, London, Chapter 7.

2. Pader, M. (1988) *Oral Hygiene Products and Practice*, Marcel Dekker, New York.
3. Schour, I. (1938) Tooth development. In *Dental Science and Dental Art*, Gordon, S.M. (Ed.). Lea & Febiger, Philadelphia, Chapter I.
4. Dawes, C. (1979) In *Proceedings, Saliva and Dental Caries*, Kleinberg, I., Ellison, A. and Mandel, I.D. (Eds). Special Suppl. Microbiology Abstracts, p. 505.
5. Bhaskar, S.N. (1981) Role of pulsating water lavage in plaque control. In *Prevention of Periodontal Disease*, Caranza, F.A. and Kenney, E.B. (Eds). Quintessence, Chicago, Chapter IV.
6. Dawes, C., Jenkins, G.N. and Tonge, C.H. (1963) The nomenclature of the integuments of the enamel surface of teeth. *Br. Dent. J.* **115** 65–68.
7. De La Rosa, M.R., Guerra, J.Z., Johnston, D.A. and Radicke, A.W. (1979) Plaque growth and removal with daily toothbrushing. *J. Periodontol.* **50** 661–664.
8. Nikiforuk, G. (1985) *Understanding Dental Caries*, Karger, Basel.
9. Lõe, H., Theilade, E. and Jensen, S.B. (1965) Experimental gingivitis in man. *J. Periodontol.* **36** 177–187.
10. Pader, M. (1992) Dental plaque. *Cosmet. Toilet.* **107** 81–89.
11. Schiff, T. and Borden, L.C. (1994) The effect of a new experimental prebrushing dental rinse on plaque removal. *J. Clin. Dent* **4** 107–110.
12. O'Mullane, D.M., Whelton, H., Phelan, J. and Gleeson, P.A. (1994) A 12-month efficacy study of a pre-brushing rinse on plaque removal. *J. Periodontol.* **65** 611-615.
13. Attström, R. (1988) Does supragingival plaque removal prevent further breakdown? In *Periodontology Today*, Guggenheim, B. (Ed.). Karger, Basel, pp. 251–259.
14. Schroeder, H.E. (1969) *Formation and Inhibition of Dental Calculus.* Hans Huber, Berne.
15. Mandel, I.D. and Gaffar, A. (1986) Calculus revisited. A review. *J. Clin. Periodontol.* **13** 249–257.
16. Van Dyke, T.E. and Zinney, W.B. (1989) Biochemical basis for control of plaque-related oral diseases in the normal and compromised host: Periodontal diseases. *J. Dent. Res.* **68** (Special Issue) 1588–1596.
17. Burt, B.S. (1988) The status of epidemiological data on periodontal diseases. In *Periodontology Today*. Guggenheim, B. (Ed.). Karger, Basel, pp. 68–76.
18. Vogel, R.I. (1975) Intrinsic and extrinsic discoloration of the dentition. A literature review. *J. Oral Med.* **30** 99–104.
19. Eriksen, H.M. and Nordbø, H. (1978) Extrinsic discoloration of the dentition. *J. Clin. Periodontol.* **5** 229–236.
20. Berman, L.H. (1985) *J. Periodontol.* **56** 216.
21. ten Bosch and Coops, J.C. (1995) Tooth color and reflectance as related to light scattering and enamel hardness. *J. Dent. Res.* **74** 374–380.
22. Rugg-Gunn, A.J. and MacGregor, I.D.M. (1978) A survey of toothbrushing behavior in children and young adults. *J. Periodontol. Res.* **13** 382–389.
23. Kleber, C.J., Putt, M.S. and Muhler, J.C. (1981) Duration and pattern of toothbrushing in children using a gel or paste dentifrice. *J. Am. Dent. Assoc.* **103** 723–726.
24. Council on Dental Therapeutics (1985) *J. Am. Dent. Assoc.* **110** 545–548.
25. Bass, C.C. (1948) *Dent. Items Interest* **70** 696.
26. Kitchin, P.C. and Robinson, H.B.G. (1948) How abrasive need a dentifrice be? *J. Dent. Res.* **27** 501.
27. Pader, M. (1993) Dentifrice rheology. In *Rheological Properties of Cosmetics and Toiletries*, D. Laba (Ed.). Marcel Dekker, New York, Chapter 7.
28. De Boer, P., Duinkerke, A.S.H. and Arends, J. (1985) *Caries Res.* **19** 232.
29. Stookey, G.K., Burkhard, T.A. and Schermerhorn, B.R. (1982) *In vitro* removal of stain with dentifrices. *J. Dent. Res.* **61** 1236–1239.
30. Hefferen, J.J. (1977) ADA dentifrice function program. *Collaborative study of clinical methodology. I. Methods to assess dentifrice cleaning function*, May, 1977.
31. Pader, M. (1984) Humectants for clear gel dentifrices. US Patent 4,435,380 (March 6, 1984).
32. Pader, M. (1988) Binders and Thickeners for toothpaste. *Cosmet. Toilet.* **103** 84–93.
33. Pader, M. (1985) Surfactants in oral hygiene products. In *Surfactants in Cosmetics*, Reiger, M.M. (Ed.). Marcel Dekker, New York, Chapter 10.
34. Pader, M. (1987) Toothpaste gels. *Cosmet. Toil.* **102** 81–87.
35. Chown, J.P. and Healey, J.M. (1976) US Patent 3,980,767 (Sept. 14, 1976).

36. Mandel, I.D. and Gaffar, A. (1986) Calculus revisited. A review. *J. Clin. Periodontol.* **13** 249–257.
37. Hodosh, M., Hodosh, S.H. and Hodosh, A.J. (1994) About dentinal hypersensitivity. *Compend. Continuing Educ. Dent.* **15** 658–667.
38. Mandel, I.D. (1994) Antimicrobial rinses: overview and update. *J. Am. Dent. Assoc.* **125** (Suppl) 25–85.
39. Bolden, T.E., Zambon, J.J., Sowinski, J., Ayad, F., McCool, J.J., Volpe, A.R. and DeVizio (1992) *J. Clin. Dent.* **3** 125–129.
40. Fischman, S.L., Truelove, R.R., Hart, R. and Cancro, L.P. (1992) The laboratory and clinical safety evaluation of a dentifrice containing hydrogen peroxide and baking soda. *J. Clin. Dent.* **3** 104–110.
41. Grigor, J. and Roberts, A.J. (1992) Reduction in the levels of oral malodor precursors by hydrogen peroxide: *in vitro* and *in vivo* assessments. *J. Clin. Dent.* **3** 111–115.

8 Perfumery

A. DALLIMORE

8.1 Introduction

The object of this chapter is to give the reader an overview of the subject of perfumery. It will cover the role of fragrance, the raw materials used, and the creative approach and construction (including technical constraints, quality aspects, and health and safety). Fragrance briefing, current trends and issues, and a short glossary of perfumery descriptors are included. A brief description of some other special additives used in perfumery, or closely related to perfumery materials, is given.

8.2 Fragrance—a definition

The terms 'fragrance' and 'perfume' are synonymous and are used interchangeably throughout the perfumery and cosmetics industries. This will also be so throughout this chapter. Fragrance (or perfume) is a blend of two or more materials characterised by having olfactive properties, and is incorporated into a preparation with the intention of imparting specific odorous characteristics to that preparation. Very occasionally, the fragrance may consist of one raw material only.

8.3 Role of fragrance

The most important function of the perfume is usually to impart a pleasant and suitable odour to the product into which it is to be incorporated. It is the job of the creative and evaluation teams to define the terms 'pleasant' and 'suitable' for each particular brief that is received. The perfume may also be required to perform other roles in the product, which may take priority over, or be additional to the primary role of imparting odour.

Most cosmetic products have a mild but detectable odour, which is derived from the raw materials in the formulation. Although this 'base' odour is usually inoffensive, it can detract from the overall aesthetic character of the product and should at least be 'masked'. This can be effectively achieved by the use of a suitable fragrance. Very pungent bases, such as permanent-wave lotions, are almost impossible to mask, even with the strongest of perfumes.

The role of the perfume in these products is therefore to produce as pleasant an odour as possible, and this can often be attained by harmonising the fragrance with the base odour.

Fragrances can be designed to give subliminal support to a particular product. For example, a fragrance may appear to make a shampoo wash the hair more cleanly, a perfumed night cream may seem more nutritional to the skin, or a perfumed body spray may appear to impart all-over freshness, all-day. In other words, the perfume may have the role of apparently enhancing product performance.

Some fragrance materials have antimicrobial activity, particularly against bacteria. Both bacteriostatic and bacteriocidal properties are evident. Consequently, fragrance blends often take on the antimicrobial activity of some of the individual components. Fragrances that have spicy and herbal notes usually exhibit some antibacterial activity owing to the presence of phenols such as eugenol (from clove oil), and thymol and carvacrol (from thyme oil). Consequently, the fragrance may also have the role of assisting the preservative system and even in extreme cases, of replacing the preservative system altogether. Similarly, the fragrance may assist a bacteriocide/bacteriostat in its effect on the skin's microflora (reducing body odour) or may be required to carry this attribute of the finished product on its own.

8.4 Perfumery raw materials

The basic components or raw materials that are used to manufacture fragrance compounds number in the region of 4000–6000. Although estimates vary from source to source, this range is widely accepted as typical. Of these materials, approximately 100 items are key ingredients, comprising the majority of most formulations by weight. Most formulations contain a significant number of these materials. Examples of key raw materials are benzyl acetate, cedarwood oil, geraniol, geranium oil, lavender oils, lemon oil, linalol, methyl ionones, oakmoss and treemoss products, phenyl ethyl alcohol, vanillin, and ylang oils.

Perfumery raw materials fall into two distinct groups: (i) natural; and (ii) synthetic. Natural components are of animal or plant origin, while synthetic materials are produced from a wide range of starting materials, using diverse reactions and synthetic pathways.

8.4.1 *Natural perfumery raw materials*

8.4.1.1 *Animal products* The use of animal products is a very topical and emotive issue, and it is quite clear that the perfumery industry is rapidly heading towards the exclusion of this group of raw materials. Traditionally, extracts of musk from the musk deer, castoreum from the beaver, civet from the civet cat, and ambergris from the sperm whale were all used in fragrance compositions, especially those created as skin perfumes. However, with modern analytical techniques and the ability of the aroma chemical industry to analyse and synthesise many of the active components, excellent substitutes are now available to the perfumer, often at much lower cost.

8.4.1.2 *Plant products* A vast range of plant extracts can be used in the creation of perfumes. The extracts are prepared in several ways, the most important of which are steam distillation and organic solvent extraction, particularly ethanolic extraction. Steam-distilled products are called 'essential oils'. Solvent-extracted products are generally referred to as 'concretes' with the alcoholic extracts having several names depending on the starting material. 'Absolutes' are ethanolic extractions of concretes, while 'tinctures' are ethanolic extracts of the raw plant material. Tinctures of animal products are also prepared. Plant resins and resinous materials may be used in the untreated states or processed by extraction. Organic solvent extraction, usually hydrocarbon based, produces the so-called 'resinoid', which may be further extracted with ethanol to give a 'resin absolute'.

Due to the presence of odorous materials in specific organs, particular parts of plants are extracted. In some instances, different parts of the plant produce olfactively differing extracts. A typical example is the bitter orange tree, of which the fruits give bitter orange oil, the flowers give neroli oil, and the leaves and twigs give petitgrain oil. Although the odour of these oils is related and many of the constituents are common, the oils have distinctive odours and composition. Some examples of commonly used plant products are rosemary, rose, lemongrass, clove, caraway and rosewood oils, oakmoss and jasmin absolutes, and olibanum and labdanum resins.

The preparation of plant extracts for use in perfumery is a worldwide industry and is subject to unpredictable factors such as climatic conditions, political changes and economic uncertainty. As a result, it is not unusual to have to deal with the problems of fluctuating crop availability, pricing and quality.

8.4.2 *Synthetic perfumery raw materials (aroma chemicals)*

The use of synthetic organic chemicals with organoleptic properties follows the advances made in organic synthesis from the middle to late nineteenth century up to the present day. A vast range of aroma chemicals is now available to the perfumer, ranging in complexity from the relatively simple esters, aldehydes and alcohols (such as *n*-hexanol) to macrocyclic molecules like cyclopentadecanolide. Most companies involved in the manufacture of aroma chemicals are actively engaged in the search for new compounds. A new fragrance chemical must have certain characteristics in order for it to become a part of the standard perfumery repertoire. First and foremost it must be safe both for the fragrance manufacturer to handle, and in its destined consumer product. Second, it must be chemically stable over a range of conditions, cost-effective in use and possess a useful odour property. Finally, it must appeal aesthetically to the creative perfumers who will use it as a part of their palate of raw materials, thus ensuring its economic viability and long-term success.

Particular emphasis is presently being placed on the development of chemicals, especially those with: (i) sandalwood characters; (ii) fruity components, especially those that are nature identical; (iii) exceptional stability in adverse media, for example media with extremes of pH or powerful oxidisers; (iv) unique odour or blending properties; and (v) versatile, economic properties.

The two main feedstock sources for aroma chemicals are turpentine oil and crude oil. Many synthetic processes are used to produce a wide range of high quality, standardised materials suitable for perfumery use. A number of examples from the hundred or so key aroma chemicals used in the fragrance industry are given below.

Benzyl acetate Characteristic jasmin note, light floral and slightly fruity; widely used; very economic.

Benzyl salicylate Warm, balsamic note, blending well with musky and floral notes, and giving body to the composition.

iso-Bornyl acetate Fresh, piney; cheap and stable; important component of all synthetic pines, in which it may be used at concentrations as high as 80%.

p-t-Butyl cyclohexyl acetate (PTBCHA) Soft woody note; stable and widely used in soap perfumery; low cost; provides good support for all other woody notes, particularly patchouli.

Cedryl acetate Sharp, woody material, particularly useful in masculine perfumes; derived from cedarwood oil.

Citronellol Main component of rose oils, therefore widely used as a key element of rose fragrances; also an effective floraliser in citrus blends; manufactured synthetically.

Dihydro myrcenol Citrus, especially lime and lavender aspects; very volatile; a valuable, stable topnote material at concentrations of 0.1% to 20%.

Geraniol Floral, rosy; found naturally in geranium, rose, palmarosa and citronella oils.

Heliotropine Sweet, floral, powdery odour; will discolour in some finished products, e.g. soap.

Hexyl cinnamic aldehyde Homologue of amyl cinnamic aldehyde, to which it is similar in character and properties, and with which it is often combined; overall more light and delicate; used extensively in modern fine perfumery in accord with methyl dihydro jasmonate.

Indole Chemically heterocyclic; floral, animalic odour; found in jasmin and orange flower absolutes; discolours readily and sublimes.

γ-Methyl ionone Woody, floral, violet notes; used in many fine perfumes for its unique effect; also supports the iris character in a composition.

Musk ketone Musky, animalic, warm, powdery; one of the nitromusks; very long lasting but can cause discoloration at high levels.

Phenyl ethyl alcohol Floral, faintly green; main component of rose water and, hence, rose fragrances; excellent blender; deceptively powerful despite being weakish in odour in the pure state.

Vanillin Sweet, powdery, ambery, vanilla; extremely powerful material that will impart softness, warmth and depth to almost any fragrance; very widely used; at concentrations as low as 0.01%, can still be effective and is a note that is almost universally liked and accepted; at high concentrations (0.5% and above), will rapidly dominate the creation if not well blended; develops as the freshly compounded fragrance matures; disadvantage of discoloration (especially in alkaline media such as toilet soap), producing a dark brown colour; discoloration is accelerated by an increase in temperature.

8.5 Development of a fragrance

The development of a new fragrance follows a clearly defined route from the customer's briefing to the production or manufacture of the selected formulation.

8.5.1 *The brief*

A perfumery company requires certain basic pieces of information in order to select a suitable fragrance from the library or to create a new, specific fragrance to satisfy the client's request. The brief should be in written form and discussed with the perfumery company's representatives, who may include a perfumer, fragrance evaluator or marketing person. The following items of information are essential to the creative team.

8.5.1.1 *Product* The application for which the fragrance is intended. Details of the formulation are useful (sometimes essential), especially if unusual, novel or reactive raw materials are being incorporated. Ideally, a quantity of the unperfumed formulation (usually referred to in perfumery circles as 'unperfumed base') should be provided for the perfumery house to conduct its testing. The intended colour for the final product should also be indicated, since the harmony between colour and odour is of utmost importance in the overall integrity of the product.

8.5.1.2 *Packaging* The proposed packaging for the new product. Although perfume/packaging interactions are infrequent, they can and do occur.

8.5.1.3 *Fragrance area* The particular fragrance type required for the proposed product. If possible, the fragrance area should be described using perfumery terms, but the use of specific raw materials, even of marketed products, can form the basis of a common language and can therefore be used as a reference point. Although the language of smell is probably the least developed for all the senses, perfumers and their clients can often broadly understand each other by using such reference points. Interim meetings between the client and perfumer during a project can provide valid direction in the creative task, and regular discussion brings a common awareness, on both sides, of what is required versus what can be achieved.

8.5.1.4 *Cost* The cost limits for the perfume submission. Pro-rata dosing should always be considered if feasible for the project, and comprises the incorporation of a more expensive perfume at a lower dosage, or a cheaper perfume at a higher dosage, both giving the same cost contribution to the final product formulation. The cost limits are usually expressed as pounds sterling per kilo, or dollars per kilo or pound but, under a pro-rata option, may be expressed as pounds sterling, or dollars, per unit weight of finished product.

8.5.1.5 *Number of submissions* The number of submissions permitted per product. The submission of two ideas per product allows the perfumery house to comply closely with the customer's request and to suggest an idea of its own—a 'flier'.

8.5.1.6 *Marketing information* The market positioning of the product. It is vital that the right type of perfume is created for the product. Consequently the proposed market positioning of the product, including such items as geographical distribution, should be made as clear as possible.

8.5.1.7 *Timescale* The deadline date for submissions. Most perfumery houses will be able to turn a brief round within two to four weeks if the submission is taken from the library and submitted as such, or perhaps modified slightly. However, the creation of a new fragrance should be considered in terms of months rather than weeks, and it is the plea of many perfumers that the fragrance should not be an afterthought but an integral part of the product at the concept stage.

8.5.1.8 *Safety* Conformity to regulations. Although conformity with guidelines set out by the Research Institute for Fragrance Materials (RIFM) and the International Fragrance Association (IFRA), and with local national legislation is standard with all reputable companies, the briefing company should make suppliers aware of any particular restrictions that they impose on materials not covered by the current recommendations. From time to time, new evidence on the safe usage level of a fragrance raw material may be produced, with the effect that current compounds may need modification to comply with the latest safety developments. It is the obligation of the perfumery house to make their customers aware of this, and the obligation of the cosmetic and toiletries houses to accept a slight change in their fragrance compounds should there be no alternative. However, in the author's experience, expert evaluators can often detect minor odour changes in a product, while most consumers, in whose interests the changes are being made, cannot.

8.5.1.9 *Submission size* The quantity of oil required, for the client's test programme. The customer should also stipulate whether the perfume submission should be demonstrated in the base, i.e. the final product, preferably the client's own.

8.5.1.10 *Testing* Stability testing and panel testing. Clients should indicate to the perfumery house any specific stability testing that is required. Most perfume houses undertake their own stability testing as a matter of course, using perfumed and unperfumed samples, with conditions of low, elevated and ambient temperature, in the light and dark, as standard. If any panel testing of support data is required, this should be discussed fully at the initial briefing so that appropriate tests can be designed.

8.5.2 *The creative process*

When creating a fragrance, the perfumer must take into account the following parameters.

(i) Odour type
(ii) Suitability for product
(iii) Performance (strength)
(iv) Cost
(v) Stability
(vi) Health and safety and other legislation
(vii) Consumer acceptability
(viii) Timescale
(ix) Possible scale-up (manufacturing) problems
(x) Availability of raw materials

Each perfumer has his or her own individual method, which is usually a combination of techniques. The combinations and techniques used vary with the type of project. The first requirement is usually the creation of the appropriate odour theme, while simultaneously taking into account the various limitations. Some constraints may be ignored during the early stages of development and attended to as the project progresses. For example, the perfumer may not adhere to the cost limits whilst exploring the main theme but, once a satisfactory odour type is achieved, then he or she may set about working it into the designated price area.

Several techniques form the basis of the methodology of creating the right odour, although it must be stated that the methods used by perfumers are diverse and individual in their nature and combination. Some of the more important of these methods will now be discussed. Using these techniques, along with skill, experience, patience and, on some occasions, considerable trial and error, a finished product will be obtained.

8.5.2.1 *Binary blending* The mixing of two components in varying proportions. As an example, consider the 10% blend of lemon oil and lime oil to give a final range of 0%–100%. The perfumer will choose the blend that appeals to him or her most. This may be the blend that has an emphasis on either the lemon or the lime, or it may be the blend at which neither component is dominant and perhaps a harmony or a novel effect is achieved. An analogy is the blending of the two colours yellow and blue across a range. Somewhere in the middle of the range an apparent 'new' colour—green—will be seen. This is the aim of the perfumer when binary blending.

Although appearing to be a basic starting procedure, binary blending may also be used at the final stage of development. For example, it may be used to obtain the optimum level of the green note that is being added at the final stage to 'finish-off' the newly created perfume.

8.5.2.2 *Formation of accords* The balancing of several components to give an interesting or novel effect. For example, a simple jasmin accord could be:

Accord 1	*Wt%*
Hexyl cinnamic aldehyde	50
Methyl dihydro jasmonate	25
Benzyl acetate	20
Dimethyl benzyl carbinyl butyrate	3
Indole	2

A floral accord could be:

Accord 2	*Wt%*
Phenyl ethyl alcohol (rose-like)	40
p-t-Butyl-(*a*)-methyl hydrocinnamic aldehyde (muguet-like)	30
Hexyl cinnamic aldehyde (jasmin-like)	20
γ-Methyl ionone (violet-like)	9
Phenyl acetaldehyde (hyacinth-like)	1

This accord is very simple, but could be made more sophisticated and more interesting by substituting the individual raw materials with representative accords.

Accord 3	*Wt%*
Rose accord	40
Muguet accord	30
Jasmin accord	20
Violet accord	9
Hyacinth accord	1

Once an accord has been made, other notes may be added to accent certain aspects of the developing fragrance. In accord 3, for example, a green topnote may be blended into the composition in the form of galbanum oil or *cis*-3-hexenol.

8.5.2.3 *Thematic construction* Using this method of creation, the perfumer selects the basic fragrance notes and builds them *in situ*. A simple example could be the creation of an aldehydic, woody, green floral fragrance. The components for each odour theme within the perfume are listed, and quantities decided.

(i) Aldehydic part representing 2% of the total
  Undecylenic aldehyde, 1%
  Lauric aldehyde, 1%
(ii) Woody part representing 55% of the total
  Cedarwood oil, 10%
  γ-Methyl ionone, 15%
  p-t-Butyl cyclohexyl acetate, 30%
(iii) Green part representing 2% of the total
  Methyl phenyl carbinyl acetate, 1%
  Phenyl acetaldeyde, 1%
(iv) Floral part representing 41% of the total
  Hexyl cinnamic aldehyde, 15%
  Terpineol, 21%
  Phenyl ethyl alcohol, 5%

Again, this is a relatively simple demonstration, since a fragrance is not limited to just four themes, or just a few ingredients within these themes. One fragrance may consist of just two items, or may contain several hundred.

8.5.2.4 *Assessing a fragrance compound* Each perfumer has his or her own way of smelling. During training it is imperative that creations are incorporated into the intended end-product and the relationship between the concentrate and how it performs in the product is understood. This enables a more predictive approach to be made to experimentation as time goes by. Fragrance concentrates should always be assessed on a smelling strip or blotter, which should be dipped to a depth of about 2–3 cm. The blotter should be held 6–10 cm from the nose and the aroma inhaled. The environment in which smelling is undertaken should be at a temperature and humidity level that is comfortable for the assessor. The fragrance should be evaluated over a period of time, possibly up to 48 hours and notes made for reference. Final assessment should always be made in the end-product and in an in-use situation.

8.5.3 *Evaluation and marketing of the new creation*

The objectives of the perfumer are: (i) to create a product that will fulfil the client's request; and (ii) to ensure that the client's product succeeds in the marketplace, with the fragrance playing its required role. Much research concerning the influence of perfume within a finished product has been undertaken. It is clear that certain perfumes can complement certain types of products, syngergise with particular colourways, subliminally support certain product attributes, and even act as a link across a product range.

Fine fragrance is marketed in such a way that, as well as perfuming the individual, the use of the scent may signify lifestyle, describe personality or alter ego, attract the desired partner or stimulate emotions in a way not thought to be possible. Once created, therefore, the fragrance is often tested by an in-house evaluation panel, which may consist of a mixture of expert 'noses' (perfumers and trained evaluators) and lay 'noses' (usually other staff, or possibly a local residents panel) to assess its suitability for the brief. The result may indicate that a reworking of the creation is needed, and the panel test may then be repeated. An outside panel may also be used to support the fragrance submission, and the client will most likely undertake their own market test regime before the product is finally launched.

8.6 The current market in fine fragrance

The current couture fragrance market for women is designer-dominated and is characterised by the launch of a profusion of fragrances that are considered fresh and natural. Predominant themes are ozonic and watery elements, light flowery notes, citrussy topnotes and fruity aspects that vary from discreet to most evident. The fruity notes are typified by blackcurrant, peach and pineapple with more tropical notes such as mango and passion fruit beginning to appear. Base combinations of musk, amber, moss and wood notes are frequently used. The overtly powerful sea themes have been replaced by more sophisticated marine blends, whilst several brands have been launched as lighter versions of their respective 1980s successes. The use of unusual or even unique natural extracts has appealed to many marketeers and the incorporation of nature-identical synthetics, especially those identified by headspace analysis, continues to grow. There is a recent trend using the vanilla note as the key theme. Already several mass market brands are in the marketplace with successful products.

The current fine fragrance market for men is also heavily influenced by designer creations, with the Fougere theme being the most in vogue. The accord of allyl amyl glycollate, dihydro myrcenol/dimethyl heptanol and absolute oakmoss continually recurs in new launches, as do the powerful woody–sandalwood aroma chemicals. An accord of sandalwood notes, patchouli oil, methyl dihydro jasmonate and a macrocyclic musk forms a very latent, diffusive base, which is used across a wide range of masculine themes. Vetivert characters, sea aspects and ozonic notes are more in evidence and may yet form the foundation of a significant stage in the evolution of the men's market.

It is clear that innovation in the fine fragrance market for men and women usually takes place in small increments and that a quantum leap in terms of 'new' odours occurs very rarely. However, once a truly innovative perfume is accepted by the market, this theme may act as an inspiration to marketing companies, designers and perfumers for a plethora of developments in which the theme itself is key.

8.6.1 *The trickle-down effect*

Fine fragrance themes act as inspiration for creative ideas for functional cosmetic products, with some odour types lending themselves to easy modification for a whole range of end uses. For example, aldehydic, woody or floral perfumes can be found in body sprays, skin-care products, hair-care products, bath products, toilets soaps and, although beyond the subject matter of this book, even in fragranced household and industrial products.

The translation of a fine fragrance theme in order to make it suitable for cosmetics and toiletries involves adjusting the formulation to ensure that it is chemically stable, performs strengthwise, is a balanced perfume, is consumer acceptable, and conforms to the lower cost limits demanded by these types of products.

8.7 Odour types

The many different odour types fall into four categories: (i) *simple notes*, which are representative of naturally occurring aromas such as fruits, herbs, spices, flowers and animal smells; (ii) *complexes*, which are combinations of singular characters—for example green floral, spicy citrus and fruity floral; (iii) *classic accords* such as Fougere, Chypre and Opoponax; and (iv) *multi-complexes*, represented by many fine fragrance types which can have up to

12 identifiable simple themes within the total structure. The definition of an odour type is very subjective. Opinions can be very disparate, but there is no real right or wrong. In fact, subjectivity reigns supreme.

8.8 Technical performance of perfumes

The custom-creation of a fragrance relates not only to the style of odour required, but also to the proposed end usage, with specific products posing specific compatibility problems. The following general comments apply: (i) oils and powders require the heavier aspects of the fragrance to be accentuated; (ii) aerosol products require less emphasis on the topnote; (iii) aqueous products require a strong, well-supported topnote; and (iv) soaps and emulsions require an evenly balanced composition.

Aerosols The perfume must be well balanced for aerosol use, topnotes being accentuated strongly by all propellants. The perfume must also mask or harmonise with any propellant odour, especially in the case of the 'ozone-friendly' hydrocarbon propellants. Interaction between the perfume and any active ingredient, e.g. an antimicrobial, must not take place. Finally, the perfume must be soluble and not precipitate, which could cause valve blockage.

Antiperspirants The acid medium of aluminium chlorhydrate based formulations is very aggressive towards some fragrance materials. Traditional citrus and lavender notes are very susceptible, but more stable versions can be made using a series of synthesised aroma chemicals. The comments made concerning aerosols also apply to antiperspirants.

Antiperspirant and deodorant sticks Performance, coloration, discoloration and reaction with the active constituents are factors to be aware of.

Body sprays and deosprays When perfuming these products, the points to note are the same as those that apply to aerosols in general, but special care should be taken to avoid suppression or potentiation of the active components.

Skin creams and lotions Discoloration and emulsion stability are potential problems. Some emulsions effectively absorb the fragrance, reducing its overall impression. If this occurs, rebalancing of the perfume is required.

Hair products

Conditioners Usually, the same compound is used to fragrance the shampoo and conditioner of a single range. Dosage in the conditioner is lower, and the conditioner base promotes different aspects of the fragrance. Careful balancing is needed.

Mousse Base stability, and performance of the fragrance are the two key aspects to be aware of.

Permanent waves The alkaline environment of 'perms' can present difficulties for the perfumer, and off-notes and discoloration are frequently encountered. Chemical stability is very important—the raw materials used in the perfume should be thoroughly tested before being used. Permanent waves also have powerful odours of their own, which are impossible to mask. Consequently, a strong perfume that will ameliorate these base odours and harmonise with them will prove to be the most acceptable answer.

Shampoos Points to consider are solubility of the perfume in the base, and acceptable performance from the bottle and in-use.

Spray Base notes from the active agents and propellants need to be harmonised with the perfume. The general comments made about aerosols also apply.

Styling gels Solubility and perfume performance are the two major sources of problems.

Bath products

Foam baths Particular problems are lack of perfume performance, insolubility, discoloration, and separation (possible precipitation) of the pearling agent where used.

Shower gels The comments are the same as those for foam baths, with the additional problem of base thinning by the perfume.

Bath oils Solubility and performance are the main technical points for the perfumer. Bath oils readily envelop fragrances, and the use of very powerful ingredients in the topnotes is one method of achieving perfume performance. Raw materials, such as fruity esters, essential oils containing pyrazines (galbanum, petitgrain and vetivert), and essential oils that are high in the terpene hydrocarbons (olibanum, nutmeg and orange), will all give beneficial results.

Talcs Points to note are the loss of volatiles and oxidation over the exceptionally high surface area, and discoloration of components like indole and vanillin, which, odourwise, are very effective in talcs.

Toilet soaps Coloration and discoloration can occur in pale-coloured and white soaps. Very volatile materials may be lost both during the soap-making process, and during storage. At high levels, fragrances may affect the finish on the final stamped soap bar.

Suncare products The perfumer must bear in mind that suncare products are usually applied to the skin when it is very warm or when it will become very warm. The profile of the fragrance should not alter drastically over these temperature ranges. Safety of ingredients is particularly important in suncare products, since some materials are photo-sensitisers.

Baby-care products Once again, safety is of paramount importance, and fragrance should convey the desired safe and 'caring' messages. Fragrances for baby care products must also be effective at low concentrations.

Lipsticks Clearly, the flavouring aspect of the fragrance must be considered along with effective base cover. Sweating can be an occasional problem, and is associated with insolubility of the perfume.

Face powders Oxidation and loss of volatiles can present difficulties. The perfumery is very similar to that for talc.

Shave products Performance and odour type are usually the main areas of difficulty, although in terms of safety, the perfumer must bear in mind that the skin is abraded during shaving.

8.9 Stability testing

A standard regimen of testing of a perfumed cosmetic or toiletry should be based around the following framework.

(i) Samples of perfumed and unperfumed base.
(ii) Storage at (1) 0°C (control); (2) ambient conditions (dark); (3) ambient conditions (window or artificial daylight); and (4) oven temperature (40°C).
(iii) Assessments at one month, two months, three months and six months.
(iv) Evaluation of samples for product stability (i.e. colour consistency, etc.), perfume character and perfume strength.
(v) Packaging to ideally be customer-provided for the specific project; otherwise an inert material (e.g. glass, lacquered cans, etc.).

As information from the stability programme builds up, the perfumer becomes more aware of potential problems and potentially unstable raw materials, and can take appropriate measures when working on a brief. However, all fragrances should be tested in the final product prior to market launch, since combination effects, both synergistic and antagonistic, can and do occur.

8.10 Compounding

Compounding is the term used by the fragrance industry for the bulk production of perfume concentrates. The compounding of quantities of perfume is basically simple, but must be carried out with the utmost precision and accuracy. Fragrance manufacture is usually a straightforward mixing procedure, therefore requires very few processing operations. However, the presence of crystals, resins and other solids or semi-solids in the formulation demands an efficient mixing regime and careful formulation by the perfumer to ensure that, once homogeneously mixed, the fragrance will not suffer from deposits of crystalline components on standing. Resinous materials are usually stored in a warm-room to keep them in a pourable state and, consequently, easier to handle. Thermolabile materials are kept refrigerated, but most other items are stored at room temperature. The ideal storage conditions for most fragrances and fragrance ingredients are those where heat, light and air are excluded or reduced to an absolute minimum. Readily oxidised materials are best kept under an atmosphere of nitrogen.

The normal sequence of addition of the formulation components is: (1) solids; (2) resins; (3) stable aroma chemicals (to act as solvents for the first two items); (4) the remaining aroma chemicals; (5) essential oils; and (6) speciality bases, sub-compounds or premixes, and highly volatile materials. High-speed stirrers will dissolve most solids and resins in the solvent part of the formula, and will give a rapid, effective mix of the total fragrance when all the ingredients have been added. Stainless steel vessels stirrers and other ancillary equipment are used universally for fragrance compounding. Once mixed the fragrance will be filtered by one of several techniques and when approved by quality control, packed into containers for shipping.

Automation and computer-controlled compounding are now being used in some of the larger fragrance houses. Traditional compounding as carried out by the majority of the trade is a highly skilled job, and full training takes at least 18 months. Knowledge, technique, accuracy and efficiency are the hallmarks of a top-grade compounder.

8.11 Quality control

The main parameters assessed during the quality control of fragrance compounds are odour, colour, physical state, density and refractive index. Occasionally, infra-red (IR) spectrographs and gas-liquid chromatographs (GLC) are used as fingerprints, and optical rotation (OR) may be important in the case of some compounds with optically active components, e.g. citrus blends. The odour of a fragrance compound should always be evaluated against a standard by a panel of at least two qualified or trained persons, and should be assessed on smelling strips, both fresh and after drying down for a minimum of six to eight hours. Since most fragrances are multi-component, a period of maturation may be expected, and all fragrances noticeably change in odour during the 24 hours following manufacture.

Although the colour of a specific perfume may vary slightly from batch to batch (often due to the variation in the natural raw materials), the clarity and mobility should not. Only in exceptional circumstances should a perfume be anything but clear. Lack of clarity suggests a non-homogeneous mix and, possibly, the presence of water.

Measurements of density (SG) and refractive index (RI) are undertaken universally, by fragrance houses and perfume purchasers alike. These physical tests give an indication that the raw materials within a specific formulation are reasonably consistent, both qualitatively and quantitatively. However, perfumery materials, especially natural products, vary from batch to batch and crop to crop. Consequently, the SG and RI of a specific perfume will vary slightly, with variations in the individual components being either compensating or cumulative. As a result, realistic ranges are set for SG and RI values. Fingerprinting by IR and GLC is less frequently used, although the very specific results achieved by a GLC assessment can be a valuable back-up when trouble-shooting problems, shown up by non-conformity of a sample to one or more of the other tests, occur.

The use of a computer 'nose' can be expected to increase. This technology is in its early stages but is progressing rapidly and is based on the principle of specific polymer receptor sites generating a response that can be interpreted and plotted by a computer. Advances in this area will take place in the structure and size of the receptors and the complexity of the computer software.

The tests described above are also carried out on all raw materials and sub-compounds being used in production, and it is at this stage that many potential problems can be eliminated. Good quality raw materials, screened by an efficient quality control procedure, are absolutely essential. In fact, the consistent quality of a company's products depends on this and many other factors. With all these factors, reputable suppliers, effective internal systems, well-trained development, production and quality control staff, and top-level management policy are crucial.

8.12 Special additives

Many essential oils used as perfumery raw materials exhibit desirable cosmetic and therapeutic properties and may, therefore, be used as 'special additives' in a cosmetic or toiletry formulation. Camomile and lemon are examples of two such oils. Several fixed oils are also incorporated as active ingredients. Aloe vera, jojoba and evening primrose oils are all currently enjoying popularity as additives that give cosmetic benefit. These fixed oils usually have relatively low odours and are often included in the formulations at low levels, presenting few problems to the formulator in terms of their effect on the overall odour of the product.

Both aqueous and glycolic extracts of plants and herbs have demonstrable beneficial properties and are used particularly in skin, hair and bath preparations. These extracts contain components that are not always present in the respective essential oils, therefore the incorporation of the essential oil may not give the desired effect. Where an essential oil does contain the required components for the cosmetic chemist, it is often possible to develop the perfume with the specific oil or oils as a part of the theme.

8.13 Glossary of odour descriptors

Aldehydic	sharp, fatty or soapy; characterised by the straight chain aliphatic aldehydes in the range C_8 to C_{12}.
Amber	sweet, warm, slightly animalic; frequently vanilla-like.
Animalic	redolent of animal odours; includes civet, musk, ambergris and castoreum.
Balsamic	warm, sweet and resinous with a faint medicinal aspect.
Camphoraceous	medicated; reminiscent of camphor and eucalyptus.
Chemical	usually harsh, aggressive and basic odours; typified by products such as amyl alcohol, acetophenone and diphenyl oxide.
Citrus	fresh, tangy and zesty; smelling of lemon, lime, orange, mandarin, grapefruit, bergamot or combinations of these.
Earthy	usually a combination of green, rooty and dank aspects.
Fatty	having the odour of animal or vegetable oils or fats.
Floral	having the odour of flowers, e.g. carnation, honeysuckle, jasmin, lily, rose, tuberose, violet or ylang.
Fresh	used subjectively depending on personal taste and experience; commonly citrus, light floral, green or fruity.
Fruity	any natural fruit note.
Green	light intense; redolent of leafy, grassy aromas.
Herbal	fresh plant odours, e.g. lavender (floral), rosemary (medicinal), camomile (fruity), basil (culinary) or coriander (spicy).
Leather	phenolic, warm, animalic.
Light	discrete; usually floral, green, citrus or combinations of these.
Medicinal	phenolic, camphoraceous, herbal; often pungent.
Mossy	earthy, woody, phenolic, green.
Nutty	sweet, oily, natural nut odours.
Pine	redolent of pine wood, needles and resins.

Powdery	soft, gentle; often balsamic, ambery and musky.
Resinous	warm, sweet, balsamic; sharp in the topnote.
Spicy	pungent, hot and culinary, e.g. bay, cardamon, cinnamon, clove, cumin, ginger, nutmeg and pepper.
Sweet	heavy, cloying; typified by vanilla and sugary notes.
Warm	typically ambery, animalic, balsamic and sweet.
Woody	natural wood notes such as sandalwood, cedar and guaiac.

Further reading

Arctander, S. (1969) *Perfume and Flavor Chemicals.* S. Arctander, Montelair, New Jersey.

Arctander, S. (1960) *Perfume and Flavor Materials of Natural Origin.* S. Arctander, Elizabeth, New Jersey.

Poucher, W.A. *Perfumes, Cosmetics and Soaps.* 8th edn: Vol. 1 (1974), Vol. 2 (1975) and Vol. 3 (1975); 9th edn: Vol. 1 (1991) and Vol. 3 (in press). Chapman and Hall, London.

Appel, L. (1982) *The Formulation and Preparation of Cosmetics Fragrances and Flavours,* Novox, NJ, USA, ISBN 1-870228-10-3.

Bauer, K. and Garbe, D. (1985) *Common Fragrance and Flavour Materials,* VCH, Weinheim, ISBN 0-89573-063-4.

Lawless, J. (1992) *The Encyclopedia of Essential Oils,* Element Shaftesbury, UK, ISBN 1-85230-311-5.

Van Toller, S. and Dodd, G. (1988) *The Psychology and Biology of Fragrances,* Chapman & Hall, London, ISBN 0-412-30010-9.

9 Personal hygiene products

M.J. WILLCOX

9.1 Introduction

The basic concept of personal hygiene is a relatively recent phenomena and, throughout history, mankind has probably been far more concerned about external appearance rather than about the general condition of the body. This chapter will consider current trends in personal hygiene products with some reference to the historical background.

9.2 Soap and other solid bathing products

9.2.1 *Toilet soaps*

Soap in its multitude of different forms would probably be regarded as the most basic material for personal hygiene. The history of soap and soap-making in its broadest sense stretches back to around 2800 BC, although in those days the products would most probably have been used for washing garments rather than for personal hygiene. Soap-making originally developed in the Mediterranean basin. Soap would have been regarded as a luxury item, therefore its availability would be quite restricted. It was not until the production of cheap caustic materials in the 19th century that any recognisable industry developed.

The basic chemistry of soap-making has hardly changed since its earliest days and involves the reaction of neutral fats or fatty acids with caustic alkali to form a soluble soap product with surfactant properties. Today, volume soap production is a highly automated process, far removed from the days of processing in open pans, although such production techniques are still widely practised by smaller manufacturers and are often considered to produce soap of a premium quality when the highest quality raw materials are selected.

Since the formulation of the basic soap does not vary greatly, the present discussion on formulation and development will be restricted to the many modifications that can be made to create a wide range of interesting and novel products. Development trials can be conducted with full-scale plant but it is more appropriate to carry out initial formulation trials using laboratory-scale equipment. This type of equipment replicates the full-sized plant and, as an absolute minimum, requires a triple roll mill and an extruder. Additional equipment such as a mixer and a stamper are ideal but not essential.

9.2.1.1 *Basic formulation requirements* Soap is a remarkably versatile medium and the choice of possible additives is almost limitless. The most basic formulation requirements will now be considered.

Soap base	to 100.00
Titanium dioxide	0.1–0.4
Colour	*q.s.*
Perfume	0.5–1.0

Soap base The initial selection of the appropriate soap base is important as it can affect colour, odour and performance. The soap base should contain an appropriate preservative system, the selection of which normally depends on the preferences of the base manufacturer. Most preservative systems are combinations of a chelating material such as EDTA to take care of any free metal ions, and an antioxidant material to prevent the possible oxidation of the fatty constituents of the base itself. If there are any doubts about the long-term stability of the base material itself, the formulator may consider it appropriate to add further EDTA and antioxidants.

Depending on the choice of fatty materials used in manufacture, the soap base can vary in colour from an almost pure white through to a dirty, creamy yellow. The odour will also vary accordingly.

Titanium dioxide This is widely used as an opacifier to provide a consistent background on which to create a wide range of different colours. Although the anatase grades have been preferred, current environmental pressures indicate that the rutile grades may become a more acceptable alternative. The latter have the disadvantage of being slightly less opaque, and of having a yellower tint.

Colour The selection of appropriate colouring materials depends on a number of factors including processing techniques, current legislation and the final market place for the product. The ideal ingredients will: (i) be readily dispersible throughout the soap mass to produce an even colour; (ii) have good tolerance to a pH of 10; and (iii) have good light stability. Inorganic pigments that fulfil these criteria include: (i) iron oxides (red, yellow and black); (ii) ultramarines (blue, pink and violet); and (iii) chrome oxides. Such pigments, however, only permit the formulator to create a limited range of final shades. Consequently, they are often complemented with organic pigments that also have good light stability and tolerance to alkaline conditions. Although these particular colours will not necessarily have been developed for use in soap (but probably more specifically for use in the plastics and paper industry), experience will have shown then to be excellent for use in soap. Advice on colours can be obtained from reputable cosmetic colour suppliers, who will share their knowledge and experience. Both organic and inorganic pigments can be incorporated into soap either as fine powders or as paste dispersions—the final decision will be determined by the style of processing.

Within the EC, colouring materials are covered by a number of positive lists, which relate to specific areas of use. It is essential to ensure the acceptability of any proposed colour system. Such considerations become even more important when export to the US and Japan is considered since regulations in both of these countries can further restrict the choice of colours (see chapter 11). For example, although soap as an individual item is not covered by the Food and Drug Administration regulations in the US, it is often considered prudent to approach formulation work as if it were. This can restrict choice to the inorganic pigments mentioned earlier, together with a limited range of certified organic dyestuffs that have poor light and product stability. Japanese regulations are very precise, and the choice of pigments and dyestuffs are quite restrictive. Certain parts of the spectrum cannot be obtained without the use of colours with extremely poor light stability.

Perfume Once an appropriate colour for a new soap formulation has been decided on, the next task is to develop a suitable fragrance. The expertise of one or other of the many fragrance supply houses may be utilised to provide a suitable perfume, at an appropriate price to satisfy the brief supplied by the marketing department, or by the customer.

As with any perfume development project, it is essential to provide the perfumer with as much information as possible regarding the base material, and with other pertinent comments. Although soap is a reasonably hostile environment to perfume, the range of perfumery raw materials with acceptable stability enables the competent perfumer to achieve most of the desired fragrance notes. In some areas, however, perfume interaction can cause coloration and discoloration problems. Perfumes with significant levels of vanilla notes suffer from particularly severe discoloration. For the majority of development projects, it is possible to eliminate or minimise discoloration problems, but more basic problems can be encountered when the soap is 'intended to form part of a product range based on a specific alcoholic fragrance. In some cases, the only option is to accept the discoloration and aim to produce a soap colour that masks or complements it. Many of the soaps produced for French couturier houses contain no colouring material, and the final colour of the product is determined by the coloration or discoloration due to perfume alone.

The level of fragrance likely to be included in the more basic formulations would probably be no more than 1%, but this is by no means absolute and normally the upper limit will be determined by cost and/or processing restrictions. Even at the upper levels, it is possible to further increase perfume performance without adversely affecting process performance. This can be achieved by the partial use of micro-encapsulation for some of the perfume dosage.

Once the initial outline formulation has been selected, pilot production trials are essential to ensure that the new formulation provides the desired characteristics for colour and perfume, both on initial manufacture and after a series of controlled stability tests. These production trials will also enable the formulator to determine how the new product behaves during manufacture. At this stage it may even be necessary to consider modifying the formulation. This may be as simple as adjusting the level of colour to correct for process scale-up, but could, under the worst conditions, require a major rethink of the basic formulation itself. Samples stored at elevated temperatures are very useful for determining whether any discoloration is likely to take place, while samples stored in daylight indicate the light-stability performance of any new formulation. It is often useful if the perfumer conducts parallel stability trials using samples produced to the final formulation.

9.2.1.2 *Variations on the basic formulation* As mentioned earlier, soap is a very versatile product and the range of possible additives is almost limitless. The only restrictions are (i) the exclusion of raw materials that are potentially harmful to the processing conditions, workers and equipment; and (ii) the exclusion of materials that degrade under alkaline conditions. All additional formulation components should be chosen so as enhance performance (either actual or perceived) and, possibly, to support specific product claims, thus giving marketing staff the selling points required to compete in the marketplace.

Superfatting agents Soap is a relatively harsh washing medium in its own right and it is therefore considered beneficial to improve the perceived performance with the use of superfatting agents. The average consumer is also much more aware of the concept of 'moisturising' since the introduction into the marketplace of cleansing/cream bars. These products have been sold in the United States for many years using well proven technology which is considered later in this chapter, but are relatively new to the European market. The formulator is often faced with the challenge of trying to reproduce the washing characteristics of these cleansing bars by the addition of cost effective superfatting systems.

The system can be based on either single products or complex cocktails of oily and emollient chemicals. Originally the most frequently used materials included mineral oil, petroleum jelly, lanolin and nut oil fatty acids at levels of 0.5–1.0%. More recently more complex molecules often based on quaternary compounds are found to be very effective at levels up to 2.0%.

However there can often be associated processing difficulties and ideally such additives should be included earlier in the soap making process prior to drying. Fatty chemicals from the oil phase of simple emulsion systems can also be included to substantiate market claims such as 'contains cold cream'. In reality, the formulator may choose to include a low level of the actual cream whilst incorporating higher levels of other appropriate fatty chemicals such as octyl hydroxystearate or glyceryl ricinoleate, which can impart the desired feel to the soap tablet and allow consumers to draw their own conclusions regarding enhanced product performance. As with perfume, it is possible to use encapsulation techniques to increase the levels of addition without the consequent production difficulties that would normally be associated with additive levels of oils and fats in excess of 2%. The formulator should never forget that products must be capable of being produced on full-scale plant, therefore plant trials on soaps containing new and exotic oily cocktails are essential.

Natural oils and extracts With current trends towards more natural products, it is possible to incorporate a wide range of natural oils such as jojoba, avocado, apricot, cherry kernel, hazelnut, etc. With these oils, it is often wise to include an appropriate antioxidant to prevent any tendency for rancidity. A wide array of natural extracts can also be used to further substantiate product description claims. In practice, these often contain very little of the true natural product but do seem to be widely used, albeit at relatively low levels, either separately or in combination.

Abrasives Oils and natural extracts can impart improvements to the feel and texture of the soap bar, but it is also possible to use soap as a vehicle to carry abrasive materials and create an astringent product that is still pleasant to use. Once again the choice of possible additives is very comprehensive and includes: oatmeal, pumice, various crushed nut kernels, seaweed, herbs and spices. The formulator is limited only by the effect that processing may have in breaking down particle size and the attractiveness of the final product. The alkaline nature of soap can also cause some added natural products, such as flower petals, to change colour but this can only really be determined by trial and error.

Deodorants In addition to formulations with aesthetic qualities it is also possible to create products that have specific treatment properties, particularly in the deodorant area. Soap is an ideal vehicle for deodorant materials, with those most frequently used being triclosan (Irgasan DP 300) and triclocarbon (TCC), either individually or in combination. Raw material suppliers can provide detailed information regarding optimum levels of addition, although certain specific combinations of the two materials that provide the best synergist effects may well be covered by patents. Triclosan and triclocarbon have significantly different kill spectrums. The choice must be balanced against the discoloration that results from the use of triclosan following exposure to light. It is also possible to substantiate general claims of a deodorant effect by the use of specific perfumes—fragrance suppliers will be able to advise on this point.

Alternative base materials Regular toilet soap is still mainly produced from raw materials based on tallow although with increasing influence of animal welfare groups, there is growing interest in the use of base materials that are totally derived from vegetable feed stocks. Such soaps are normally produced by the substitution of palm oil or palm oil fatty acids for the tallow or tallow fatty acids in the regular material. The resulting bases can range in colour from white to cream but, as far as the formulator is concerned, they should be treated as separate raw materials. All of the previous comments on raw materials remain relevant.

9.2.2 *Shaving soaps*

Many of the comments and observations concerning the formulation of toilet soaps (section 9.2.1) also apply to shaving soaps. Shaving soap base differs from toilet soap mainly by the addition of caustic potash at the saponification stage. This change, in combination with the correct choice of fatty acids, creates a softer base material with enhanced lathering characteristics. To facilitate processing, the moisture content of the base material is normally reduced to around 8%. The control of free alkali is crucial, and it is not uncommon to finish shaving soaps with a slight excess of free fat to avoid the possible risk of reaction to free alkali during use.

The choice of possible options for the formulator are much more restricted than for toilet soap, with the principal requirements being fragrance and skin-feel. The earlier comments made on such items as opacifiers, colours, sequestering agents and antioxidants also apply here, and the general philosophy with regard to fragrance selection remains constant, although the levels of incorporation are often 10–20% higher than one would normally expect to use for the equivalent toilet soap.

The use of superfatting systems in shaving soaps improves the skin-feel after shaving, although it is important to keep the levels of oily constituents at around 1% to prevent any significant effect on lathering.

9.2.3 *Transparent/translucent soaps*

The recent trend towards 'natural' products has seen a marked interest in 'clear' products of all types, and soap is no exception.

9.2.3.1 *Transparent soaps* True transparent soaps can be produced in a number of different ways. One of the oldest methods involves dissolving soap in alcohol with gentle heating to form a clear solution, which is then coloured and perfumed. The colour of the finished bar depends on the choice of starting materials and, unless a good quality soap is selected, may well be quite yellow. This presents difficulties with certain colours.

The traditional process involves partial removal of alcohol by distillation, and casting of the liquid soap into blocks. The blocks are left to condition for up to three months before being moulded and packed into their final presentation. This process, by its very nature, is expensive and is restricted to some familiar products that have been in the marketplace for many years. It is unlikely that the formulator would need to deal with this particular style of formulation, since alternative, cheaper methods have been developed using castor oil and sugar in addition to the regular soap-making raw materials. With these methods, it is possible to produce transparent/-translucent soaps directly from the constituent raw materials, without any need to pre-prepare the soap as an intermediate stage in the process. A typical basic formulation for this type of transparent soap is shown below.

	Wt%
Tallow fatty acids	27.00
Coconut oil	7.00
Castor oil fatty acids	5.00
Alcohol	10.00
Sodium hydroxide	6.20
Sugar	15.50
Glycerine	9.00
EDTA	0.25
Water	to 100.00

The selection of raw materials particularly the fatty ones, will have a significant effect on the colour of the finished product. As always, there has to be a commercial balance between the most refined, and therefore most expensive materials if one is to achieve a realistic end cost for the product. Castor oil is usually considered to play an important part in the formulation in achieving optimum clarity, although it does have the disadvantage of giving the finished tablet a distinct yellow colour. Other alkaline materials (e.g. triethanolamine) may be used to effect saponification, although the resulting bars are often much softer than those produced with caustic soda.

The method of production involves the melting of the fat phase and the preparation of a water phase in which the sugar, glycerine and preservatives are dissolved. These two phases are reacted with an alcoholic solution of caustic soda under controlled heating. Some simple refluxing is an advantage. Having allowed the reaction to subside, the soap mass is checked to ensure complete saponification with a very slight free caustic finish (less than 0.1%). This soap mass is then available for colouring and perfuming.

The choice of perfumes, additives and colouring materials is more restricted because of the process conditions and it is essential that none of these additives have any adverse effect on the transparency of the finished bar.

It is very important that the choice of perfume should take account of process heat (up to 4 hours at 70°C) and it is essential to avoid the more volatile perfumery materials along with those which may be prone to change colour/degrade at these elevated temperatures. Natural extracts, oils and vitamins may also be added, always bearing in mind the comments regarding transparency. Organic dyestuffs are probably the best choice for colouring the bars but as they can have poor light stability it is often prudent to include an ultraviolet screen to minimise fading problems.

Following colouring and perfuming, the finished soap is poured into separate moulds or cups and allowed to set before packing. The process can be scaled up for bulk production without any major difficulty, although the systems for dosing and solidification become relatively sophisticated.

The use of alcohol can be considered something of a problem as it does necessitate the provision of controlled production areas and equipment to avoid any risk of explosion. More recently transparent soaps are being produced using combinations of soap and detergent which result in products with excellent colour and clarity. An outline formulation for this type of transparent soap is shown below:

	Wt%
Stearic acid	15.00
Coconut fatty acid	6.00
Propylene glycol	18.00
Glycerine	8.00
Sodium hydroxide	4.20
Sugar	10.00
Sodium laureth sulphate	16.00
Sodium lauryl sulphate	12.00
EDTA	0.20
BHT	0.20
Water	to 100.00

As one is able to select virtually water-white raw materials the resulting bar has a very pale colour and consequently it is possible to obtain more delicate pastel shades. The method of production is similar to the method given above with the surfactants and the propylene glycol being combined with the fatty phase. With the high level of volatiles it is essential that the finished bars are wrapped in a film with adequate barrier properties and this also has the added benefit of enhancing the visual appearance of the bar.

9.2.3.2 *Translucent soaps* A recent development has been the appearance of translucent soaps either tallow-based or produced totally from vegetable raw materials. Unlike transparent soaps, these are produced on regular soap-making equipment by extrusion. This means that they are capable of being moulded into a wide range of shapes, which tend to be simple distortions of oval or rectangular blocks.

It has long been recognised that soap passes through a translucent phase during the processing of the basic soap base. The ability to control the production of this translucent phase by the addition of glycerine or other polyol-type materials, linked with processing modifications, has allowed the soap manufacturer to develop a wide range of novel presentations. The formulator need not be concerned with the formulation of the soap base itself, and the following comments relate to the options available for the development of the finished formulation.

(i) As with transparent soap, the most important consideration is to ensure that nothing is done to affect the translucency of the finished soap tablet. It should also be recognised that the translucency of these soaps is often enhanced during storage.

(ii) The selection of colours will follow the same general consideration mentioned previously although, as the level of colour required is normally very low, it is possible to use organic pigments in addition to organic dyestuffs. One disadvantage of the low level of incorporation is the possible reduction in colour stability on exposure to light. This is particularly relevant to translucent soaps which are most often presented unwrapped or with a clear film overwrap.

(iii) The selection of perfume is critical since certain perfumery raw materials affect translucency. The competent perfumer will be able to provide a wide range of interesting fragrance alternatives once the limitations of the final presentation are made known, and provided he or she is able to work with the relevant soap base material.

The formulation for a translucent soap is normally quite simple and straightforward, adding the following blend to the chosen base.

	Wt%
Perfume	1.0–1.5
Colour	*q.s.*
Preservative	*q.s.*

Experience suggests that the level of perfume should be no higher than 1.5% to ensure optimum clarity.

Despite the need for optimum translucency, it is possible to incorporate low levels of natural extracts and natural oils to enable claim validations to be made for certain soaps. The limitations are normally established by process experience but, for an acceptable product, a total additive level of 2.5% is a likely maximum. Although soap bases are normally fully preserved it may be prudent to add a small quantity of BHT (butylated hydroxy toluene) to ensure good product stability. The translucent appearance may be exploited further by the use of solid material or pearl additives, both of which can provide novel appearances to further enhance the soap tablets.

As with regular toilet soap, it is important to carry out stability tests on prototype samples over an extended period to ensure that no deterioration occurs. This is particularly important with perfumes, as any discoloration effect is readily apparent.

9.2.4 *Synthetic detergent/combination bars*

The development of modern detergent systems has provided raw materials that can be used singly or in combination to produce a solid bar with a much lower pH than regular soap. Indeed some of the undoubted success of recently introduced cream / cleansing bars has been the ability to demonstrate that products mimic the pH of skin.

Although many of the basic raw materials are more expensive than regular soap the added benefits do allow a higher retail price for such products. When these products are produced from the most basic raw materials, specialised equipment is required and indeed many of the processes may be covered by patent restrictions. However, synthetic detergent/combination bars can be processed on regular soap-making equipment although it is essential that the temperature is carefully controlled to ensure that the detergent mass is maintained in a suitable plastic condition without becoming too mobile. The product is also difficult to handle on pilot plant equipment.

Formulations for synthetic detergent/combination bars tend to be relatively simple and because of the less aggressive conditions, it is possible to use them as vehicles for specific treatment chemicals that would degrade at higher pH levels. The lower background odour means that perfume levels can be reduced and, since good colour distribution through the bar is often difficult to achieve, the products are very often white or off white.

Combinations of synthetic detergent systems and regular soap can offer the opportunity to reduce cost but on the down side the pH is only reduced slightly to around 9.0–9.5. These formulations can pose some production problems and it is again essential to maintain close control of process temperatures.

9.2.5 *Bath salts/bath crystals/bath cubes*

Solid bath additives are generally intended to soften the water and produce a delicate fragranced atmosphere for bathing. It is also possible to include other additives to further enhance the bathing experience.

9.2.5.1 *Bath salts* The basic formulation of bath salts is shown below.

	Wt%
Sodium sesquicarbonate	97.50
Mineral salt	1.50
Perfume	1.00
Colour	*q.s.*

This product could be produced by simply using a ribbon mixer to blend together the solid components, after which the perfume and colour would be dispersed on to the surface to create the desired end result. The most commonly used mineral salt is sodium chloride, although combinations of other salts can be used to simulate seawater.

Starting with the basic outline formulation, it is possible to include solid detergents to enhance water softening and to create an illusion of foaming. One of the sodium dodecylbenzenesulphonates would be appropriate. Fatty acid ethanolamides and fatty esters are also added to improve skin-feel. As the level of additives is increased, the mass becomes quite sticky. It is necessary, therefore, to include a suitable drying/flow agent to ensure that the product remains a free-flowing powder. The process of replacing the sodium sesquicarbonate with other components has no real limit apart from cost. In practice, the total additives could probably account for as much as 40% of the total formulation; however, the fatty/liquid components would account for no more than 10%.

The product can be tinted using pigments dispersed on to a suitable extender, and the product can be further enhanced by the use of a solid, water-soluble dyestuff, which will colour the bath water. This can be a totally different colour to the product itself. Although in the laboratory the formulator is able to approximate process conditions by simple mixing, it is essential to check the levels of colour dispersion and the amount of drying required for any individual formulation by means of a full-scale trial. A ribbon mixer with a separate high-speed mixing blade to ensure good dispersion of all raw materials is ideal, and equipment for sieving the finished product to remove any large undispersed particles is advisable.

9.2.5.2 *Bath crystals* Bath crystals are normally formulated in a similar manner to bath salts but using raw materials with a definite crystal structure, which enhances the appearance of the finished product. Sodium sesquicarbonate is available as fine, needle-like crystals. Sodium chloride can be obtained in a number of different crystalline forms, both regular and in lumps. Sodium thiosulphate has also been used. With these basic constituents it is only possible to surface coat the crystals, and the formulation requires little more than the addition of a liquid colour and perfume.

9.2.5.3 *Bath cubes* Bath cubes are basically bath salts combined with suitable binders and compressed to form solid tablets. A typical formulation is as follows.

	Wt%
Sodium sesquicarbonate (powder)	92.00
Sodium sesquicarbonate (crystals)	2.50
Maize starch	2.00
Talc	1.50
Mineral oil	1.00
Perfume	1.00

The formulation of bath cubes is very difficult to simulate and is probably best left to suppliers. As with solid bath additives, all perfumes should be selected with the end use in mind. One recent innovation has been that of 'fizzing' cubes. These incorporate suitable alkaline and acidic materials into the cube so that the product dissolves with the formation of carbon dioxide, and generates bubbles.

9.3 Liquid bathing and showering products

The market for bathing and showering products has seen tremendous growth in recent years and there has been a significant move away from the solid bathing products mentioned in the previous section. The range of newer products includes foam baths, shower gels, bath oils and after-bath preparations. This section will consider the formulation options available, based on surfactants materials.

Regular soap is an excellent surfactant that can be produced from readily available raw materials. Its only major disadvantage is its poor performance in hard water, and this feature prompted the development, from vegetable oils, of sulphonated derivatives for use in textile processing. The technology developed gradually, with the production of the first surfactants from petroleum-based feedstocks around the time of World War I. The basic technology for the production of surfactants developed between the two world wars, and the expansion in the petrochemical industry made available large volumes of reasonably priced raw material. Today, a wide range of alternative raw materials are available to the formulator.

9.3.1 *Foam baths*

Foam baths, in either liquid or gel form, are probably the most widely used bathing preparations currently available in the marketplace. A bewildering array of clear and pearlised products, in a rainbow of colours, can be seen on supermarket shelves.

The principle function of bathing, in the 'western' sense of the word, is the cleansing of the body, although the therapeutic effect of relaxing in a hot bath should not be ignored. A foam bath will help to soak off dirt and body oils and ensure that they are left suspended in the bath water. It will also ensure that no tell-tale 'bath ring' is deposited, as is the case when soap is used. Although cleansing is the principle function of a foam bath, good perfuming and a capacity to impart some skin conditioning properties is desirable. As already mentioned, the total bathing experience should create a pleasant, relaxed atmosphere.

9.3.1.1 *Basic formulation requirements* When formulating foam baths, a number of fundamental attributes should be considered.

(i) Good foaming characteristics
(ii) Non-irritant to eyes, mucous membrane and skin
(iii) Reasonable cleansing power without adverse skin effects
(iv) A clean, fresh and attractive effusive perfume.

The selection of the principle raw materials will have a significant bearing on the final product.

The main ingredient, even in the most basic formulations is the foaming agent and selection of this surfactant should be made with the basic attributes in mind. Surfactants are available in four principle types.

(i) Anionic—with the polar group having a negative charge
(ii) Non-ionic—with the polar group having no charge
(iii) Cationic—with the polar group having a positive charge
(iv) Amphoteric—with both positively and negatively charged polar groups.

Anionic surfactants are probably the most widely used, as they are reasonably priced and available as very pale coloured/water-white variants, which allows the formulation of very pale/tinted products. Among of the most popular forms of anionic surfactant used in foam bath formulations are the fatty alcohol ether sulphates. The sodium form based on lauryl/myristyl alcohols containing either two or three moles of ethylene oxide forms the basic constituent of many formulations. More recently, the equivalent magnesium forms of the same basic structure have become available. Although more expensive, these have a much lower irritancy potential. The formulation for a typical, moderately priced foam bath is shown below.

	Wt%
Sodium lauryl ether sulphate, 28% active	50
Coconut diethanolamide	3
Perfume	1
Sodium chloride	1.5
Preservative	*q.s.*
Colour	*q.s.*
Water	to 100

Surfactant The combination of the basic anionic agent with the non-ionic diethanolamide imparts improved foam stability and can help with the creation of a suitably viscous product. These materials can also assist with the solubilisation of perfume and other skin-care ingredients where needed.

Perfume The choice of perfume is crucial to the development of a good foam bath, and a reasonable amount of time should be spent in the evaluation and panel testing of this vital component. The selected perfume should mask any odour from the raw materials and give an impressive initial impact in the head space of the container—a critical aspect most likely to stimulate purchase by the consumer. The perfume should have a good 'lift' from hot water to create the ideal bathing environment, and tenacity for the skin. The level of perfume in the product will be determined by cost constraints, and the formulator is under increasing pressure to obtain fragrances with high impact at minimal cost. Fragrance levels of up to 5% are chemically feasible, although some additional solubilisation may be necessary. Reasonable shelf-life is an obvious requirement, and adequate product stability testing is vital as perfumes can have adverse effects on colour, viscosity, clarity and even preservative performance.

Sodium chloride This is widely used as a viscosity modifier and the level of addition will need to be determined at the development stage once the total formulation is available. The level of salt in the basic surfactant will vary from batch to batch, therefore the level of viscosity modifier will need to be adjusted according to analysis of the incoming bulk raw material. On occasions, reduction in the viscosity of products that have become too thick may be required. This can be achieved by the addition of alcohol, propylene glycol or hexylene glycol.

Preservative Preservation of foam-bath formulations is vital to prevent product degradation by moulds and bacteria. The basic surfactants may contain preservatives, but addition of a selected preservative will almost certainly be required. Validation must be carried out by challenge testing with an approved method, normally involving the use of a range of specified organisms. It is also essential to use an 'in-house' cocktail of organisms that are specific to the individual manufacturing location. The choice of preservative will be determined by the ultimate market for the product, since many countries—and indeed specific customers—have approved lists of preservatives. The whole question of preservation is highly volatile, since proven products can be de-listed almost overnight. With increasing demands from the consumer for 'safer' products, the levels of preservative may well be reduced, since they can contribute to irritation difficulties for some consumers. In these situations the importance of good manufacturing practice is essential to ensure that there is no unnecessary overload on the preservative system by poor 'in-house' techniques.

Colour The selection of the colour system will be determined mainly by the legislative requirements of the final market, and it is common practice to use a range of US certified colours to create an appropriate colour palette. This can prove problematic, since specific colours may be de-listed at any time. Other markets, such as Japan, can be even more restrictive in the choice available. Organic colorants may give rise to colour instability problems, especially when pale-coloured products are required. Light stability must be carefully tested.

Simple foam baths are relatively easy to manufacture and require no specialised production equipment other than a suitable vessel equipped with a large paddle mixer. The ability to control the speed of the mixer is vital if unnecessary aeration is to be avoided.

9.3.1.2 *Variations on the basic formulation* The basic outline formulation for a foam bath can be used by the formulator as the starting point for a whole series of foam baths. Within reason, the active content can be reduced to produce inexpensive formulations but these will obviously not have the performance attributes that were identified earlier in this section. Possibly of more interest is the addition of other materials to improve both the appearance and performance of the product; here the only determining factor will be the final cost objective for the finished product.

Opacifiers One of the most frequently used methods of enhancing the appearance of the product is to include an opacifier to impart a pearl effect. This can be achieved by the addition of alcohols (such as stearyl alcohol and cetyl alcohol), ethylene glycol monostearate or glyceryl stearate. A disadvantage of using these particular raw materials is the necessity to heat the formulation (less perfume) to around 70°C, and then allow the bulk to cool slowly with stirring to ambient temperature to develop the pearl. An alternative method not requiring heating is the use of a commercially available pearl agent (opacifier–detergent blends) incorporated at a level of between 2.5 and 5.0%. In any opacified/pearl formulation it is vital to maintain optimum viscosity to ensure that the pearl remains in suspension. Stability checks are the only way to establish adequate shelf life and, in particular, to determine the effect of fragrance on pearl systems.

Amphoteric surfactants Although the sodium lauryl ether sulphates are reasonably mild products in their own right, the formulator may elect to substitute a proportion of the ether sulphate with one of the amphoteric surfactants mentioned earlier. Amphoteric surfactants exhibit excellent mildness and very good foaming characteristics but, because of their higher cost, they are normally used in conjunction with anionic surfactants. Typical amphoterics are alkyl imidazoline betaines or alkyl amino betaines, which, when incorporated at the appropriate ratio with the selected anionic, can also be used to enhance formulation viscosity. Amine oxides are often claimed to have improved foam boosting properties when compared with alkanolamides and may well be preferred by the formulator.

Emollients Even when carefully formulated, foam baths can possibly cause harmful effects on the skin, and the addition of emollients can be very beneficial to product performance, and in helping to leave the skin soft and smooth. A wide range of emollients is available. Although it may be difficult to substantiate some of the moisturising claims that are made for many foam baths, they do have a perceived effect on the skin and are very popular.

Taking all of the preceding comments together, an outline formulation for an improved milder formulation with good afterfeel can be arrived at.

	Wt%
Sodium lauryl ether sulphate, 28% active	40
Cocamidopropyl betaine	5
Opacifier—detergent blend	2.5
PEG-7 glyceryl cocoate	2.5
Cocamidopropylamine oxide	1.5
Perfume	1.5
Sodium chloride	1
Preservative	*q.s.*
Citric acid	*q.s.* to pH 7
Colour	*q.s.*
Water	to 100

Although the comments made in this section are restricted to certain specific anionic, non-ionic and amphoteric surfactants, the formulator will soon recognise that there are many other surfactants in each particular category that offer different properties. All surfactant suppliers will readily provide alternative outline formulations to enable a wide choice of alternatives to be created.

It is often difficult to describe foam baths as natural products, even though the prime building blocks for the raw materials do have natural sources. However, it is not uncommon for similar formulations to be marketed under a 'natural' banner by the addition of small quantities of natural extracts. It is best left to the individual to make their own decision on the ethics of such marketing platforms.

9.3.2 *Bath oils*

It may be appropriate to question whether bath oils should really be included within the general area of personal hygiene products since they do not contribute to cleansing. Their principle function is skin lubrication, with a secondary function of applying fragrance to the skin.

9.3.2.1 *Foaming bath oils* Many products in the marketplace are described as foaming bath oils but are best considered as foam baths with increased levels of emollient oils. They foam well, do not leave any bath ring, and normally have higher levels of fragrance than regular foam baths. Consequently, they may require additional materials to solubilise the fragrance.

An outline formulation for such a bath oil is shown below.

	Wt%
Sodium lauryl ether sulphate, 28% active	55
Cocamidopropyl betaine	7.5
Coconut diethanolamide	2.5

Cocamidopropylamine oxide	2.5
PEG-7 glyceryl cocoate	1.5
PEG-6 caprylic/capric glycerides	1.5
Perfume	1.5
Preservatives/colour	*q.s.*
Water	to 100

As with regular foam baths, formulation variants would follow a parallel route.

9.3.2.2 *Soluble bath oils* If the level of emollient oils are increased significantly, the foaming characteristics of these products will be drastically reduced and they cannot be considered as foaming bath oils. They are more correctly categorised as a form of soluble bath oil, although they impart a significant level of skin lubrication, which is not necessarily the case with regular soluble bath oils.

Soluble bath oils are normally formulated to deliver high levels of fragrance to the bath water but with few additional benefits such as emollient properties. They are relatively simple formulations, consisting of little more than anhydrous mixtures of perfume, surfactant and colour. These products are quite expensive and water may be included to reduce cost. When choosing the surfactant, the level of fragrance and the desired viscosity of the final product must be considered. Perfume solubilisation can be achieved using an appropriate hydrophilic surfactant. An outline formulation would be as follows.

	Wt%
Perfume	10
Polysorbate 20/polysorbate 80	20
Preservatives/colour	*q.s.*
Water	to 100

These products are very easy to manufacture and require no heating. The perfume is mixed with the surfactant prior to dilution and the addition of minor additives.

This type of formulation can be an ideal vehicle for producing 'aromatherapy' bath oils by the simple substitution of individual or blended oils for the perfume. The choice of the oils imparts the different mood conditions that such products are reputed to create.

9.3.2.3 *Dispersible bath oils* These are designed to be emulsified in the bath water, creating a milky bloom. The emollient oils are balanced with appropriate levels of surfactant to create a product that has no significant oily feel and does not leave a bath ring. In many respects these products are closely related to the foaming bath oils mentioned earlier, although the latter products do not create a blooming effect.

Dispersible bath oils are relatively easy to manufacture and, as always, the cost of the finished formulation will determine the choice of raw materials. In its simplest form, the product consists of an oil and perfume with an appropriate surfactant. Mineral oil is a relatively inexpensive oil base, but vegetable oils can be used for partial substitution. Other emollient oils will also be required to aid perfume solubilisation and impart greater skin-feel to the product. Polyethylene glycol ethers are most commonly used as a surfactant and a reasonably priced formulation is outlined below.

	Wt%
Mineral oil	60
Corn oil	5
Isopropyl myristate	15
Oleth-3	12.5
Perfume	7.5
Colour	*q.s.*

Balancing the levels of the various components is required to achieve adequate solubilisation for the chosen fragrance and to achieve the desired amount of bloom when the product is added to the bath.

9.3.2.4 *Non-dispersible/floating bath oils* As their name implies, these are designed to float on top of the bath water. On emergence from the bath the bather's skin is coated with a thin oily film. Non-dispersible bath oils provide a distinctly different after-bath sensation and, as the fragrance in the product is concentrated in the thin surface layer, the release is greatly enhanced by hot water—creating an acceptable fragranced atmosphere in the bathroom. Their disadvantage is the oily ring left around the bath—if the bather is also using a regular soap in an area of hard water, the effect is even more unsightly because of the additional soap scum.

The ideal non-dispersible bath oil will cover the surface of the bath completely and only deposit a light oily film on the skin. This requires a careful balance between the oily and emollient constituents and the chosen surfactant to ensure that the oil spreads rapidly across the water surface. Laboratory trials using water at an appropriate temperature are essential, since the spreading characteristics of systems are greatly affected by temperature.

As with the dispersible oils, mineral oil forms a significant part of the formulation, but its greasy feel may be modified by combination with isopropyl myristate or one of the newer emollients such as the polypropylene glycol ethers. These materials also have much greater solubilisation powers. The choice of other emollient oils is almost limitless, and vegetable oil can also be used to achieve differing skin-feeling characteristics. Materials such as the condensation products of oleic acid esters of sorbitol have excellent solubility and spreading properties. The choice of surfactant is also important to ensure that emollient additives are soluble in the chosen oils.

Perfume is the most critical cost element in the formulation and the optimum level will depend on the final cost target for the formulation. A typical floating bath oil formulation is outlined below.

	Wt%
Mineral oil	43.5
PPG-15 stearyl ether	45
Corn oil	5
PEG-40 sorbitan dioleate	1.5
Perfume	5
Colour	*q.s.*

Product stability and performance testing is essential and should include a careful evaluation of the proposed packaging. Significant proportions of oily materials can interact with some plastics used in packaging. Although glass containers may always be preferred, both for appearance and for lack of any interaction, cost probably excludes their use for all but the most expensive presentations. Glass is also considered unsuitable for bathroom use because of the risk of breakage and injury.

These products also serve as excellent vehicles for 'aromatherapy' products as indicated above in section 9.3.2.2.

9.3.3 *Shower gels*

Shower gels are, in the most simple terms, higher viscosity variants of foam baths, often containing higher levels of active materials. The increase in viscosity may simply be achieved by increasing the active content and/or electrolyte levels, or by adding an alternative thickening agent to the formulation.

Most of the formulation criteria discussed in section 9.3.1 are relevant for shower gels. However, as the method of use involves direct application to the body, some attention needs to be paid to the mildness of the proposed formulation. Consequently, a higher level of amphoteric surfactant is often used. With careful adjustment of the levels of the two principal surfactants, a range of viscosities up to an almost solid gel can be obtained.

An outline formulation for a moderately priced product is given below.

	Wt%
Sodium lauryl ether sulphate, 28% active	45
Cocamidopropyl betaine	15
Coconut diethanolamide	2.5
PEG-7 glyceryl cocoate	2.5
Perfume	1.5
Propylene glycol	1
Preservatives/colour	*q.s.*
Water	to 100

Recent formulation innovations have included the addition of materials such as hydrobeads to create a more interesting visual appearance and these beads can contain additional emollient chemicals. Even solid particles such as loofah can be incorporated into gel formation if the viscosity of the product is sufficient to prevent separation/settling.

9.3.4 *After-bath and after-shower conditioners*

After-bath and after-shower products are intended to replace some of the natural skin oils that are removed during bathing or showering. Many of these products are simple oil-based systems and may be considered unsatisfactory because of their oily nature. More popular are low viscosity body lotions (see chapter 3).

A more novel approach is to apply a product such as a liquid talc, or even a body conditioner, which attempts to replicate the well established effect of hair conditioning.

Liquid talc products are suspensions of talc or starch derivatives suspended in a light oil or alcohol base. These products are rubbed gently on to the skin to produce a fresh, cool sensation, and to leave an emollient lubricity —softening the skin.

10 Antiperspirants and deodorants

R. GIOVANNIELLO

10.1 Introduction

Based on an historical perspective, antiperspirants and deodorants are relatively new personal care products, covering a 100 year span of development and use. One can appreciate the technological advancements of these compositions, considering that over 300 generations have passed during which some kind of cosmetology was crudely practiced. Botanically extracted oils, fragrances and metallic-based ointments were extensively used for skin care, eye and facial decor.

In the early 20th century the antiperspirant active choices available to consumers were few. This undeveloped cosmetic science, together with poor product delivery forms applied to the axillary body region was considered nothing less than crude in the extreme.

The earliest known marketed deodorant product was a zinc oxide-based cream composition [1] introduced within the United States in 1888. Through the years zinc in various salt forms has been evaluated for its antiperspirant and deodorant properties. The incorporation of zinc halides into basic aluminum and zirconium chloride complexes has resulted in salt complexes described as useful antiperspirants possessing good alcohol solubility [2, 3]. Zinc metal complexes however were never adopted into the Category I antiperspirant active listing. Other zinc-based compounds known to have deodorant potential were zinc peroxide, zinc salicylate, zinc sulfocarboxylate and zinc sulfide.

In the early 1960s zinc phenolsulfonate was introduced in aerosol form and was coupled with hexachlorophene. The effective antimicrobial action of the two ingredients quickly led to high consumer acceptance and some cannibalization of the decade-old roll-on antiperspirants. Today zinc phenolsulfonate is one of the few clinically accepted zinc salts that can be found in deodorant compositions. Hexachlorophene however was banned in the mid-1970s due to potential skin penetration and neurotoxicity.

The earliest antiperspirant was a non-formulated product consisting of a dilute aqueous solution of aluminum chloride. The application was simply one of cotton swabbing the axillary area and waiting for the sting and water to disappear in that order. Aluminum chloride despite its strong acidity is perhaps the most effective active to this day. It can be found in pharmacies in a regulated dosage of not greater than 15% on a hydrated basis in non-aerosol form.

Since the early 1900s, aluminum salts were primarily the only active ingredients recognized as effective in reducing sweat and controlling odor. Efficacy was not quantified until 1916, the first published report appearing in the *Journal of the American Medical Association.* At the same time promotional campaigns were under way leading to national popularity for aluminum chloride solutions.

Esthetics were clearly lacking from the simple solution forms of antiperspirant but by 1930, cream compositions were introduced containing aluminum sulfate in wax bases. Compared with current standards the esthetics may appear to have been derived from poor cosmetic science, but were in fact elegant to the consumer of that era. By 1945, 88% of the US antiperspirant market was in the form of creams, the greatest share of an antiperspirant form ever to exist.

In the 1950s hand squeezed aerosol and roll-on antiperspirants made their way onto the market, representing unique delivery forms. The package technology was not perfected however, leaving the consumer with convenient but often faulty delivery systems. The attractive features were the fact that the antiperspirants could now be applied without the use of fingers. Popularity grew slowly but as early as the next decade, roll-ons dominated the market over creams for a short period of time.

During the same era, phenomenal changes occurred in the way aerosols were designed. Deodorants and antiperspirants were incorporated into pressurized propellant systems [4] in metal containers leading once again to high consumer acceptance. Valve perfection coupled with anhydrous suspension compositions dramatically reduced the rate of clogging. By 1973 aerosol forms achieved greater than 80% market share. The huge success of aerosols sparked interest in many countries. Currently, approximately two thirds of the UK antiperspirant and deodorant market is in the form of aerosols [5].

The mid-1970s in the United States can be remembered as an era of growing environmental safety awareness which hit hard on aerosol compositions. Both fully halogenated chlorofluorocarbon propellants and aluminum–zirconium complexes were prohibited from use in aerosols and sprays leading to the fastest market decline in an antiperspirant/deodorant category ever witnessed.

The greatest formulary ingredient ever recognized for esthetic improvements came about at the same time that aerosols were under fire. The generic name for this miracle additive was cyclomethicone. It possessed high esthetic and lubricative properties which led to rapid technological advancements in suspension roll-ons and sticks. These anhydrous suspension systems have become consumer favorites and today more than 80% of all antiperspirant formulations contain cyclomethicone. In Europe and other parts of the

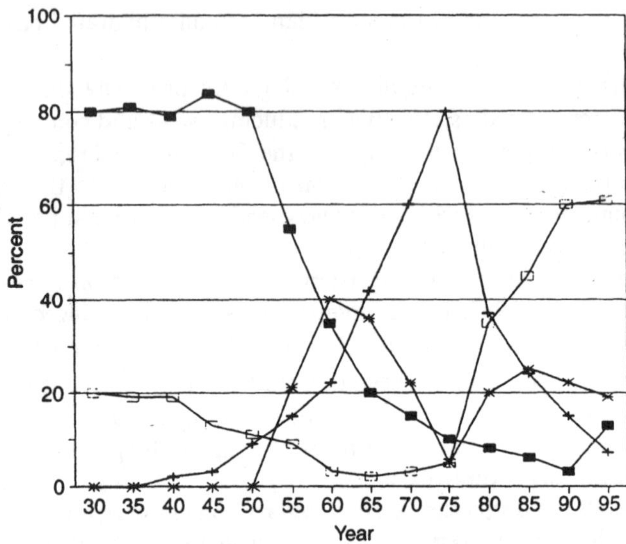

Figure 10.1 A/P-DEO product form market history. US market = 1.8 billion \$ = 80% AP + 20% DEO. Key: —■— semisolids, —+— sprays, —✳— liquids, —⊟— solids.

world, aqueous based roll-ons and sprays dominate suspensoid stick compositions. The existing antiperspirant/deodorant market comprises roll-ons, sticks, aerosols, gels and creams. Since 1985 consumers have had the choice of conventional and enhanced performance types. The US market history of antiperspirant and deodorants is graphically illustrated in Figure 10.1.

10.2 Regulations

Unlike most cosmetic products, antiperspirants have been regulated more stringently by the US Food and Drug Administration, both with respect to the active raw materials and the finished product compositions. Up until 1938 there were no regulations relative to cosmetic materials. The Food Drug and Cosmetic Act which was enacted in 1938 by the federal government clearly stipulated that deodorants which did not alter a bodily function were viewed as cosmetics but antiperspirants which affected the operation of the sweat glands were considered drugs. This physiological concept is interpreted in basically the same way today throughout the world, however labeling and testing requirements for cosmetic drugs do vary from country to country. The

United States has initiated the most specific labeling requirements and Canada probably leads in the most conservative testing programs.

Antiperspirants are classified as over-the-counter (OTC) drugs and deodorants are designated as cosmetic products. The regulatory control tightens specifically when claims are made by the manufacturers of such products.

The proposed rules for classifying OTC drugs were drawn up in the Federal Register in 1972 by the FDA. The last final tentative drug monograph CFR 21 part 350 was published in 1982 and since then it has been regarded as the only document to spell out rules regarding safety, clinical testing and active categorization. Labeling of OTC drugs is discussed in CFR 21 part 201.

In addition to the tentative final document, Category I antiperspirant actives which were effective in January 1995 were adopted by United States Pharmacopeia (USP) as drug compounds, assuming identity standards and test methods consistent with USP requirements. Currently, additions and modifications are underway to provide full coverage of Category I actives. The specifications will include the total range of elemental atomic ratios, solvents and concentrations. The changes are consistent with the industries use or intended use within the specifications already permitted in CFR 21 part 350. For commercial considerations, only those actives listed in Category I should be of interest to the formulator as shown in Table 10.1.

Table 10.1 US FDA over-the-counter Category I antiperspirant actives

	Atomic ratio range	
Active ingredient	Metal chloride	Al:Zr
Aluminum chlorohydrate	2.1–1.91:1	—
Aluminum dichlorohydrate	1.25–0.9:1	—
Aluminum sesquichlorohydrate	1.9–1.26:1	—
Aluminum chlorohydrex PG	2.1–1.91:1	—
Aluminum dichlorohydrex PG	1.25–0.9:1	—
Aluminum sesquichlorohydrex PG	1.9–1.26:1	—
Aluminum chlorohydrex PEG	2.1–1.91:1	—
Aluminum dichlorohydrex PEG	1.25–0.9:1	—
Aluminum sesquichlorohydrex PEG	1.9–1.26:1	—
Aluminum-zirconium trichlorohydrate	2.1–1.51:1	2.0–5.99:1
Aluminum-zirconium tetrachlorohydrate	1.5–0.9:1	2.0–5.99:1
Aluminum-zirconium pentachlorohydrate	2.1–1.51:1	6.0–10.0:1
Aluminum-zirconium octachlorohydrate	1.5–0.9:1	6.0–10.0:1
Aluminum-zirconium trichlorohydrex GLY	2.1–1.51:1	2.0–5.99:1
Aluminum-zirconium tetrachlorohydrex GLY	1.5–0.9:1	2.0–5.99:1
Aluminum-zirconium pentachlorohydrex GLY	2.1–1.51:1	6.0–10.0:1
Aluminum-zirconium octachlorohydrex GLY	1.5–0.9:1	6.0–10.0:1
Aluminum chloride	—	—
Aluminum sulfate buffered	—	—

10.3 Mechanism of sweating

Antiperspirants are intended for use in the axilla region of the human anatomy. This area contains aprocrine, eccrine and sebaceous glands. These glands generate fluids and chemical substances which lead to the development of body odor. The eccrine glands average about 200 per square cm and produce the majority of sweat. The eccrine secreted fluid is composed of a hypotonic solution of sodium chloride, urea, lactates and other metabolic wastes. The sebaceous gland is a sub-surface gland evidenced by a hair follicle. It excretes lipids and fatty acid substances which are responsible for the generation of coryneform bacteria. The apocrine gland, also a sub-surface gland with a protruding hair follicle secretes lipids, cholesterol and steroids. The biological degradation of certain steroids by coryneform bacteria is mainly responsible for the development of odiferous compounds. The total mechanism is still not fully understood.

The production of sweat by the eccrine glands is initiated by both emotional, thermal and sensory stimuli. The actual mechanism is somewhat complicated in the sense that water is not just pumped through the sweat duct to the surface. The transfer of sweat fluid results from the enzymatic degradation of Na^+/K^+-ATPase within the basal membrane. The movement of Na^+ ions across the luminal membrane in the secretory coil causes an osmotic pressure gradient. The relief of this gradient condition is accomplished by the movement of water across the cells of the sweat gland secretory coil into the lumen. This movement of water continues under either stimuli, increasing luminal hydrostatic pressure. The excess fluid carries over to the absorptive duct in the secondary coil and proceeds to the pore opening on the skin.

There are a number of theories about how antiperspirants affect the sweat gland to diminish the flow of fluid. One of the earliest investigations by Shelly and Horvath [6] suggested that protein plugs were created when aluminum chloride was used to treat the axilla. This was challenged as being more of an injurious effect causing a cell transformation to restrict flow. Papa and Kligman [7] theorized that aluminum chloride damaged the sweat duct causing cell erosion, thus allowing for sweat to leak from the inner duct into interstitial spaces rather than flow to the skin surface. Others postulated that the pores swell shut when contacted by antiperspirant materials.

The most respected theory which has been substantiated by much scientific evidence is the hydroxide plug development by aluminum and other metallic salts. This theory best coincides with the hydrolysis chemistry of aluminum chloride and its basic salt complexes. These complexes are polymeric in nature and possess a high degree of cationic charge relative to the average molecular weight of the polymer distribution. Aluminum chloride has the highest charge and lowest molecular weight compared to the lower charges of higher molecular weight basic aluminum halides. Table 10.2 illustrates the charge-to-weight relationship in polymer development. One should

Table 10.2 Charge/mass ratios for Al polymers with formula: $[Al_n(OH)_{2n-2}(H_2O)_{2n+4}]^{n+2}$, where charge/mass $= n+2/3n$

n Al	Formula	$n+2/3n$ Decreasing charge/mass ratios
Straight chains		
1 Al	$Al(H_2O)^{3+}$	3/3
2 Al	$Al_2(OH)_2(H_2O)_8{}^{4+}$	4/6
3 Al	$Al_3(OH)_4(H_2O)_{10}{}^{5+}$	5/9
4 Al	$Al_4(OH)_6(H_2O)_{12}{}^{6+}$	6/12
5 Al	$Al_5(OH)_8(H_2O)_{14}{}^{7+}$	7/15
6 Al	$Al_6(OH)_{10}(H_2O)_{16}{}^{8+}$	8/18
Cyclic		
6 Al	$Al_6(OH)_{12}(H_2O)_{14}{}^{6+}$	6/18

conclude that the smallest polymer size species with high charge will enter the sweat duct by some migratory movement perhaps assisted by charge attraction. During this journey in a very diluted state, neutralization followed by hydrolysis occurs causing the hydroxide plug to form in the intracorneal and intraepidermal ducts thereby restricting flow. These plugs in fact have been confirmed using electron microscopy and aluminum fluorescence techniques to detect the presence of the gel.

Studies by Quatrale [8] have been conducted on the antiperspirant action of aluminum chloride, aluminum chlorohydrate (ACH) and aluminum zirconium glycinate (AZG) during application. By performing cellophane stripping of the applied axilla region it was demonstrated that hydroxide plugs from aluminum chloride were found to be the deepest and most difficult to remove. The restoration of the affected sweat glands back to normalcy was slow. ACH hydroxide plugs were next easiest to restore followed by AZG. One would have expected AZG hydroxide plugged glands to be deeper and restore slower than ACH because AZG typically demonstrates 40% relative higher sweat reduction than ACH. It should also be recognized, however that under dilute conditions as would be the case in underarm applications, polymer growth for most AZG compounds exceeds that for ACH. This faster, more extensive superficial hydrolysis at an earlier stage of migration into the duct makes AZG an effective antiperspirant.

10.4 Antiperspirant active properties

10.4.1 *Basic aluminum chloride*

The basic aluminum chlorides (BAC) defined in the FDA-OTC monograph include aluminum dichlorohydrate (ADC), aluminum sesquichlorohydrate

(ASC) and aluminum chlorohydrate (ACH). Aluminum chlorohydrate is manufactured from aluminum metal, water, hydrochloric acid and/or aluminum chloride according to the following chemical equations of synthesis:

$$10Al + 2AlCl_3 + 30H_2O \longrightarrow 6Al_2(OH)_5Cl + 15H_2$$
$$2Al + HCl + 5H_2O \longrightarrow Al_2(OH)_5Cl + 3H_2$$
$$2Al(OH)_3 + HCl \longrightarrow Al_2(OH)_5Cl + H_2O$$
$$4AlCl_3 + 10H_2O \longrightarrow 2Al_2(OH)_5Cl + 5Cl_2 + 5H_2$$

Quite often ACH will also be referred to as 5/6 basic aluminum chloride which pertains to the fully basic most polymerized form of the salt. Synonymous names for ACH that industry has become familiarized with are aluminum chlorhydrate, aluminum chlorohydroxide and aluminum hydroxy chloride.

ASC and ADC are more acidic, less polymerized forms of basic aluminum chloride. They may be prepared by using a stoichiometric excess of acid according to the following equation:

$$6Al_2(OH)_5Cl + 2HCl \longrightarrow 6Al_2(OH)_{4.666}Cl_{1.333} + 2H_2O$$

From a commercial perspective, ACH and other basic aluminum salts prior to 1980 were not polymerically or clinically differentiated. It was discovered that polymer alterations could boost sweat reduction performance. The performance benefits were first disclosed by Gosling in a 1977 Canadian patent application [9].

Currently there are numerous polymeric versions of ACH and AZG with enhanced or activated properties. Most types of the enhanced salts are offered in dry powder form, but there are aqueous and propylene glycol complexes that are stable as well.

It is advisable for the formulator to have a polymer fingerprint for each particular batch of an active to ensure that the physical properties are consistent with those used during the developmental stages of formulation. A chromatographic analysis of an enhanced active is the only means of verifying potential performance. Figure 10.2(a)–(f) illustrates various polymer distributions of the aluminum complexes. The formulator should also be familiar with other physical properties of the active found during the research and development stage so that scale-up can be better understood. Process variables can impart changes to the active itself.

The hydrolysis of trivalent aluminum is extremely complicated. The formulator who incorporates these polymer blends should be aware that solvated forms of BAC will in fact reach a different polymeric equilibrium under certain conditions of use. Such conditions would include the dilution of active with polar solvents, freezing, process temperature, time and the use of pH altering additives. Most marketed forms of concentrated basic

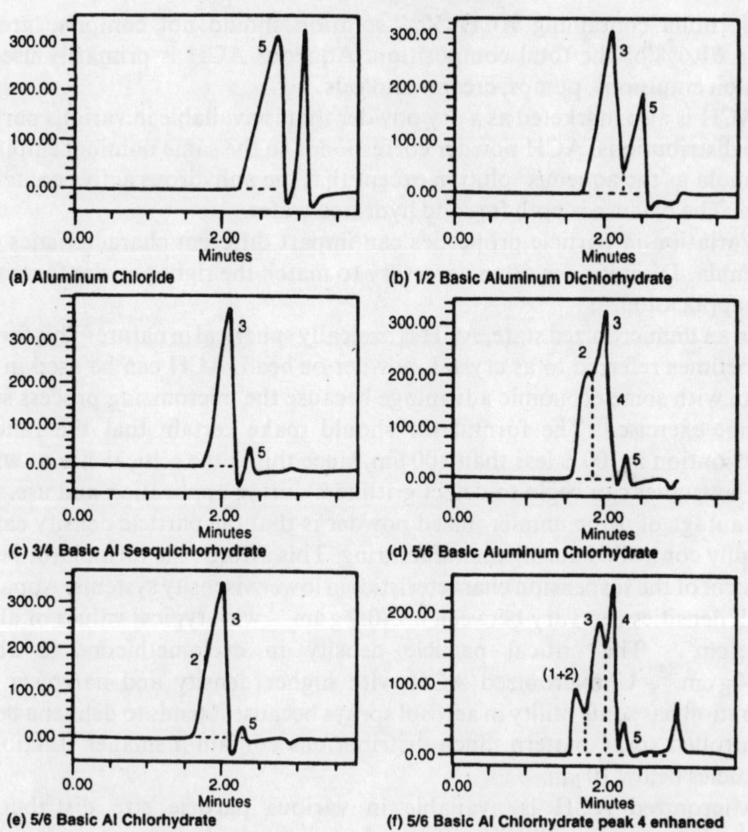

Figure 10.2 Typical polymer distributions of aluminum antiperspirant actives.

aluminum chlorides will show a shift both in the molecular size distribution and structural coordinates during the formulation process. This can lead to reduced efficacy.

Conventional aqueous ACH is the most popular form of BAC, which is normally supplied as a 50% hydrated salt solution. It is fully miscible with anhydrous ethanol, propylene glycol, glycerine and most low molecular weight polyols. The typical polymeric distribution for ACH 50% aqueous solution shows 35% peak 2 polycations of about 8000 g m^{-1}, 60% peak 3 intermediate polycations of about 5000 g m^{-1}, 3% peak 4 Al_{13} polymers of about 3000 g m^{-1} and 2% peak 5 oligomers and monomers of less than 2000 g m^{-1}.

The nominal empirical formula for ACH is $Al_2(OH)_5Cl.2\frac{1}{4}H_2O$ and the anhydrous active content of a 50% solution is 40.5%. BACs are permitted in all types of formulations not to exceed a maximum anhydrous dosage of 25%.

A formula containing ACH 50% solution should not comprise greater than 61.6% of the total composition. Aqueous ACH is primarily used in roll-on emulsions, pumps, creams and gels.

ACH is also marketed as a dry powder that is available in various particle size distributions. ACH powder corresponds to the same nominal empirical formula as the aqueous solution except that the anhydrous active content is 81%. The balance is both free and hydrated water.

Variation in particle properties can impart different characteristics to a formula. Discretion is often necessary to match the right powder form with the application.

In an unmicronized state, ACH is basically spherical in nature. This form is sometimes referred to as crystal, powder or bead. ACH can be used in this form with some economic advantage because the micronizing process stage is not exercised. The formulator should make certain that the random distribution is 100% less than 100 μm, since this is the critical size at which the consumer can begin to detect grittiness during application and use. One advantage of using unmicronized powder is that the particle density can be readily controlled during manufacturing. This affords the formulator better control of the suspension characteristics in lower viscosity systems. Apparent bulk densities can vary between 0.2–0.9 $g\,cm^{-3}$ with typical values of about 0.5 $g\,cm^{-3}$. The critical particle density in cyclomethicone is about 0.24 $g\,cm^{-3}$. Unmicronized ACH with higher density and narrower distribution has some utility in aerosol sprays because it tends to deliver a better controlled spray pattern. Such distributions contain a smaller fraction of particles below 10 μm.

Micronized ACH is available in various particle size distributions depending on the application. The industry standard grade more often than not will be found suitable for use in most suspension systems. The standard particle size specification for controlling distribution is 97% minimum less than 325 mesh. The mean particle size is about 10 μm. Extra fine ACH is sometimes required for ultimate suspension capability. This grade generally contains 80% minimum less than 10 μm, and a mean particle size of 5 μm.

The pH of a 15% ACH solution is 4.3 compared with 3.9 and 3.5 for a nominal metal/chloride ratio ASC and ADC, respectively. Since pH is a logarithmic function of acidity/basicity, it is likely that ADC and the lower range of ASC will show more irritation at the same dosage level as ACH. This is the predominant reason why highly basic ASC and ACH nearly always have an exclusive presence in the marketplace. A basicity index range for antiperspirant actives relative to their metal/chloride specifications is shown in Figure 10.3.

ACH can be polymerically optimized with higher peaks 3 and 4 to impart a greater efficacy potential over the conventional version. These salts may be referred to as enhanced or activated because they can reduce sweat by 30–50% relative to the conventional type. As those experienced in the field

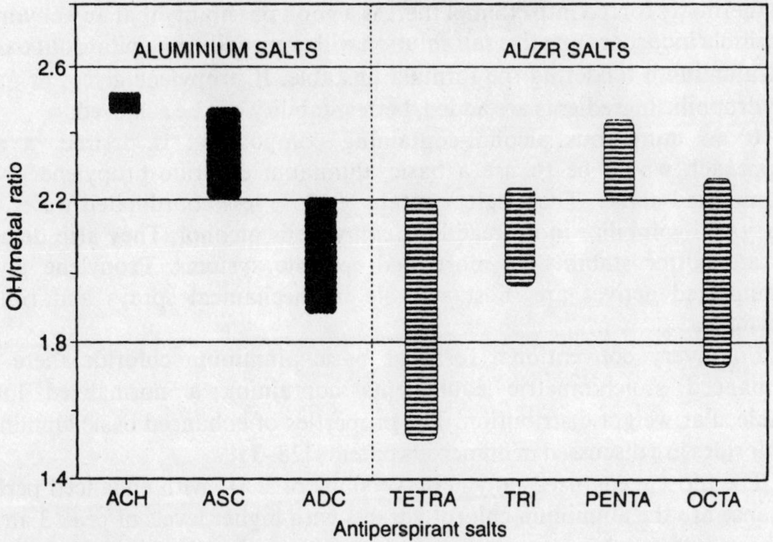

Figure 10.3 Basicity index (OH/metal ratio). OTC mongraph A/P salts.

of antiperspirants have sometimes recognized, other formula ingredients do not always behave in the same way in the presence of enhanced salts as they do with conventional forms. These interactions can impart rheological changes to the formula as well as diminish the potential difference of the anticipated clinical result. It is the writers contention that some non-ionic surfactants, in particular those with strong emulsifying characteristics, will selectivity seek out the most available coordinated water in anhydrous formulations. This water can be found in the lower molecular weight bridged polycations of the enhanced polymer group. Once this bound water is drawn into phase equilibrium with the emulsifier, migratory movement of the enhanced species to the sweat duct is impaired.

Solubility restrictions are also encountered by the encapsulating effects of fatty alcohols, waxes and emulsifiers. A partial or delayed release of active over time leads to lower efficacy. Ingredient interaction with enhanced salts has plagued formulators for nearly two decades. It is advisable to conduct an in-house clinical evaluation first whenever enhanced salts are being considered for a performance geared project.

Basic aluminum chloride powders demonstrate varying degrees of solubility in anhydrous ethanol and propylene glycol. The degree of solubility depends on the amount of hydrated water available in the complex as well as the temperature of the dissolution step. As a general rule ACH containing 47% aluminum oxide will dissolve in 180 proof ethanol. ASC and ADC will dissolve in 190 proof ethanol and 200 proof, respectively. If the salt is thermally forced into ethanol there is a good possibility that an anhydrous formula incorporating this salt solution will eventually precipitate out oxides of aluminum rendering the formula unstable. If propylene glycol or other hydrophilic ingredients are added, better stability can be achieved.

If an anhydrous alcohol-containing composition is desired, a safe approach would be to use a basic aluminum chloride propylene glycol complexed active. These salts contain 50–75% less coordinated water, yet they will solubilize more readily in anhydrous alcohol. They also demonstrate better stability in more hydrophobic systems. Propylene glycol complexed actives are most suitable in mechanical sprays and roll-on formulas.

For every conventional form of basic aluminum chloride there are enhanced stoichiometric equivalents containing a normalized lower molecular weight distribution. The properties of enhanced basic aluminum chlorides are discussed in numerous patents [28–35].

The most recognized polymeric versions of BAC with enhanced performance are the aluminum chlorohydrates with higher levels of peak 3 and 4. These salts contain an intermediate polymer distribution with reduced amounts of peaks 1 and/or 2 polycations. Higher efficacy has been associated with smaller polymer size and increasing cationic charge. It is clearly logical to anticipate that greater amounts of peak 4 species present in BAC will demonstrate higher sweat reduction performance than peak 2 polymer-enriched conventional actives. It is also logical to assume that peak 3 enriched BAC will also demonstrate better performance than peak 2 enriched conventional actives although less than peak 4 enriched BAC. Al^{27} NMR studies reveal that 5/6 basic aluminum chloride containing excess peak 4 can undergo structural change to diminish the Al_{13} polymer-activated species without changing the polymer content of peak 4. For this reason gel permeation chromatography techniques are not always adequate for the evaluation of enhanced active polymer properties. The technique however is useful as a quality control method to determine if the specified polymers are present.

Conditions of very high basicity, low concentration, high temperature and extended time can have a degradative effect on the preferred Al_{13} polymer with the subsequent generation of lower charged peak 1 polymers. In today's competitive environment, the demand exists for various physical polymeric forms of antiperspirant compositions.

Enhanced BAC can be manufactured in particle forms similar to those discussed for conventional BAC. The predominant use of peak 4 enhanced BAC is in anhydrous formulations. These salts lack peak 4 stability in aqueous media. Formulas containing glycol solubilized peak 4 enhanced actives with less than 5% added water will result in substantial peak 4 preservation. Peak 3 enriched actives demonstrate aqueous and hydro-alcoholic polymer stability. Every cosmetic chemist dedicated to this field should have a basic understanding of the polymer network of the active raw material ingredients.

10.4.2 *Aluminum zirconium complexes*

The aluminum zirconium complexes (AZC) defined in the FDA-OTC monograph include aluminum zirconium tetrachlorohyd*rate* (*drex-gly*), aluminum zirconium trichlorohyd*rate* (drex-gly), aluminum zirconium octachlorohyd*rate* (drex-gly) and aluminum zirconium pentachlorohyd*rate* (drex-gly). All AZC are comprised of a basic aluminum chloride and a zirconium chloride polymer which are combined in aqueous media with or without glycine to form Al–Zr oxo bridged complex polymer structures.

The following empirical equations represent some starting materials which when combined in aqueous media form various basic aluminum zirconium active complexes, ZC is zirconium oxychloride and ZHC is zirconium hydroxychloride.

$$2Al_2(OH)_5Cl + \underset{ZC}{ZrOCl_2} \xrightarrow{aq.} \underset{\text{Al/Zr tetrachlorohydrate}}{Al_4(OH)_{10}Cl_2 \cdot ZrCl_2}$$

$$2Al_2(OH)_5Cl + ZrOCl_2\ NH_2CH_2COOH \xrightarrow{aq.} \underset{\text{Al/Zr tetrachlorohydrex-Gly}}{[Al_4(OH)_{10}Cl_2 \cdot ZrOCl_2]\ NH_2CH_2COOH}$$

$$2Al_2(OH)_5Cl + \underset{ZHC}{Zr(OH)Cl} \xrightarrow{aq.} \underset{\text{Al/Zr trichlorohydrate}}{Al_4(OH)_{10}Cl_2 \cdot Zr(OH)Cl}$$

$$2Al_2(OH)_5 + Zr(OH)Cl + NH_2CH_2COOH \xrightarrow{aq.} \underset{\text{Al/Zr trichlorohydrex-Gly}}{[Al_4(OH)_{10}Cl_2 \cdot Zr(OH)Cl]\ NH_2CH_2COOH}$$

$$4Al_2(OH)_{4.5}Cl_{1.5} + Zr(OH)Cl \xrightarrow{aq.} \underset{\text{Al/Zr octachlorohydrate}}{Al_8(OH)_{18}Cl_6 \cdot Zr(OH)Cl}$$

$$4[Al_2(OH)_{4.5}Cl_{1.5}] \cdot NH_2CH_2COOH + Zr(OH)Cl \xrightarrow{aq.} \underset{\text{Al/Zr octachlorohydrex-Gly}}{[Al_8(OH)_{18}Cl_6 \cdot Zr(OH)Cl] \cdot NH_2CH_2COOH}$$

$$4Al_2(OH)_5Cl + Zr(OH)Cl \xrightarrow{aq.} \underset{\text{Al/Zr pentachlorohydrate}}{Al_8(OH)_{20}Cl_6 \cdot Zr(OH)Cl}$$

$$4[Al_2(OH)_5Cl] \cdot NH_2CH_2COOH + Zr(OH)Cl \xrightarrow{aq.} \underset{\text{Al/Zr pentachlorohydrex-Gly}}{[Al_8(OH)_{20}Cl_4 \cdot Zr(OH)Cl] \cdot NH_2CH_2COOH}$$

The incorporation of glycine in any portion of the aqueous synthesis will form the aluminum zirconium chlorohydrex-gly salt. Aluminum zirconium polynuclear complexes are extremely complicated in their structure and the methods of their manufacture play a key role in their physical molecular makeup.

The most popular aluminum zirconium complexes are the aluminum zirconium tetra- and tri-chlorohydrex-gly. The industry standard for the aluminum/zirconium atomic ratio is about 3.5:1. The metal/chloride atomic ratios however are significantly different, the tetrachlorodrex-gly being 1.4:1 and trichlorohydrex-gly being 1.7:1. The metal/chloride atomic ratio is a critical stoichiometric characteristic of the formulation because this ratio dictates the structural properties of the aluminum zirconium oxobridged coordinates. Overall the higher metal/chloride ratio actives are more formulator friendly for several reasons. The pH is higher, in this case, 4.0 (15% w/w aq. at 25°C) for the trichlorohydrex-gly and 3.8 for the tetra-chlorohydrex-gly.

Chromatographic analysis reveals that the free acid content of trichlorohydrex-gly is roughly half that of tetrachlorohydrex-gly. Acidity is an inherent disadvantage for antiperspirant actives in general because it limits the number of compatible solidification agents, gel stabilizers and thickeners that can be incorporated into antiperspirant compositions. In essence the pH window for these products is 3.5–5.0 compared with a neutral–mildly alkaline environment for deodorant preparations.

Another possible advantage of tri- over tetra-chlorohydrex-gly in anhydrous formulations is that the amount of coordinated water that is present under comparable conditions of active manufacture is less for tri- than for tetra-chlorohydrex-gly. The differences arise from the inverse proportional relationship between lower coordinated water and increasing basicity. Greater amounts of glycine introduced into the active followed by drying also lead to lower coordinated water as a result of direct substitution by glycine in the complex. It cannot be overemphasized that the active component in an anhydrous recipe will contribute the greatest amount of inherent water to the formula. This can make or break the objective stability goals for anhydrous semi-solid and solid products.

Another consideration when choosing the active type is the effect of stability on the fragrance in the formula. The amount of free acid whether in the form of hexa-aquo aluminum or hydrochloric acid will act unfavorably toward certain unsaturated fragrance compounds. Generally fragrance-based ketones, esters and nitriles exhibit good stability. Aldehydes and alcohols with moderate unsaturated components are not as stable in acidic media and at elevated process temperatures.

Since 1985 aluminum zirconium chlorohydrex-gly powders have been available in polymerically optimized forms, giving the formulator active performance options. Analogous to the basic aluminum chlorides, for any given stoichiometric arrangement of aluminum zirconium glycinate complex, there are numerous polymeric versions of the salt. Figure 10.4 (a) – (h) illustrates typical polymer variations in Al/Zr complexes.

The conventional aluminum zirconium glycinate typically contains 10% peak 2 polycations, 65% peak 3 intermediate polycations, 5% peak 4 Al_{13}

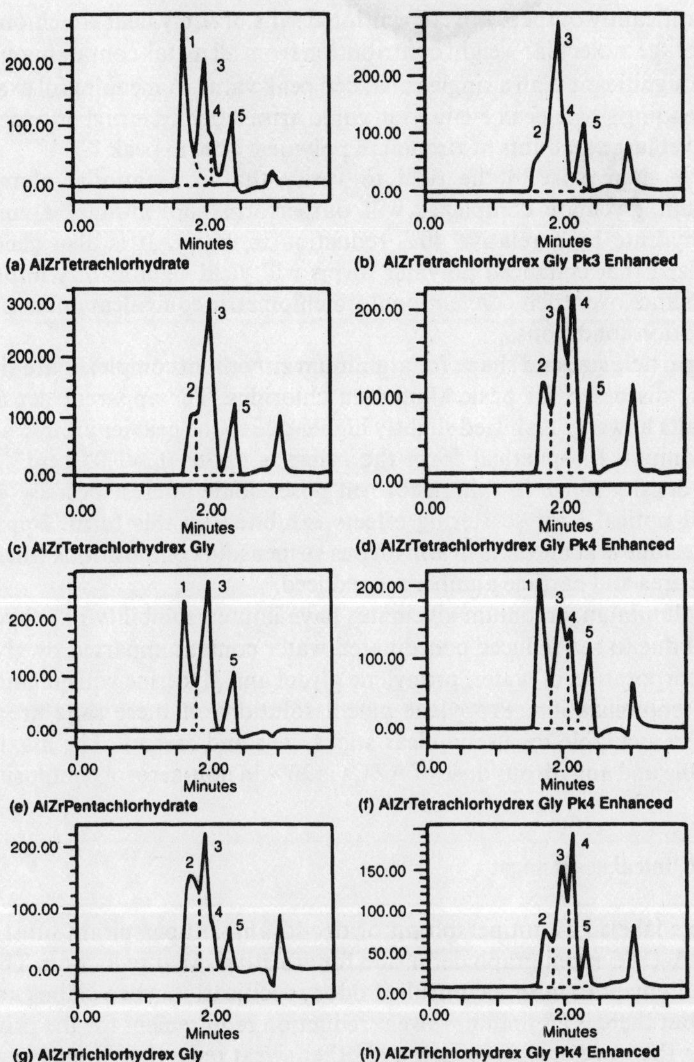

Figure 10.4 Typical polymer distributions of aluminum / zirconium A / P actives.

polymer and about 20% monomer and oligomers. Peak 4 enhanced aluminum zirconium glycinates typically contain 10% peak 1 large zirconium polymers, 40% peak 3 species, 40% peak 4 species and 10% peak 5 species. Peak 3 enhanced aluminum zirconium glycinates typically exhibit 10% peak 1, 65% peak 3, 5% peak 4 and 20% peak 5.

Abiding by the concept that lower molecular weight polymeric distributions are increasingly effective given non-interfering effects of other formula ingredients, one can assume that peak 4 and peak 3 enhanced salts will significantly outperform conventional salts of equivalent stoichiometry. The average molecular weight contribution from all metal-containing species is more significant than a single enhanced peak value. A meaningful example of misleading single peak evaluation could arise if peak 4 is higher relative to peak 3 yet large amounts of zirconium polymers exist in peak 1.

Ample data exist in the field to verify that conventional aluminum zirconium glycinate complexes will outperform conventional aluminum chlorohydrate by a relative 40% reduction in sweat. It is also generally recognized that enhanced polymer forms will yield significantly improved performance over their conventional stoichiometric equivalent under proper formulation conditions.

The particle size and shape for aluminum zirconium complexes are similar to those discussed for basic aluminum chlorides. The apparent density of these salts however is shifted slightly higher due to the heavier atomic weight of zirconium. In spherical form the range is about 0.3–1.05 $g\,cm^{-3}$. The higher density range in spherical form poses some interest because of the reduced optical light scattering effects exhibited by this form. Improved lower residue is achievable in anhydrous suspensoid compositions where the surface area and particle number are reduced.

The aluminum zirconium glycinates have limited solubility in anhydrous ethanol due to the reduced coordinated water content imparted by glycine. The incorporation of water, propylene glycol and glycerine will enhance the rate of solubilization. Propylene glycol solutions of these salts are commercially available for use in clear sticks, gels and creams. The maximum allowable and anhydrous dose of AZCs is 20% in non-aerosol compositions.

10.5 Clinical assessment

Products labeled as antiperspirant or deodorant–antiperspirant must meet the FDA-OTC monograph definition for effective sweat reduction. There is no requirement to meet a controlled odor specification when claims are not made, but there is a minimum sweat reduction requirement for the privilege of using the word antiperspirant and other sweat reduction language which appears on the label.

The term deodorant is an inherent property of antiperspirant products and it may appear on the label as well. The minimum efficacy requirement for antiperspirants is 20% sweat reduction in 50% of the users as demonstrated under specific test guidelines. Sweat rate evaluations dating back to the 1920s were essentially visualization techniques using starch and indicators to identify the quantity of sweat droplet development. Some sensory techniques were also developed by researchers such as thermographic and gravimetric analysis, however this work has remained as a laboratory curiosity and has not been adopted as a standard clinical procedure.

The modern clinical efficacy test accepted by both government and industry is the gravimetric measurement technique that was developed after 1950 and has since been perfected. The most recent method which has been adopted with minimal modifications is described by Majors and Wild [10]. Thorough analytical efficacy assessment is discussed by Murphy and Levine [11].

The regulatory guidelines set forth by the OTC drug review panel and adopted into the 1982 tentative final monograph are the current procedural test requirements used to qualify antiperspirant drug products. These rules are:

1. Panelist screening to obtain representative subjects.
2. 17 day antiperspirant and deodorant abstinence by test subjects prior to sweat study.
3. (a) Hot room conditions of 100 ± 2°F, relative humidity of 35–40% or
3. (b) Ambient environment allowing daily routines for subjects during collection period.
4. Use of control test formula for each subject, equally but randomly dividing subjects so that they receive equal but opposite axilla treatment following reversed axilla treatment of the other formula.
5. The application must represent a typical normal dose.
6. A 40 minute warm-up period is required.
7. Two 20 minute collection periods are made using pre-weighed cotton pads with gravimetric collection during one of the two periods.

If no pretreatment evaluation is made it is necessary to make a right to left sweat ratio adjustment by dividing the raw milligram moisture measurement of the right axilla. The sweat reduction calculation is made using the Wilcoxon rank sum test method. An error is not allowed in more than 5% of the cases where a 20% minimum number is not achieved.

Clinical efficacy studies are costly to run and generally are instituted when new or modified formulas are encountered. It is to a researcher's benefit to have a budget sufficient to use these studies in developmental screening of trial and error bench preparations. It is helpful to consult experts in the industry who administer such tests routinely so that the objectives can be met.

The application of antiperspirant composition to the skin is generally recognized as safe with few negative side effects. As is the case with any dermatological product, safety pre-screening new or modified formulas is essential taking into account normal mode of use and possible misuse. Potential side effects to consider might be dermal irritancy and contact sensitivity.

Dermal irritation of aluminum and aluminum zirconium complexes can vary considerably depending on the basicity of the complex. The maximum allowable dose defined in the OTC drug monograph is a built-in safety factor to help prevent adverse reactions. Active complexes with metal:chloride atomic ratios of greater than 1.3:1 have a long reputation of being dermalogically compatible. Unlike the aluminum chloride and aluminum sulfate precursors, they have less affinity for and reaction with the epidermal keratin of the skin. Formulations containing Category I actives and cosmetic ingredients with non-primary irritation scores are generally subjected to a dermal evaluation on the axillary region of human subjects. A more conservative approach might involve their backs using an occlusive or semi-occlusive patch test. In every instance a control sample is used.

Antiperspirants and deodorants are regarded as being non-sensitizing. Other components in the formula such as perfumes and oils are more likely to be sensitizing agents and to be photo-sensitizing. Comedogenicity and irritation of skin-care ingredients is discussed by Fulton [12]. A re-evaluation is most advisable if ingredient changes are made. It should be noted that there will always be a fair percentage of consumers who will abstain from antiperspirant use entirely because of irritation effects. A safety assessment of aluminum compounds is described by Lansdown [13].

10.6 Formulary considerations

10.6.1 *Performance*

Antiperspirant products are strictly designed to control wetness and odor in the axilla region of the body. Not all finished products will deliver the same degree of wetness protection, so there should be some target goal before commencing a project. Odor control is less of a concern because antiperspirants are good deodorants. As a conservative measure, sometimes an anti-microbial agent will be employed in small quantities for insurance purposes.

In nearly every instance, formulators design the antiperspirant product to deliver a 25% or better sweat reduction if it is to remain within the competitive performance arena of current products. Efficacy is diminished to some extent by hydrophobic non-volatile ingredients. Given proper selection and quantity used, 25% reduction should be achievable. The active concentration in the composition should be at least 12% calculated on an anhydrous non-buffered basis. The customary range of anhydrous active in non-aerosol compositions is 15–20% and 7–10% for aerosols. The delivery dose to the axilla under normal application should be between 200–500 mg. If the dose is too small, performance objectives might not be met. Product stability also meets a greater challenge due to extended shelf-life.

As a general rule, a conventional active in an aqueous-based formulation will yield better clinical performance over a silicone-based liquid suspension which in turn will edge out wax/silicone-based suspensoid solids. The reason for this order is solely due to the rate of availability of the active in a

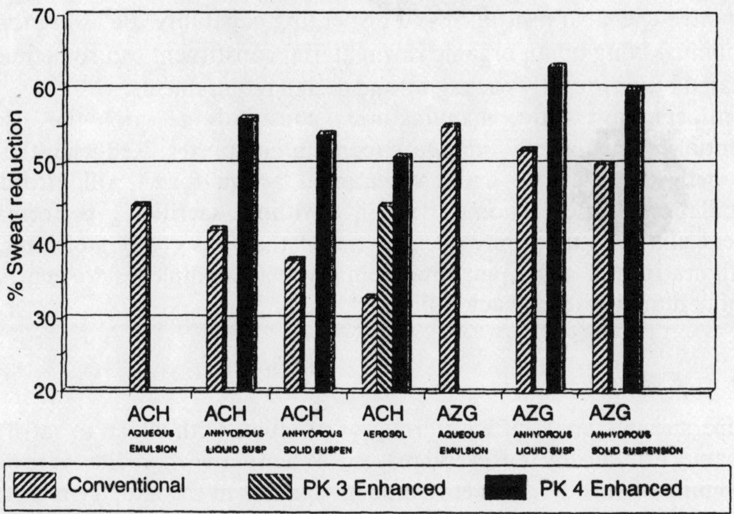

Figure 10.5 Clinical efficacy expectations of A/P actives in specified drug form.

solubilized state after application. Aerosols are the weakest contender because of the failure of the spray to deliver the full dose to the axilla. Enhanced ACH is often employed to boost performance in aerosols to help offset the delivery problem. Figure 10.5 [26] provides a guide to clinical expectations for the customary active dosage in specified drug form.

10.6.2 *Cost*

Antiperspirants can be formulated from hundreds of ingredient choices. It will be the responsibility of researchers and marketers to weigh the cost benefit ratio of expensive ingredients in terms of their esthetic or product integrity contribution.

The esthetic demands of the United States consumer run high and this has driven the total raw material cost higher. Anhydrous product forms are more expensive, where water has been replaced by silicones to improve feel. In other parts of the world, esthetics is also important, but there is greater tolerance of aqueous compositions. A cost/esthetic balance is also achieved where oil-in-water or water-in-oil emulsions are used.

The benefits of aqueous emulsions are that they are easy to prepare, they comprise a less expensive ingredient mix and they generally meet the efficacy criteria. They will always survive because they provide an economical alternative to the consumer. If the formulator wants total performance esthetic benefits, then the anhydrous enhanced active compositions are preferable.

A large cost of antiperspirant/deodorants can be the package container itself. Container and ingredient compatibility must be studied carefully to guarantee chemical resistance and dispensing capability. Be aware that an economic saving on an organic raw material constituent can sometimes be defeated by more costly packaging and design requirements.

Antiperspirant active manufacturers constantly research the efficacy potential of enhanced aluminum zirconium complexes. Reducing the cost gap between enhanced and conventional active forms will afford the formulations some economic flexibility without sacrificing performance. Recent clinical studies have demonstrated that less costly stoichiometric configurations of aluminum zirconium glycinate complexes are achievable without diminishing efficacy [14].

10.6.3 *Esthetics*

Antiperspirants are a difficult group of products with which to satisfy the consumer esthetically for several reasons. When active ingredients are placed in a limited hydrodynamic environment such as in the axillary region, the active will exhibit some degree of tack. Emollients with lasting properties can easily offset tack, but there are limitations to the amounts that can be employed in the formula. Excessive amounts will inhibit the solvation rate of powdered actives, thereby interfering with the pathway of the active into the viable epidermis region of the sweat duct.

Volatile silicones have tremendously improved the esthetic quality [15] without negative performance drawbacks. They carry the ingredients to the skin with a dry feel and good spreadability. Their high vapor pressure allows them to depart quickly leaving behind the active and smaller amounts of less volatile oils. If tack is experienced in an underarm application for more than several minutes it should be minimized by formula adjustments, as it will probably have negative consequences for the consumers.

Aqueous-based products will have more pronounced tack profiles than anhydrous suspension products. With the proper emollient balance, they will generally not draw strong criticism. Large amounts of humectants such as glycols and ethoxylated derivatives can also impart a long tackier drying process.

Good spreadability is an important attribute in antiperspirant applications because thick films adhering to the skin and hair follicles will give the perception of greasiness. Smooth glide is another application benefit because women in particular will associate glide with elegance. Hard, high, particulate-containing suspensoid sticks sometimes possess more drag due to the increased levels of stearyl alcohol and suspending agents. Formulas with poor glide may also have an irritating effect on the skin due to the frictional effects of solid particulates.

The underarm is a sensitive area and fast evaporation from water and alcohol can cause an undesired cold sensation during the drying process. In such cases the cooling rate can be lowered with oil phase ingredients.

An esthetic appearance of the contents is always of interest to the consumer. Be aware that high levels of alcohol can evaporate faster than any other ingredient from the system. This could change the product appearance and consistency. Content shrinkage in time is inevitable in solid products containing large amounts of alcohol. Propylene glycol may in part be a good substitute for alcohol in formulas where solvent power is not required.

The color of a formula that can be viewed by the consumer is of significance. Opaque compositions should be white or off white, while clear formulas should be as colorless as possible. A yellow appearance, particularly against a white package, is subconsciously viewed as poor quality although the functionality of the product could still be the same.

Product clarity is the most recent marketing campaign used by antiperspirant deodorant manufacturers. Clear cosmetics express purity and beauty. Clear gels have been fairly successful in promoting the no product residue concept. The flaky or after whitening residue is a wide concern addressed mostly by the female antiperspirant user. Low residue products are quickly becoming an important segment of most product lines. Several formulation approaches will lead to reduced residue. These include microemulsion and suspension technology.

A functional esthetic enhancer in anhydrous formulas comes from the use of encapsulated fragrances that are capable of providing time or demand release of fragrance. Micro-encapsulation is specialized technology and the procedure is often sent out to experts in the field who can provide the proper encapsulate for the formulator. Some common coating materials include sucrose, polysaccharides, polyurea and maltodextrins. Micro-encapsulation is thoroughly discussed elsewhere [16, 27].

10.7 Formulations

10.7.1 *Roll-on products*

The majority of roll-on products in the US and UK markets are antiperspirant/deodorants. Both aqueous- and silicone-based preparations are quite popular. The predominant forms in the US are silicone emulsions and suspensions whereas aqueous emulsions take precedence throughout Europe and other countries. Aqueous versions usually contain greater than 50% water and incorporate a thickening agent such as magnesium aluminum silicate, carbomer, hydroxypropyl cellulose or hydroxyethyl, cellulose. Other components might include 2–20% emollients such as myristyl, palmityl or stearyl esters and emulsifiers such as alkoxylated fatty ethers/esters and glycerol stearate.

Formulas containing ample quantities of ethyl alcohol are regarded as hydroalcoholic systems. Clear stable formulas can be achieved using cellulose polymers. The polymer is often prepared in an aqueous concentrate and deaerated before adding the active and emollients. A pre-solubilized fragrance can be added followed by a final dilution with water. When paraben preservatives are used, some heating will be required to bring them into solution but the main mix need not be heated.

The general principles of emulsion technology [17] apply to water-in-oil (w/o) and oil-in-water (o/w) emulsions. This involves separately heating both phases with miscible surfactant emollients and adding the water phase to the oil phase with stirring until emulsification at a lower temperature is achieved. Homogenization is sometimes needed to reduce the internal phase droplet size to increase emulsion stability. Water-in-oil microemulsion formulas can be clear and require an experimentally fine-tuned hydrophilic/lipophilic balance (HLB) to maximize formula stability. Clear emulsions can also be produced by matching the refractive indices of the two phases at use temperature. This is sophisticated formulation. Two or more non-ionic surface active agents are recommended to accomplish this balance. The viscosity can be regulated by altering the weight ratio of the phases. Oil-in-water emulsions are more common in lotion formulas. There is a greater esthetic advantage of w/o type micro-emulsions because of the dry, lower tack feel upon application. Fragrances are easy to incorporate in the last stage of process because of the solvent power of the external phase. Very stable w/o emulsions have been achieved with organosilicone emulsifiers [18]. The enhanced stability arises from the highly absorptive phase boundaries resulting from the alkyl side chains and the polyglycol groups attached to the silicone polymer.

The recommended emulsification procedure involves adding an ambient temperature water phase to an 80°C oil phase at a 3:1 weight ratio with vigorous mixing. The cooled mixture is lightly homogenized for several minutes.

Dry suspension roll-on compositions have few hydrophilic characteristics because they contain no water. They can be thickened with bentonite clays or high surface area silica to moderate viscosity levels of several thousand centipoise. Between 20 and 25% of powdered antiperspirants are suspended in a base of cyclomethicone or other light oils. Emollients might include diesters, polydecene, dimethicone, alkoxylated ethers and soluble waxes. Dry suspension roll-ons exhibit an excellent dry, non-greasy esthetic quality when applied. High levels of sweat reduction can be achieved with enhanced aluminum zirconium glycinate antiperspirant actives.

Typical formulation of an emulsion roll-on

Part	*Ingredient*	*Wt%*
A	Water	48.0
	Magnesium aluminum silicate	1.5

B	Lanolin alcohol	0.5
	Peg 40 stearate	2.0
	Cetyl alcohol	1.5
	Lanolin	2.0
C	Glycerine	2.0
	Mineral oil	3.0
	Phenyl trimethicone	3.0
D	Fragrance	*q.s.*
E	Aluminum chlorohydrate (50% aq.)	36.50

Procedure: Disperse A using high shear mixing. Heat A to 85°C and add B. When melted add C, reduce heat to 40°C and add D and E.

Typical formulation of a clear micro-emulsion roll-on

Part	*Ingredient*	*Wt%*
A	Polydecene 364	21.0
	Oleth 2	12.0
	Oleth 20	7.0
	Cetyl dimethicone copolyol	4.0
B	Aluminum zirconium tetrachlorohydrex-gly (35% aq.)	49.0
	Dipropylene glycol	5.0
	PPG 14 butyl ether	2.0
C	Fragrance	*q.s.*

Procedure: Heat A and B separately to 85°C. Slowly add A to B while stirring. Continue stirring until cooled to 45°C. Add C.

Typical formulation of a dry suspension roll-on

Part	*Ingredient*	*Wt%*
A	Octamethylcyclotetrasiloxane	40.0
	Decamethylcyclopentasiloxane	25.0
	Quaternium 18 hectorite	2.5
B	Dimethicone (50 cps)	5.2
	SDA 40 alcohol	2.0
C	Fumed silica	0.3
	Aluminum zirconium tetrachlorohydrex-gly	25.0
D	Fragrance	*q.s.*

Procedure: Mix A with a high speed agitator. Add B and continue mixing. Add C and D and continue mixing for 15 minutes. Transfer to a homogenizer and homogenize for 3 minutes.

10.7.2 *Stick products*

A large amount of research time has been dedicated to the solid and semi-solid antiperspirant/deodorant product forms. The introduction of volatile silicones as the base solvent and carrier over two decades ago has led to high consumer loyalty particularly in the United States, Canada and the United Kingdom.

The solid suspensoid antiperspirants are comprised of a blend of waxes, silicone, emollients, couplers and antiperspirant powder. The benefits of solid formulas are those of controlled delivery and good esthetics. They also have a good reputation for exhibiting long lasting chemical stability. The packaging components are manufactured from various plastic materials such as polyethylene, polypropylene or polyethylene terephthalate. It is important to match the chemical compatibility of the plastic used in barrel construction to the oils chosen in the formulation. Some emollients can be aggressive solvents when in contact with plastic particularly on longer term stability. The distortion temperature of plastic should also be above the process fill temperature. The general procedure used in preparing suspensoid solid products is to dissolve the fatty alcohol waxes and couplers in the silicone at elevated temperature. The melting point of the castor wax will generally dictate the working process temperature needed to achieve a homogenous molten state. Sometimes a higher molecular weight wax such as paraffin or behenyl alcohol is used in small quantities to fine tune stick hardness. The materials to be suspended such as the active powder, talc, silica or other fillers are then added. The temperature is lowered to several degrees above the solidification temperature, fragrance is added and the contents are poured. A carefully controlled cooling stage is critical to the final physical appearance of the stick. High viscosity pours will have less cavitation after solidification. It is sometimes helpful before final cool down to subject the stick to a brief period of infrared heating to smooth the top and reduce blemishes.

Stick quality is evaluated by a number of procedures 24–48 hours after solidification. Chemical analysis at various vertical depths is done to assess active homogeneity. Hardness is measured by taking penetration readings using a penetrometer. Sticks designed for the male market are generally harder with readings of 7–10 mm of penetration compared with sticks designed for the female market which penetrate 10–13 mm. Other specifications might apply to horizontal bond strength, color and posting failures.

With some minor chemical adjustments, suspensoid compositions can be formulated with lower residue properties without departing from the solid stick concept. Consumer residue complaints are not all warranted but sometimes are the fault of the user based on how the product was applied. The real residue issue pertains to the amount of white material that occurs on the skin. Hours after application white flakes may fall off onto darker garments without much notice. The residue occurs either from the encapsulating effects of fatty alcohol on the active and other inert ingredients or from the white crystalline active after the volatile silicone has evaporated. Certain emulsifiers that impart creaminess can also prevent the solubilization of active leading to a white protective barrier of residue particulates.

Branched esters and diesters when incorporated into these waxy based formulas can significantly reduce residue properties. Low surface area antiperspirant actives can also reduce residue because of a lower number of encapsulated particles observed. These actives can exhibit reduced light scattering characteristics by virtue of special manufacturing techniques. Softer stick preparations containing C_{16} and C_{18} alcohols are lower in residue than more crystalline C_{18} by itself based on comparable dosage evaluation.

Semi-solid gels and creams have sparked great consumer interest because of their low residue properties and esthetic value. The package design is quite similar in appearance to solid formulas, but delivery through a well designed applicator head becomes an additional package component necessity. Controlling dosage is paramount so that the product can be applied without mess.

Gels are w/o or o/w micro-emulsions and they have the advantage of being very clear if both phases can be combined with matching refractive indices of less than 5×10^{-4} units. The w/o type has a dryer feel and can be prepared in a similar fashion to that discussed previously in roll-on preparations. A thickening agent such as a dimethicone copolyol is useful in bringing the phase to a high viscosity level of at least 50 000 cps. The formulas are quite stable at high viscosities because the semi-solid characteristics help to immobilize the dispersed phase in the external phase.

Aluminum zirconium complexes might be the active of choice in w/o micro-emulsions in order to boost the efficacy while overcoming the encapsulating effects of the external oil phase.

Creams can be of the emulsion type containing about 50% water, the balance consisting of emulsifiers, gel thickeners, emollients and active. They are similar to lotions to prepare except that a higher level of thickening agent is required. An esthetically rewarding feel can be achieved from anhydrous creams that are made by suspending fine particulate active powder in an organic base of volatile emollients, esters, emulsifiers and surface active suspending agents. High surface area clays and silica are useful suspending agents and stabilizers. In anhydrous formulas such as these, very small amounts of polar additives such as glycerine and propylene carbonate can assist in charge development of the surface active particulate.

Solid clear gel antiperspirants represent another approach to preparing no residue products. The composition consists of a gel agent, propylene glycol complexed active, solvent glycols, emollients and couplers. A useful gellant is dibenzyl sorbitol which is dissolved in polar glycols at temperatures of about 80 to 120°C. Some emollients can be added to this phase but care should be taken to maintain a low enough viscosity to maintain the solubility of the gellant. The other phase should contain the active and any remaining ingredients at some lower elevated temperature between 50 and 80°C. A pH adjusting agent is needed to maintain a formula pH of 4.5–5.0 to prevent the dibenzyl sorbitol from degrading in an acidic environment. The final phase blending should be done in a micro-batch or continuous blend sequence with adequate mixing at a temperature of 70–90°C. The solution should be poured promptly in order to avoid gellation in the process line. Solidification time can be 1–20 minutes depending on the viscosity and the amount of gellant. The entire process is more challenging than most, therefore a reliable automated line is recommended in scaling up for production quantities.

Typical formulation of an antiperspirant stick

Part	*Ingredient*	*Wt%*
A	Decamethylcyclopentasiloxane	50.0
	Stearyl alcohol	13.5
	Hydrogenated castor oil (70°C mp)	3.5
	PPG 14 butyl ether	3.5
B	Aluminum zirconium trichlorohydrex-gly (80% < 10μm)	24.0
	Talc	5.5
C	Fragrance	*q.s.*

Procedure: Heat A to 75°C with mixing until clear. Lower temperature to 65°C and add B and disperse. Lower temperature to 56°C, add C and fill at 54°C.

Typical formulation of a low residue antiperspirant stick

Part	*Ingredient*	*Wt%*
A	Decamethylcyclopentasiloxane	41.0
	Diisopropyl adipate	6.0
	Isocetyl octanoate	3.5
	Stearyl alcohol	10.0
	Cetyl alcohol	5.0
	Hydrogenated castor oil (80°C mp)	3.0
	PEG 8 distearate	1.0
	Ceteareth 20	1.0
B	Hydroxylated milk glycerides	1.5
C	Aluminum zirconium tetrachlorohydrex-gly (enhanced)	25.0
	Talc (90% 325 mesh)	1.0
D	Silica	1.0
E	Polysaccharide/fragrance	*q.s.*

Procedure: Heat A to 85°C with mixing until clear, lower temperature to 62°C, add B and dissolve. Add C and disperse. Add D and disperse. Lower temperature to 58°C, add E and fill at 56°C.

Typical formulation of a clear antiperspirant stick

Part	*Ingredient*	*Wt%*
A	Aluminum zirconium pentachlorohydrex-gly 30% pg sol/ with zinc glycinate	44.0
	Glycereth 7 benzoate	2.8
	Glycereth 4.5 lactate	0.9
	PPG 2 isodeceth 4	1.0
	Myristyl lactate	3.3
	Diisopropyl adipate	3.5
A/B (30/70)	Propylene glycol	16.2
A/B (30/70)	Dipropylene glycol	26.0
C	Dibenzyl sorbitol	2.3
D	Fragrance	*q.s.*

Procedure: Using a micro-batch process, heat A to 68°C with mixing. Add A portion of glycols. Separately heat B portion of glycols to 110°C with good mixing and slowly add C by increments while increasing temperature to 125°C until dissolved. Immediately lower to 118°C and add to A. Immediately add D and pour at 78°C within 3 minutes.

Typical formulation of a clear w/o gel stick

Part	*Ingredient*	*Wt%*
A	Cyclomethicone and dimethicone copolyol (DC 3225C)	8.0
	Dimethicone	5.0
	Decamethylcyclopentasiloxane	7.0
	Dioctyl succinate	10.0
B	Aluminum zirconium tetrachlorohydrate (50% sol.)	50.0
	Alcohol, SDA 40B	9.0
	Methyl propanediol	8.0
	Glycerine	3.0
C	Fragrance	*q.s.*

Procedure: Combine A ingredients and mix. Combine B ingredients and mix. Measure refractive index (RI) of each phase at 25°C. Adjust RI of B up or down to match A to within 0.0005 units by adjusting glycerine or water, respectively. Add B to A with gentle mixing. Add C. Homogenize briefly to reach desired thickness.

10.7.3 *Spray products*

Aerosols are a major product category in the UK. They have dramatically declined in popularity in the US due to an environmental campaign to phase out volatile organic carbon (VOC) chemicals, the heart of the propellant system. Due to the lack of suitable non-VOC substitute propellant, research has remained low key in this area, awaiting the final call by the environmental regulatory agency. The California Air Resource Board (CARB) has been instrumental in developing the rules for reducing VOC emissions, where antiperspirant and deodorant products have sustained more criticism than other aerosol product categories (see section 13.8).

Most antiperspirant aerosol formulations are variants of suspensions of active salts in different oil and solvent blends. Aluminum chlorohydrate powders of micronized or controlled particle size grades are predominantly used in spray systems. Both aluminum zirconium complexes and aluminum chloride are prohibited from use based on health concerns stemming from the less controlled delivery form.

Typical ingredients useful in aerosols are isopropyl myristate, isopropyl palmitate, dibutyl phthlate, volatile silicone, dimethicone, silica, suspension clays, propylene carbonate and ethyl alcohol. The propellant is liquified isobutane, or blends of isobutane and propane.

A thick concentrate is prepared by shearing a bentonite or hectorite clay with the emollient in the presence of a polar additive such as ethyl alcohol or propylene carbonate. The active is slowly added and the final blend is screened to remove any agglomerates. The formula is added to a corrosion resistant pressure rated aerosol, and propellant is charged at approximately a 3:1 weight ratio of propellant:concentrate, to satisfy California requirements of 60% maximum high volatility organic compounds (HVOC).

The success behind an antiperspirant aerosol with good delivery characteristics is based on low viscosity and good valve design. The vapor–tap valve and mechanical break up buttons are essential for reduced particle size sprays. The emollient oils used in the formula also act to reduce the bounce effect of the spray hitting the axilla surface in order to land a greater percentage of active at the site. A dispensing rate of 0.08 $g\,s^{-1}$ for 2 s of active should be sufficient to exceed the minimum efficacy requirements of 20% sweat reduction in half the users. The handling of hydrocarbon propellants is hazardous and special precautions must be taken to avoid fire and explosion.

Mechanical pump sprays are also available on the market but are less popular. Pump sprays consist of low viscosity solutions or emulsions of solubilized basic aluminum chloride, volatile silicones, ethyl alcohol, water and various emollients and couplers. The mean diameter droplet size for pump sprays should be between 50–100 μm. This may require inserts ranging in dimension of 0.010″ to 0.014″ × 0.010″ shallow. The output per stroke will be about 120 or 140 μl. The performance of mechanical pump sprays is largely dependent on the design of the pump mechanics. Exhaustive trial and error may be needed to optimize the functional components.

Stability testing for pump sprays should be done daily and involves two sequential strokes per day for multiple units accurately to assess the rate of failure.

The greatest formulation challenge will arise where VOC restrictions might apply. It will be necessary to reduce ethyl alcohol to a low percentage and still maintain low viscosity and a good evaporative drying rate. The following are typical guide formulations for aerosols and pump sprays:

Typical formulation of a low VOC w/o antiperspirant pump

Part	*Ingredient*	*Wt%*
A	50% Aluminum chlorohydrate	44.0
	Water	27.0
B	PPG 10 butanediol	16.0
	Lauryl dimethicone copolyol	8.0
	Laureth 2 octanoate	5.0
C	Fragrance	*q.s.*

Procedure: Combine A, combine B and add to A with propeller mixing. Add C.

Typical formulation of a suspension aerosol

Part	*Ingredient*	*Wt%*
A	Aluminum chlorohydrate powder (enhanced)	11.0
B	PPG 3 myristyl ether	2.0
C	Isopropyl myristate	11.0
D	Alcohol SDA 40	0.8
E	Organofunctional clay	0.8
F	Isobutane/propane (80/20)	74.4
G	Fragrance	*q.s.*

Procedure: Disperse E in B and C with a high shearing homogenizer for 10–15 minutes. Add D and G with continuous high speed mixing until a viscous gel is obtained. Slowly add A at reduced shear speed by incremental additions to ensure homogeneity. When addition is complete continue mixing for at least 15 minutes. Filter concentrate through a 50 mesh sieve and homogenize at 6000 psi. Fill the concentrate into corrosion resistant lined aerosol containers and add F at an 80:20 ratio, F:(A–E).

Typical formulation of a hydroalcoholic antiperspirant pump

Part	*Ingredient*	*Wt%*
A	Alcohol (190 proof)	45.0
B	Aluminum sesquichlorohydrate powder	20.0
C	Water	5.0

D	Octamethylcyclotetrasiloxane	2.5
	Laureth 2 octanoate	5.0
	PPG 15 stearyl ether	12.5
	PEG 20 isohexadecyl ether	4.0
	Laureth 7	5.5
	Laureth 23	0.5
E	Fragrance	*q.s.*

Procedure: Heat A to 50°C, add B and dissolve. Add C followed by D and E.

10.8 Deodorants

Deodorant products are designed specifically for the control of malodor and cater in the US more for the male market. In Latin America and Europe, deodorants dominate market share. Deodorants date back to many centuries of use when they were most likely to be natural fragrance oils and aromatic compounds applied to all areas of the body.

Continental Europe and most parts of the world favor deodorant products to control odor in lieu of antiperspirants, however there is a gradual decline of that ratio every year. In the US, approximately one in every four people use deodorants instead of antiperspirants. A basic advantage of deodorants is that they tend to be less irritating than antiperspirants, promoting loyalty among their users.

10.8.1 *Odor control*

There are three approaches to odor control:

1. Topical application of anti-microbial agents to suppress the bacterial degradation of aprocrine sweat secretions.
2. Topical application of antiperspirants to master a dual function of sweat reduction and odor prevention by the anti-microbial function of aluminum.
3. Other miscellaneous means such as the systemic administration of anti-cholinergic agents to diminish perspiration, enzyme inhibition to prevent odor formation, olfactory competition (fragrance) and complexation of odoriferous compounds.

This section primarily focuses on antimicrobials incorporated into formulations.

The simplest approach to deodorizing the axilla is in the use of masking agents applied neat or in formulation. Any fragrance will work as such, however there is no long term control mechanism in place to maintain a stable low odor plateau for extended periods.

The best masking agents are deodorant fragrances. These fragrances are designed to complement the body odor but at the same time not to reinforce it. According to Sturm [19] it is possible to select certain deofragrances that not only mask but have some anti-microbial action on bacteria. This may require considerable evaluation in formula development and perhaps may not be practical.

Odor absorption is an old concept for malodor control and it has been revisited within the last several years. Sodium bicarbonate is just starting to be used in antiperspirant and deodorant products to absorb odors by the chemical neutralization of odorous short chain fatty acids. The use of sodium bicarbonate in suspension systems at a pH of 7–8.5 is feasible. Process temperatures above 50°C should be avoided since sodium bicarbonate begins to loose carbon dioxide above this temperature.

Sodium bicarbonate is not stable in aqueous antiperspirant deodorant compositions. It is stable and more effective if it is encapsulated in anhydrous preparations. Recent innovative technology [20] teaches hydrophilic polymer encapsulation of bicarbonate compounds. The deodorization of bicarbonates is discussed by Lamb [21].

Zinc glycinate [22], zinc oxide and metallic carbonates also demonstrate deodorizing capability.

Antimicrobial agents [23] that are currently being used are very effective in destroying gram positive and some gram negative micro-organisms. Two that have withstood extensive safety testing are 3,4,4′-trichlorocarbonilide (triclocarbon) and 2,4,4′-trichloro-2′-hydroxydiphenyl ether (triclosan). Their structural formulas are given below.

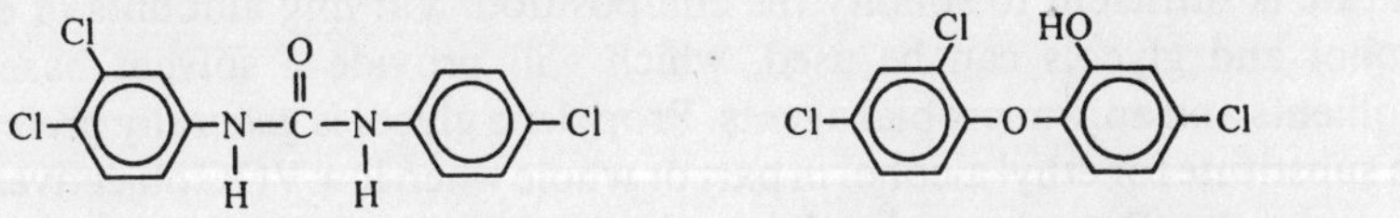

3,4,4' - Trichlorocarbonilide

2,4,4' - Trichloro-2'-hydroxydiphenyl ether

Triclosan is more readily used in underarm deodorant preparations at about a 0.05–0.3% level. It is not necessary to exceed these levels because the contact time in underarm topical applications is hours long. This gives better protection than deodorant soaps which can be more readily washed away. The effect of triclosan on bacteria is discussed by Meinike *et al.* [24].

10.8.2 *Clinical assessment*

Deodorants are cosmetic and a deodorant claim by itself is a cosmetic claim. There are instances where claims are made in relation to an antiperspirant or an antimicrobial [25] which require that a clinical study be made to substantiate the claim.

Some *in vitro* test procedures can be conducted to determine the effect of compounds in deodorant products but they are not reliable enough to indicate the true potential for malodor control. The best evaluation should involve a panel study. There are many rules to follow by the panelists and the odor judges. Clinical testing houses are best suited to conduct this type of study. A direct evaluation is performed by several judges who will give a rank odor score from 0–10. The lowest tabulated scores will indicate the best deodorant efficacy. A less critical evaluation can be done in house by volunteers who agree to wear the deodorant and provide a personal assessment.

10.8.3 *Formulations*

The majority of underarm deodorant products are in the form of aerosols or sticks. The preparations for deodorants are similar to antiperspirant aerosols except that the anti-microbial is used in place of the antiperspirant.

Various plant extracts such as lichen and alpine herb offer anti-microbial benefits. The following formulations can be used as a guide to prepare deodorant products.

10.8.3.1 *Stick products* Deodorant stick products are gelled in a different manner to antiperspirant sticks by using sodium stearate in lieu of C_{16}–C_{18} fatty alcohols. Deodorant sticks are relatively easy to prepare. The gelling agent is normally pre-formed by the supplier of the raw material, however the formulator can prepare it from aqueous sodium hydroxide which is added to an alcoholic solution containing stearic acid. About 5–8% of sodium stearate is sufficient to solidify the composition. Varying amounts of ethyl alcohol and glycols can be used, which will provide a solvent base for emollients and anti-microbial agents. Propylene glycol is generally preferred as a substitute for ethyl alcohol in part or whole when low VOC objectives are being sought. Propylene glycol is quite popular because it can act as a plasticizer in the stick, yet it will be absorbed and is non-greasy after application. Most deodorant sticks are translucent in appearance. The clarity and feel can be improved with branched esters and alkoxylated ethers such as PPG 3 myristyl ether. Other ingredients might include dimethicone, silicones, fatty acid amides and plant extracts. Deodorants in the form of micro-emulsion clear gels may also be of interest to today's consumer. The technology is similar to that discussed in the antiperspirant section.

10.8.3.2 *Spray products* Deodorant sprays can be in the form of an aerosol or pump. These formulas can be made to contain deofragrances with or without anti-microbial agents. If triclosan is used it must be dissolved in the organic additives when preparing the concentrate. Since antiperspirant actives are good deodorants, aluminum chlorohydrate powder can be utilized at about 3–4% of the level of conventional ACH. In conjunction with an anti-microbial, it makes a very effective product. Enhanced aluminum chlorohydrate can also be utilized in even less quantities to obtain the desired performance.

Typical formulation of a deodorant stick

Part	*Ingredient*	*Wt%*
A	Sodium stearate	7.0
	Propylene glycol	52.0
	PPG-15 stearyl ether	20.0
	Ethyl alcohol	15.0
	Water	5.0
	Triclosan	0.2
B	Fragrance	*q.s.*
	Color	*q.s.*

Procedure: Dissolve the solids in A by heating to 85°C, cool to 80°C, *q.s.* with B, and fill into molds.

Typical formulation of a microemulsion deodorant gel

Part	*Ingredient*	*Wt%*
A	DEA oleth-3-phosphate	6.8
	Oleth 3	4.1
	Oleth 5	2.7
	Light mineral oil	13.6
	2-Ethyl-1,3-hexanediol	3.4
B	Diazolidinyl urea	1.0
	Propylene glycol	0.4
	Deionized water	68.0
C	Fragrance	*q.s.*

Procedure: Heat A to 85°C. Heat B to 85°C with good mixing add B to A. Add C with mixing before filling. Fill at 56°C.

Typical formulation of a deodorant aerosol

Part	*Ingredient*	*Wt%*
A	Aluminum chlorohydrate powder	4.0
B	Dimethicone (50 cps)	5.0
C	C_{12}–C_{15} Alcohol benzoate	15.0
D	Alcohol, SDA40	1.0
E	Organofunctional clay	0.8
F	Triclosan	0.1

G	Fragrance	*q.s.*
H	Isobutane/propane (80/20)	76.1

Procedure: Disperse E in B, C and F with a high shearing homogenizer for 15 minutes. Add D and G with continuous high speed mixing until a viscous gel is obtained. Slowly add A in small increments to ensure homogeneity. When addition is complete, continue mixing for at least 15 minutes. Filter concentrate through a 50 mesh sieve and homogenize at 6000 psi. Fill the concentrate into corrosion resistant lined aerosol containers and add H in an 80:20 ratio.

References

1. Carson, H.C. (1981) *Household Personal Products Industry* **18** 33.
2. Jones, J. and Rubino, A. (1968) US Patent 3,405,153.
3. Jones, J. and Rubino, A. (1975) US Patent 3,928,545.
4. Alexander, P. (1987) Refreshing antiperspirant formulation developments. *Manufacturing Chemist.* **58** 51–52 (June).
5. Ross, S. (1994) An overview of the aerosol market in Europe. *Spray Technol. Marketing* **4** 16–25 (November).
6. Shelley, W.B. and Horvath, P.N. (1950) Experimental miliaria in man, II. Production of sweat retention anhidrosis and miliara crystallina by various kinds of injury. *J. Invest. Dermatol.* **1** 9–20.
7. Papa, C.M. and Kligman, A.M. (1967) Mechanisms of eccrine anhidrosis, II. The antiperspirant effect of aluminum salts. *J. Invest Dermatol.* **49** 139–145.
8. Quatrale, R.P., Waldman, A.H., Rogers, J.G. and Felger, C.B. (1981) The mechanisms of antiperspirant action by aluminum salts, I. The effect of cellophane tape stripping on aluminum salt-inhibited eccrine sweat glands. *J. Soc. Cosmet. Chem.* **32** 67–73.
9. Gosling, K., Jackson, N. and Leon, N. (1982) Canadian Patent 1,115,930.
10. Majors, P.A. and Wild, J.E. (1974) The evaluation of antiperspirant efficacy: influence of certain variable. *J. Soc. Cosmet. Chem.* **25** 139.
11. Murphy, T. and Levine, M. (1991) Analysis of antiperspirant efficacy from test results. *J. Soc. Cosmet. Chem.* **42** 167–197 (May/June).
12. Fulton, J.E. (1989) Comedogenicity and irritancy of commonly used ingredients in skin care products. *J. Soc. Cosmet. Chem.* **40** 321–333 (November/December).
13. Landsdown, A.B.G. (1947) Aluminum compounds in the cosmetic industry. *Soap Perfumery Cosmet.* **47**(5) 209.
14. Westwood Chemical Corporation, Antiperspirant Study at Hilltop Research Inc., unpublished clinical study, January 1994.
15. Abrutyn, E. and Bahr, C (1993) Formulating enhancements for underarm applications. *Cosmet. Toilet.* **108**(7) 51–54 (July).
16. Kroschwitz, J. (1987) *Encyclopedia of Polymer Science and Engineering, Microencapsulation,* 2nd Edition. Volume 9, Wiley, New York, pp. 724–745.
17. Morey, E. (1991) Emulsification, emulsion preparations and the HLB system facts and hands-on-tips. *For Formulation Chemists Only* F_2CO **1** 10–13.
18. Hameyer, P. (1991) Comparative technological investigations of organic and organosilicone emulsifiers in cosmetic water-in-oil emulsion preparations. *Happi.* **28**(4) 88–94 (April).
19. Sturm, W. (1979) Deosafe fragrances: fragrances with deodorizing properties. *Cosmet. Toilet.* **94** 35–48 (February).
20. Murphy, R. and LuJoie, S. (1994) Antiperspirant-deodorant cosmetic products, US Patent 5,376,362.
21. Lamb, J.H. (1946) Sodium bicarbonate: An excellent deodorant. *J. Invest. Dermatol.* **7** 131–133.
22. Charig, A., Froeke, C., Simone, A. and Eigen, E. (1991) Inhibitor of odor producing axillary bacterial exoenzymes. *J. Soc. Cosmet. Chem.* **42** 133–145 (May/June).

23. Department of Health and Human Services, Food and Drug Administration (1979) OTC topical antimicrobial products, over-the-counter drugs generally recognized and safe, effective and not misbranded. *Federal Register*, part II. January 6.
24. Meinike, B.E., Kranz, R.G. and Lynch, D.L. (1986) Effect of irgasan on bacterial growth and its adsorption into the cell wall. *Microbios* **28** 133–147.
25. Shehadeh, N.H. and Kligman, A.M. (1963) The effect of topical antibacterial agents on the bacterial flora of the axilla. *J. Invest. Dermatol.* **40** 61–71.
26. Reheis Inc. (1990) Antiperspirant efficacy guide, published literature, May.
27. Buttery, H. (1994) Encapsulation of fragrances. *Cosmet. Toilet. Manufacture Worldwide* 165–171.

Miscellaneous patent literature

28. Fitzgerald, J.J., Phipps, A.M., McClean, J. and Wu, M.S. (1980) Aluminum chlorhydroxide and preparation thereof. UK Patent Application 2,048,229A.
29. Gosling, K., Mulley, V.J. and Baldock, M.J. (1982) Inhibition of perspiration. UK Patent 2,027,419B.
30. Giovanniello, R. (1994) Method for preparing basic aluminum halides by reacting aluminum halide with aluminum. US Patent 5,356,609.
31. Giovanniello, R. (1994) Method for preparing basic aluminum halides and product produced therefrom. US Patent 5,358,694.
32. Murray, R.W., Nelson, R.E. and Rubino, A.M. (1993) Enhanced efficacy aluminum chlorhydrate antiperspirant and method of making same. US Patent 5,234,677.
33. Callaghan, D., Phipps, A. and Provancal, S. (1992) Antiperspirant composition with improved efficacy. US Patent 5,114,705.
34. Giovanniello, R. and Howe, S.M. (1989) Antiperspirant composition and method of preparation. US Patent 4,871,525.
35. Shin, C.T., Slade, M.T. and Nersesian, A. (1988) Antiperspirant composition containing aluminum chlorhydrate, aluminum chloride and an aluminum zirconium polychlorhydrate complex and method of use. US Patent 4,774,079.
36. Brewster, D.A., Kuznitz, M. and Faryniarz, J.R. (1992) Clear cosmetic sticks. US Patent 5,128,123.
37. Angelone, P., Karassik, N. and Grace, W. (1992) Clear gel composition. WO Patent 9,205,767.
38. McCrea, A.D. and Diulus, M.D. (1994) Stable anhydrous topically active composition and suspending agent. US Patent 5,292,530.
39. Orr, T.V. (1991) Antiperspirant creams. US Patent 5,069,897.
40. Luebbe, J.P., Tanner, P.R. and Farris, D. (1988) Antiperspirant gel stick. US Patent 4,781,917.

11 Regulation of cosmetic products

E.G. MURPHY and P.J. WILSON

11.1 Historical development

The modern day mass market toiletry and cosmetic industry began in the US in the early years of the 20th century, and particularly between the two world wars. Whilst the industry record for in-use safety of its products was good, this rapid expansion in consumer exposure gave rise to calls for action to pre-empt possible problems that could arise. Consumer pressure and advances in the science of toxicology added to the general concern.

In the US, the Pure Food and Drugs Act of 1906 [1], enacted to extend formal federal controls over foods and drugs, did not include cosmetics, reportedly as the result of a political compromise [2]. Not until passage of the Food, Drug and Cosmetic Act in 1938 [3] were cosmetics included in the general prohibitions against manufacturing and marketing adulterated or misbranded products. The 1938 Act also made it possible for products previously unregulated to come under Food and Drug Administration (FDA) [4] control as drugs if they were intended not only for medicinal and therapeutic purposes, but also if they were intended to affect the structure or function of the body. The 1938 Act has since been amended to control the safety of cosmetic colors. Regulations under both the FDC Act and the Fair Packaging and Labeling Act (FDCA), administered by FDA, restricted or prohibited cosmetic ingredients found to be unsafe and required extensive informative labeling of cosmetic packaging.

Other national and regional governments have likewise sought to govern cosmetic products. In Japan, cosmetics are included in the Pharmaceutical Affairs Law administered by the Ministry of Health and Welfare. The law requires that cosmetic manufacturers and importers be licensed and that each cosmetic placed on the Japanese market either be in conformance with the monographs that make up the Comprehensive Licensing Standards of Cosmetics or have obtained direct individual approval from the Ministry.

In Europe prior to 1976 there was no comprehensive control of cosmetics and local legislation varied widely. Some countries, e.g. Germany, already had a highly developed regulatory system, whilst others relied almost solely on rather antiquated rules for poisons control. With the development of the then European Economic Community, later the European Community and now the European Union (EU), with its twin aims of free movement of goods between member states (MS) and consumer safety, this situation was unacceptable. In 1976 the then MS agreed a Directive on Cosmetic Products which would be applicable to the Community as a whole. In the pursuit of adaptation to technical progress, this Directive has, in the meantime, been amended no less than 24 times.

Other countries have developed their own regulatory styles, but in the main these are based on those of one of the three major markets: Japan, USA or Europe.

11.2 Self-regulation

In many areas of the world, and especially in the US, the cosmetic industry has embarked on voluntary programs aimed at improving the safety and effectiveness of cosmetics and at the same time, at avoiding more intrusive government regulation. These activities have included codes of good practice and self-regulatory programs such as voluntary registration of cosmetic manufacturing establishments, cosmetic products and ingredients, and voluntary reporting of adverse reaction experiences. In addition, in the US, the industry has undertaken a program [the Cosmetic Ingredient Review (CIR)] to assess the safety of cosmetic ingredients through an independent panel of scientific experts which meets, deliberates and publishes its findings. CIR reports on specific cosmetic ingredients are used by industry and government to establish the safety of these materials. In addition, the perfumery industry set up the international organizations IFRA (International Fragrance Research Association) and RIFM (The Research Institute for Fragrance Materials) to examine and review perfumery materials, and to control their use. The aerosol industry published a code of practice for manufacture of aerosol products, while the European Chemical Industry set up ECETOC as a review body to control information on the safety of chemical substances. COLIPA (Comité de Liaison des Associations Européennes de l'Industrie de la Perfumerie des Produits Cosmetique et de Toilette), the representative body for trade associations within the EC, published '*Advisory Notes to Manufacturers*' together with a number of other advisory documents covering the safety of cosmetic and toiletry products. These included the '*Education and Training of Personnel*', '*Product Safety Verification During Marketing*' and '*Avoidance of Nitrosamine Formation in Cosmetic Products*'.

Although much of the voluntary activity described eventually led to legislation, there still remains a wealth of voluntary activity, and the legislation imposed has been maintained at realistic levels. Industry continues to adjust its voluntary controls in the light of consumer pressures, environmental concerns and legislation in other areas that impinges on the products manufactured and sold as cosmetics and toiletries. Constant vigilance and awareness of these issues is imperative for all scientists aspiring to create new approaches and new products in this area.

11.3 Regulation in the United States

11.3.1 *Federal regulation of cosmetics*

The Federal Food, Drugs and Cosmetic Act (FDC Act) [1] is the primary food and drug law in the United States. The FDC Act prohibits [2] the distribution or importation of cosmetic products that are adulterated [3] or misbranded [4], terms that have been interpreted by Congress, the Food and Drug Administration (FDA), and the courts to include certain characteristics. However, in general, the term 'adulterated' includes products that are defective, unsafe, filthy or produced under insanitary conditions. A cosmetic may be adulterated if it or its container is composed of a deleterious substance that may render it injurious to users under customary or labeled conditions. A cosmetic may also be adulterated if it contains any 'filthy, putrid or decomposed substance' or if it has been produced, packed or held in insanitary conditions under which it may have become contaminated. Finally, a cosmetic may also be adulterated if it contains an unapproved color additive.

The term 'misbranded' refers to statements, designs or pictures in cosmetic product labeling that are false or misleading or which fail to provide required information, and to the use of deceptive or improper containers. A cosmetic may be misbranded if it fails to bear all of the information required under both the FDC Act and the provisions of the Fair Packaging and Labeling Act (FPLA) [5], which sets forth the content and placement of information required on all consumer packages.

The FDC Act also requires that companies maintain certain records [6], and authorizes FDA to inspect cosmetic products and establishments, including manufacturing and packaging plants, warehouses and other storage facilities, vehicles used for transporting cosmetic products and containers of imported goods held at US ports of entry [7].

The Food and Drug Administration, an agency within the Department of Health and Human Services, administers and enforces both the FDC Act and the FPLA. FDA conducts routine and regular inspections of cosmetic establishments in the United States to determine that products produced, packed and stored are neither adulterated nor misbranded. The FDA has not issued good manufacturing practice (GMP) regulations for cosmetics, although cosmetics that are also drugs are subject to the pharmaceutical GMPs. In practice, the same inspection team generally conducts inspections at both over-the-counter drug and cosmetic facilities. In addition to its routine inspection activities, FDA will also investigate products and establishments on the basis of consumer complaints or because a company has filed a trade complaint concerning a possible violation of the law by one of its competitors.

Persons and firms responsible for violations may be subject to civil and criminal sanctions. Products found to be in violation may be seized and their further manufacture or distribution may be enjoined by orders obtained by FDA in the Federal court [8]. If the offense is egregious, deliberate or repeated, FDA may bring a criminal suit against the company and its responsible officials [9]. Imported products that appear to be in violation of the law may be detained at the port of entry by FDA inspectors and if found to be in violation, may be destroyed if they cannot be brought into compliance or are not re-exported [10].

Although FDA brings many formal cases against violators and products each year, it also works with domestic companies toward voluntary recall or correction of illegal goods and practices. This in part is due to the scarce resources of the FDA and because the FDA has the burden of proving that a product is either adulterated and/or misbranded. The first step in the FDA's enforcement procedures is therefore usually a warning letter to the responsible company or individual from the FDA charging a violation of the Act and regulations and providing a limited time, usually 15 days, for the company to correct the violation.

Despite the formidable enforcement tools at the disposal of the FDA, cosmetics remain the most lightly regulated products within the FDA's jurisdiction, in large part because more stringent restrictions have not been needed. Cosmetic products may be placed on the US market without FDA premarket clearance as long as the products do not contain any prohibited or unsafe ingredients (not adulterated) and are properly labeled and promoted (not misbranded). There exists no Federal requirement for cosmetic establishment registration or for listing cosmetic products or ingredients with the FDA. The regulations do include procedures for voluntary registration and listing of cosmetic establishments, products and ingredients and many industry members participate in this voluntary program [11].

11.3.2 *Cosmetic composition*

11.3.2.1 *Safe ingredients* Each cosmetic manufacturer is responsible for using only safe and suitable ingredients in its products and for substantiating the safety of the finished product [12]. Products whose safety has not been substantiated must bear a warning statement on the label [12]. Although to this author's knowledge, no cosmetic product has ever been so labeled, in the event that a product is shown to be unsafe and is marketed without such a warning, FDA could charge the responsible company with failure to warn under the provision. This use of the 'misbranding' provisions of the statute, together with the adulteration prohibitions in the statute, encourages voluntary testing of cosmetic products and ingredients prior to marketing [13].

The FDA has published a short list of restricted or prohibited substances, including items such as bithionol [14], hexachlorophene [15], mercury compounds for all but preservative use in eye area products [16], vinyl chloride [17], halogenated salicylates [18], zirconium in aerosols [19], chloroform [20], methylene chloride [21] and chlorofluorocarbon propellants [22].

11.3.2.2 *Color additives* The FDA has also published regulations stipulating the color additives that may be used in cosmetic products, including the product categories in which they may appear, any restrictions in the concentration level of the colors and any required label warning statements [23]. The list of color additives is divided into those that must be certified by the FDA prior to their sale (all 'FD&C' and 'D&C' colors) [24] and so-called 'natural' colors for which no certification is required.

In contrast to some other color additive regulatory schemes, most synthetic coal tar colors (the colors identified by 'FD&C' and 'D&C' nomenclature, such as FD&C Red 6) have not been authorized for use in products intended for the eye area. The exceptions are FD&C Yellow 5, FD&C Green 5, FD&C Blue 1 and FD&C Red 40.

11.3.2.3 *Hair dyes* The FDA's prohibition against the use of 'poisonous or deterious substances' in cosmetic products [25] does not apply to coal tar hair dyes as long as the statutory warning statement appears on the product containers. In addition, coal tar hair dye colors are not subject to the regulatory requirements for cosmetic color additives [26]. The warning statement required by the FDC Act is as follows:

> Caution—This product contains ingredients which may cause skin irritation on certain individuals and a preliminary test according to accompanying directions should first be made. This product must not be used for dyeing the eyelashes or eyebrows; to do so may cause blindness. [27]

11.3.3 *Cosmetic labeling*

The FDA has issued detailed rules for the labeling of cosmetic products*. These rules are issued under both the FDC Act and the FPLA. Because the FPLA applies only to the outermost containers of retail cosmetic packages, some labeling requirements such as the ingredient declarations, product identity statements and placement and typesize of net content declarations, only apply to the outermost retail container. Cosmetics must be labeled with the information detailed in sections 11.3.3.1 to 11.3.3.6.

11.3.3.1 *Product identity* [28] This statement should appear on the principal display panel (PDP) of the outer container in bold type and in lines parallel to the bottom of the package. The law does not require a product

*The term 'labeling' as defined in the Federal Food, Drug and Cosmetic Act (FDC Act) includes not only the product label, containers and package inserts, but also material that accompanies the product such as booklets, catalogs and even advertising.

identity statement on the inner container of cosmetics that are packaged in an inner and outer container. However, the inner containers of most products are labeled with a product identity statement to avoid consumer confusion and potential misuse that could result in a product liability suit.

The name used may be the common name of the cosmetic, a descriptive name or fanciful name (if the nature of the cosmetic is otherwise obvious) or the cosmetic may be identified by an appropriate illustration of the cosmetic's intended use.

Products sold to beauty salons for professional use by beauticians, gifts or free samples, testers or demonstrators and theatrical make-up for professional use only are exempt from product identity labeling requirements. However, in practice, these products generally are labeled with an identity statement to avoid user confusion and potential misuse that could result in user injury.

11.3.3.2 *Name and address of the manufacturer, packer or distributor* [29] The name and address of the manufacturer, packer or distributor must appear on both the inner and outer containers. The street address may be deleted if the company appears in the local telephone or city directory. If the name that appears on the container is not the manufacturer, it must be prefaced by an explanatory phrase such as 'Distributed by', 'Manufactured for', 'Packaged by' or a similar description. The name and address must be placed conspicuously on the product, although specific type size and placement are not required. Labeling the bottom panel is not generally considered to be 'conspicuous'. The actual corporate name of a corporation must be used. Partnerships, individuals or associations may be identified by the name under which they do business.

11.3.3.3 *The net content in American (ounces/fluid ounces) and metric measures* [30] This statement should appear in American units within the lower 30% of the PDP in lines parallel to the bottom of the package. If the PDP is less than 5 inch^2 (32 mm^2), the declaration may appear anywhere on the PDP. If there is more than one PDP, the declaration must appear on each. The required typesize for the net quantity of contents varies with the size of the PDP.

If a cosmetic is sold in an inner and outer container, the inner container must also bear a net quantity of contents statement. However, the typesize and placement requirements for the net quantity of contents do not apply to the inner container as long as the statement appears in a conspicuous location.

11.3.3.4 *A statement of ingredients in the product in descending order of predominance* [31] The ingredients are only required on the outer container and generally are listed in order of predominance in the product. The minimum required typesize for product ingredient statements is $\frac{1}{16}$ inch (1.6 mm), except that $\frac{1}{32}$ inch (0.8 mm) may be used if the total surface area available for labeling is less than 12 inch^2. Placement must be on an appropriate information panel (bottom panel labeling is not generally considered 'appropriate').

Products sold to beauty salons for professional use by beauticians, gifts or free samples, testers or demonstrators and theatrical make-up for professional use are exempt from ingredient labeling requirements. Special ingredient labeling rules exist for certain cosmetic products, such as lines of shaded products, assortments, products sold by mail order and products displayed in cases and display units.

11.3.3.5 *Any warnings required by regulation* [32] (*such as the warnings required for aerosol products* [33] *and bath foam products* [34]) Warning statements must be conspicuous and prominently displayed on both the inner and outer containers and in typesize no smaller than $\frac{1}{16}$ inch (1.6 mm). Warning statements have been prescribed in the US for foaming bath products, hair dyes, aerosol products, feminine deodorant sprays and for products without adequate safety substantiation.

11.3.3.6 *Country of origin, if imported* [35] US Customs Service regulations require that cosmetic product containers be marked with the country of origin of their contents. Expiration dating and lot or batch codes are not required by specific regulation. Most companies do include an expiration date if the product is not likely to be sold and used within its period of stability. Lot and batch codes are commonly added to cosmetic product labeling to facilitate recall and identification of products in the marketplace, when necessary [36].

11.3.4 *The relationship of cosmetic products to drugs*

The distinction between cosmetic and drug products is based upon the definitions of these articles in the FDC Act. The FDC Act defines cosmetics as '(1) Articles intended to be rubbed, poured, sprinkled, or sprayed on, introduced into, or otherwise applied to the human body or any part thereof for cleansing, beautifying, promoting attractiveness, or altering the appearance, and (2) articles intended for use as a component of any such articles; except that such term shall not include soap' [37].

The FDC defines drugs in pertinent part as: '(B) articles intended for use in the diagnosis, cure, mitigation, treatment or prevention of disease . . .; and (C) articles . . . intended to affect the structure or any function of the body. . . .' [38].

Thus, the representations made for a product in its labeling determine its legal status. The definitions are not mutually exclusive and a product can be both a drug and a cosmetic, in which case the product must meet all the applicable requirements. As drugs may only be marketed if the FDA determines that they are generally recognized as safe and effective for their labeled use, classification as a drug will raise the regulatory burden.

11.3.4.1 *Cosmetics and anti-aging claims* Over the years, the FDA has regularly challenged representations for cosmetic products it considers either false and misleading or representations that fall within the FDC Act drug definition. As most companies faced with misbranding or new drug charges choose to modify product labeling rather than to develop the data required to support drug approval, only a few judicial decisions address the cosmetic–drug distinction in any detail. The chief cases stem from the 1960s, at which time the FDA initiated civil seizures of three skin care products which were widely promoted to smooth, reduce or prevent wrinkles.

'Line Away Temporary Wrinkle Smoother' was advertised as being able to 'visibly smooth out fatigue lines, laugh lines, worry lines, frown lines, tiny age lines and crows feet, while discouraging new lines from forming'. Advertising for Line Away contained statements that 'Line Away is an amazing protein liquid. Contains no hormones or harmful drugs. It's the only wrinkle smoother packaged under biologically aseptic conditions in the pharmaceutical laboratories of (the manufacturer)' (*United States v. An Article ... 'Line Away'*, 283 F. Supp.* 107 (D.Del, 1986), *aff'd*, 415 F.2d 369 (3rd Cir. 1969)).

'Sudden Change' was labeled and advertised as the 'new and improved, dramatically different wrinkle-smoothing cosmetic. By simple, dynamic contraction, it lifts, firms, tones slack skin ... smooths out wrinkles, lifts the puffs under the eyes, leaving your contours looking beautifully defined. It acts noticeably, visibly ... not a hormone or chemical astringent, Sudden Change is a concentrated purified natural protein ....' Sudden Change was also advertised as a 'face lift without surgery! ... (it) is the one cosmetic that can make you look years younger for hours ... although it cannot eliminate wrinkles permanently ... Just smooth it on and watch it smooth away crows' feet, laugh and frown lines—even under-eye puffiness ... Sudden Change ... does not change the structure or function of your skin in any way' (*United States v. An Article ... 'Sudden Change'*, 288 F.Supp. 29 (E.D.N.Y. 1968), *rev'd.*, 409 F.2d 734 (2d Cir. 1969)).

'Magic Secret', the third product, was advertised to 'smooth away wrinkles in minutes ... keep them away for hours', to 'firm the skin', to 'tighten and moisturize tired skin', to 'smooth away crows feet, puffy under eye circles, laugh, frown, and throat lines in just minutes'. Users would 'feel an astringent

* F. Supp., Federal Supplement; D. Del, District of Delaware; 3rd Cir., Third Circuit; aff'd, Affirmed; E.D.N.Y., Eastern District of New York; F. 2d, Federal Reporter 2nd Edition; D. Md., District of Maryland; 2d Cir., Second Circuit.

sensation ... gently firming and toning the skin' (*Unites States v. An Article ... 'Magic Secret'*, 331 F.Supp. 912 (D. Md. 1971)).

Only the intended uses for 'Magic Secret' were held to be within the Act's cosmetic definition [39]. The courts in 'Line Away' and 'Sudden Change' found these products within the FDC Act's 'drug' definition [38] based upon their promoted ability to affect the structure and function of the body by 'lifting' the skin, by discouraging new wrinkle lines, and in the case of 'Line Away', by the therapeutic implications created by statements that the product was produced in a 'pharmaceutical laboratory' and packed under 'biologically aseptic conditions'.

Since these cases were decided, the FDA has taken administrative action against products represented for uses the Agency believes are intended to affect the structure or function of the user's body. In a letter to cosmetic skin care companies in 1987, the FDA stated that,

> We consider a claim that a product will affect the body in some physiological way to be a drug claim, even if the claim is that the effect is only temporary .... Therefore, we consider most of the anti-aging and skin physiology claims ... to be drug claims. For example, claims that a product "counteracts," "retards." or "controls" aging or the aging process, as well as claims that a product "rejuvenates," "repairs," or "renews" the skin, are drug claims because they can be fairly understood as claims that a function of the body, or that the structure of the body will be affected by the product.
>
> For this reason, also, all the examples ... (that) allege an effect within the epidermis as the basis for a temporary beneficial effect on wrinkles, lines or fine lines are unacceptable. A claim such as 'molecules absorb ... and expand, exerting upward pressure to lift wrinkles upward' is a claim for an inner, structural change. [40]

The FDA further stated:

> ... we would not object to claims that products will temporarily improve the appearance of ... outward signs of aging. The label of such products should state that the product is intended to cover up the signs of aging, to improve the appearance by adding color or luster to skin, or otherwise to affect the appearance through physical means .... We would consider a product that claims to improve or to maintain temporarily the appearance or the feel of the skin to be a cosmetic. For example, a product that claims to moisturize or soften the skin is a cosmetic. [40]

In the same letter, the FDA stated that products represented for OTC drug purposes, such as sun protection or protection from harmful or annoying stimuli would be regulated as OTC drug sunscreens or OTC drug skin protectants. Other representations that the FDA has considered to be either misleading cosmetic claims or unapproved new drug claims include anti-cellulite claims and skin renewal claims for which mode of action is to affect the structure function of the body. Most recently, the FDA has investigated the safety and mode of action of α-hydroxy acids in cosmetic products.

11.3.4.2 *Cosmetics and OTC drugs* Many products considered to be cosmetics in other countries are regulated in the US as over-the-counter (OTC) drugs. These products include antiperspirants, anti-dandruff shampoos, sunscreen products, skin lighteners, anti-acne products, anti-cavity and anti-plaque dental products and skin protectants. A product that is represented as intended for one of these uses will be regulated by the FDA as a drug. If the product is also labeled for cosmetic uses, such as for cleaning, beautifying or moisturizing, it will be considered a cosmetic as well as a drug product, in which case it must bear both cosmetic and drug labeling. Products that do not comply with FDA monographs (final rules) for OTC drug products may be considered to be unapproved new drugs that may only be marketed subject to an approved new drug application.

11.3.4.3 *Future cosmetic–drug overlap* The 'cosmetic' and 'drug' definitions have not changed since the FDC was enacted in 1938. In the meantime, cosmetic products have become more sophisticated and the line between cosmetic and drug product has often blurred. Although the FDA has recognized in the OTC drug review procedures that limits for materials used in drug products do not apply when used in cosmetic products [41] the presence of some ingredients in a cosmetic product is likely to cause its regulation as a drug by the FDA. These ingredients tend to be those for which no cosmetic use is known (for example, penicillin) or which have been heavily promoted by the manufacturers for therapeutic purposes (for example, certain UV absorbers). The response of the FDA to date to technology advances such as fluoride toothpaste, antiperspirants and sunscreens, has been to designate and regulate the products as drugs under the more stringent drug premarket review and approval procedures. It remains unclear whether the FDA will take similar steps to control the use of ingredients such as the α-hydroxy acids, which have been shown to have a physiological effect, or whether the FDA and industry will find other methods of assuring the safety of these innovative cosmetic ingredients and products. One solution would be to use the industry-funded Cosmetic Ingredient Review (CIR) program to assess the safety of cosmetic ingredients. The CIR, which has a permanent staff headquartered in Washington, D.C., commissions independent expert panels of scientists to evaluate the available published and unpublished literature on the safety of specific cosmetic ingredients. The FDA, industry and consumer organizations may each appoint a non-voting representative to attend the expert panel meetings.

11.3.5 *Regulation of cosmetics by other federal agencies*

11.3.5.1 *Federal Trade Commission* The Federal Trade Commission (FTC) and the Food Drug Administration (FDA) have overlapping and concurrent jurisdiction over the advertising and labeling of foods, drugs, medical devices and cosmetics. The FTC regulates 'deceptive' or 'unfair' advertising, including promotional claims that appear on packages or in the media and is chiefly concerned that the consumer has reliable, substantiated information upon which to base purchasing decisions.

The FDA regulates 'false or misleading' statements made in product 'labeling' and, although its jurisdiction extends to economic deceptions, the FDA is primarily concerned that the public health and safety is not endangered through consumer reliance on unsubstantiated claims to safety and effectiveness.

Although in regulating product claims, the FTC focuses chiefly on consumer deception and the FDA chiefly on health and safety concerns, the responsibilities of the two agencies overlap and nothing in the law would prevent simultaneous FTC and FDA prosecutions against deceptive, unfair and misleading labeling and promotional claims. Therefore, to avoid unwanted duplicative proceedings, the FDA and the FTC have entered into a liaison agreement (Memorandum of Understanding between the FDA and FTC, 36 Fed. Reg. 18,539 (1971)), in which the FTC agrees to take primary responsibility for regulating 'advertising' and the FDA agrees to take primary responsibility for regulating 'labeling'. In practical terms, this means that questionable representations made only in advertising are handled by the FTC, while questionable representations made in labeling or in labeling and advertising are generally handled by the FDA. As part of the agreement between the two agencies, the FTC and FDA exchange information about enforcement proceedings as needed.

The FTC regulates advertising claims through industry-wide rulemakings, through administratively adjudicated cease-and-desist orders entered against specific companies, through issuance of FTC Policy Guidelines and through prosecution of specific cases in the federal courts. An FTC investigation of advertising claims may be prompted by requests made by consumers, competitors, state officials or the Commission itself.

Advertising claims may be considered deceptive because they contain a 'material' representation, omission or practice that is likely to mislead consumers 'acting reasonably under the circumstances'. A representation or omission is 'material' if it affects the consumer's choice of product, such as statements about the product's ingredients or the nature of risks involved in product use. (*Cliffdale Associates, Inc. v. FTC*, 103 FTC 110 (1984)). In addition, representations about a product can be considered deceptive if the advertiser lacks substantiation that demonstrates there is a reasonable basis for a claim. In fact, inquiries into claim substantiation are the primary basis for FTC investigations and challenges to advertising.

An FTC request for substantiation is non-public and is 'directed to individual companies via an informal access letter or, if necessary, a formal civil investigative demand'. Requests may be directed to one company or to several companies within the same industry. Unless the claim made in advertising is substantiated, the advertiser may be held to have violated the prohibition Section 5 of the Federal Trade Commission Act prohibiting 'unfair or deceptive acts or practices in or affecting commerce' [42].

11.3.6 *Cosmetics and the Consumer Product Safety Commission*

Cosmetics are explicitly excluded from regulation under the Consumer Product Safety Commission (CPSC) [43]. However, in the absence of FDA jurisdiction, the CPSC has regulated such items as soap. The CPSC also has authority to issue regulations establishing special packaging requirements under provisions of the Poison Prevention Packaging Act [44].

11.3.7 *Regulation of cosmetics by the states*

Each state has the authority to regulate products and to impose additional requirements that do not conflict with federal regulation of cosmetics. Most states have adopted 'mini FDC Acts' that in general are identical to the federal statute. Some states, however, go beyond the FDC Act to require registration of cosmetic establishments and products.

Furthermore, state environmental legislation has tended to include cosmetic products and ingredients. For example, at the time of this writing, at least eight states (California, New Jersey, New York, Oregon, Rhode Island, Connecticut, Massachusetts and Texas) have either proposed or issued regulations restricting the levels of volatile organic chemicals in cosmetic products such as deodorants, fragrances, and/or hair products and similar regulation by the federal agency, the Environmental Protection Agency (EPA), is expected. California's Proposition 65 [45] requires a warning statement on products that contain ingredients known to the state to be carcinogenic or reproductive toxins. Other states are beginning actively to regulate the composition and disposal of plastic and other cosmetic containers.

11.3.8 *Conclusion*

Cosmetics are the least regulated of the products subject to FDA jurisdiction. Cosmetics may be placed on the US market without premarket clearance as long as they are neither adulterated or misbranded and may be distributed in all the states as long as they comply with individual state regulations, including environmental legislation affecting cosmetic products. The chief challenge to regulators in the future will be the control of innovative cosmetic products and ingredients to protect the public health and safety, while the chief challenge for industry will be to keep these cosmetics out of the drug definition and away from regulation as drugs. The other major concern for the next decade will undoubtedly continue to be regulation of cosmetic product composition and packaging under state and federal environmental laws and the subsequent challenge to industry to develop environmentally safe products.

11:4 Regulation in Europe

Since 1976, the size of the EU has steadily increased, with currently fifteen members and more applications pending. In each of these, the Cosmetic Directive (76/768) [46] as amended, now forms the basis of local legislation with dual aims of enabling the free movement of goods between member states (MS) and ensuring consumer safety. In all, the 1976 Directive has been adapted on 24 occasions, the most recent major revision being the Sixth Council Amendment (93/35) [47]. Agreeing changes to the Directive is a long and tedious process, involving industry, all MS, the Commission, the European Parliament and the Council of Ministers. As a result of this complex procedure, the Sixth Amendment was under discussion for no less than seven years.

Following these changes, the principal features of EU cosmetic regulation now include:

1. A definition of a cosmetic product
2. A requirement that cosmetic products should be safe for their intended purpose under normal and foreseeable conditions of use
3. The establishment of an inventory of cosmetic ingredients
4. Pack labeling to include a full listing of ingredients
5. The requirement for manufacturers to maintain comprehensive product information, including a formal safety assessment.

In addition, Annexes specify prohibited and controlled materials, as well as listing the only permitted coloring agents, preservatives and UV-filters. These latter 'positive lists' may well be extended to other classes of ingredient in the future. Other proposals will eventually forbid the use of ingredients and products which have been tested on animals.

National regulations in MS reflect the EU Directive, but manufacturers should study these carefully in order to identify any local variations. Compliance with the Cosmetics Directive and its national versions is mandatory but facilitates the free movement of products throughout the EU. Imports from outside the European Union are subject to similar controls.

Other European countries, outside the EU and particularly in the East, are rapidly developing their consumer-based industries. It can be expected that their regulatory regimes for cosmetics will closely parallel those of the European Union in anticipation of eventual application for membership of that group.

In addition to the rules specifically governing cosmetics, other more general Directives exist, which may have some impact on the manufacture and sale of cosmetics within the EU. These are listed below. It should be remembered that European legislation is continuously developing and that any of these may well have been the subject of more recent amendment.

- Medicines Directive (65/65/EEC)
- Prepackaged Products Directive (78/891/EEC)
- Prescribed Quantities Directive (80/232/EEC)
- Aerosol Directive (75/324/EEC)
- Dangerous Substances: Classification, Packaging and Labelling (67/548/EEC)
- Solvents Directive (73/173/EEC)
- Product Liability Directive (85/374/EEC)

Other general Directives cover such matters as advertising and health and safety at work and apply equally to the cosmetics industry as to others.

11.5 Regulation in Japan

Cosmetics are regulated in Japan under the authority of the Pharmaceutical Affairs Law (PAL) which requires that the manufacturer or importer of cosmetics be licensed and that all products be either licensed with the facility or approved individually. A simplified licensing system known as the Comprehensive Licensing Standards of Cosmetics (CLS) has been instituted within the past decade to permit easier marketing access to cosmetic products that are formulated in accordance with the standards. Ingredients that may be used for cosmetic products in Japan are identified in several sources, including the CLS monographs, the Japanese Standards of Cosmetic Ingredients (JSCI), and the industry publication *Japanese Cosmetic Ingredient Dictionary* (JCID). Portions of the PAL and implementing regulations have also been published in English in *Principles of Cosmetic Licensing in Japan* [48].

11.6 Regulation in other countries

Most countries regulate the content and labeling of cosmetic products and in many countries, the manufacturer or importer of cosmetics must be licensed by the national health agency. Therefore, if a product is to be sold in several markets, the initial formulation must take into consideration the rules governing permissible cosmetic ingredients in as many potential markets as possible. Details of individual national laws and regulations are available from local health authorities and trade associations, local branches of the Society of Cosmetic Chemists, and in-country Embassy commercial officers.

11.7 General considerations

Formulating a cosmetic or toiletry product requires not only technical knowledge, but also familiarity with the formidable array of regulations, laws, restrictions and consumer preferences in each market. Being aware of these and other factors will help the cosmetic scientist to advise the marketing staff on subjects such as product claims, safety, acceptability, and product liability concerns.

11.7.1 *Environmental impact*

Laws covering the pollution of air, water and land exist both nationally and internationally, and are constantly under review. The consumer interest in environmental issues, together with trend towards green politics and consumerism has developed public opinion to a degree where products may be accepted or rejected according to their environmental performance. Controversy concerning the ozone layer and the use of aerosol products is a particular example. Marketing edge may be achieved by developing products that can be shown to be harmless or beneficial to the environment. In this respect, ingredients must be chosen with due care. Packaging, labeling, natural materials and use of energy all require careful consideration and, without doubt, new issues will arise in the future. Vigilance and current awareness is essential.

11.7.2 *Animal protection and rights*

The problems encountered in this area are associated not only with the use of animals for laboratory testing, but also with the use of materials and ingredients derived from animal sources. The use of animals for cosmetic testing has decreased substantially over the past decade and continues to decline worldwide. However, international regulations still require safety data based on animal studies, therefore new materials not used for other industries such as plastics, food, chemicals or pharmaceuticals, would require animal testing. Fortunately, consumer pressure is now aimed at the regulators of international law (e.g. the EC Commission) and ultimately these requirements will be modified to make the situation much clearer and easier to cope with.

Due to the growing pressure from vegetarians, religious groups and animal rights groups, this area of concern is constantly changing. The labeling of products as 'non-animal tested,' 'contains no animal ingredients,' 'kosher', etc. requires careful consideration.

11.7.3 *Drug or cosmetic? Borderline products*

The description 'borderline' is applied to those products which, by their composition or claims, may fall within the regulatory regime of either cosmetics or medicines.

The development of more effective, and in some cases therapeutic, products based on modern cosmetological research and development makes a consideration of the borderline area essential. Authorities are vigilant in ensuring that claims, composition and modes of action of intended cosmetic products do not place them in the field of drugs. Thus it is essential for the formulator to be fully aware of where the boundary lies, as the penalties for straying across it are severe.

Whilst the amount of pre-marketing administration for cosmetics is generally relatively straightforward in most world markets, the same cannot be said for medicines. Here, the requirement to demonstrate benefit, efficacy, safety and quality in order to obtain the necessary licences demands significant resources, even for a relatively simple product making modest claims. In addition, the time delay in bringing a new product to market may be commercially damaging, as may restrictions on advertising or distribution.

The approach to this question varies from one jurisdiction to another. In the US and Japan there are three categories of product—cosmetics, prescriptions, drugs and an intermediate group (over the counter drugs in the US; quasi-drugs in Japan). In the EU there is no third category and products may only be classed as cosmetics or medicines. The producer on the international scale must therefore be fully aware of the local conditions in his intended markets.

It is beyond the scope of this book to discuss the borderline situation in any detail. For a fuller description, the reader is referred to the following reviews; by Wilson and Adams for Europe [49], for Japan [50] and for the US [51].

11.7.4 *Product liability*

The situation with regard to product liability has always been much more strict in the US then elsewhere. In Europe the law is under considerable debate and is moving toward the US situation based on the 'deepest pocket' principle. However, in Europe it is still necessary to prove that the product was defective and that it caused the damage complained of, before a successful suit for damages may be concluded. This situation is under review and remains an issue of great debate. As a result, change is inevitable, and it is necessary to keep abreast of current developments when considering ingredients for use in the products under consideration, the product design and method of application.

References

1. 21 United States Code (USC) §§ 301 *et seq.*
2. 21 USC § 321.

3. 21 USC § 361.
4. 21 USC § 362.
5. 15 USC §§ 1451 *et seq.*
6. 21 USC § 373.
7. 21 USC § 374.
8. 21 USC § 332, 334.
9. 21 USC § 333.
10. FDA Investigations Operations Manual (1994) US Department of Health and Human Services, 265, 270 (October).
11. 21 Code of Federal Regulations (CFR) Parts 710, 720, 730.
12. 21 CFR § 740.10(a).
13. O'Reilly, J.T. (1993) *Food and Drug Administration.* Chapters 17–25, p. 26.
14. 21 CFR § 700.11.
15. 21 CFR § 250.250.
16. 21 CFR § 700.13.
17. 21 CFR § 700.14.
18. 21 CFR § 700.15.
19. 21 CFR § 700.16.
20. 21 CFR § 700.18.
21. 21 CFR § 700.19.
22. 21 CFR § 700.23.
23. 21 CFR pts 73, 74.
24. 21 CFR pts 74, 80.
25. 21 USC § 361.
26. 21 USC § 361(e).
27. 21 USC § 361(a).
28. 21 CFR § 701.11.
29. 21 CFR § 701.12.
30. 21 CFR § 701.13.
31. 21 CFR § 701.3.
32. 21 CFR pt 740.
33. 21 CFR § 740.11.
34. 21 CFR § 740.17.
35. Tariff Act of 1930, as amended (19 USC § 1304).
36. In contrast, FDA requires that non-prescription (OTC) drug products bear expiration dating unless the drug has no dosage limits and has shown to be stable for at least three years and that all drugs bear lot or batch codes. 21 CFR §§ 201.17, 201.18 and 211. The National Drug Code (NDC) may appear, but is not required, on OTC drug product packaging. The NDC consists of a 10-digit number made up of the labeler code assigned by FDA and the product code and packaging code assigned by the manufacturer, 21 CFR § 201.2.
37. 21 USC § 321(i). This definition is incorporated in the FPLA at 15 USC §§ 1454, 1456, 1459(a). The soap exemption from regulation under the FDC Act has been interpreted narrowly in regulations issued by FDA at 21 CFR § 701.20 as applying only to products in which the bulk of the non-volatile matter consists of an alkali of fatty acids and the detergent properties are due to the alkali-fatty acid compounds and then only if the product is labeled, sold, and represented as a soap (without cosmetic representations).
38. 21 USC § 321(g).
39. 21 USC § 321(i).
40. Letter from John. M. Taylor, Associate Commissioner for Regulatory Affairs, to Stuart Friedel, Davis and Gilbert, re: *Cosmetic Regulatory Letters*, November 19, 1987.
41. Federal Register (FR) 6822 (Feb. 15, 1988).
42. 15 USC § 45(a) (1) (supp. 1987).
43. 15 USC 2051 *et seq.* The Consumer Product Safety Act provides that 'consumer products' do not include cosmetic products as defined by the FDC Act.
44. 15 USC 1471 *et seq.* The Poison Prevention Packaging Act includes 'cosmetic' within the Act's definition of a 'household substance'.
45. Proposition 65 (1995) *The Safe Drinking Water and Toxic Enforcement Act of 1986*, codified at California Health and Safety Code, section 252495 *et seq.*

46. (1976) *Official Journal of the European Communities.*
47. (1993) *Official Journal of the European Communities,* **L151** 32.
48. The sources of Japanese Cosmetics law referred to are available from YAKUJI Nippo Ltd, 1 Kanda Izumicho, Chiyoda-ku, Tokyo 101, Japan.
49. Wilson, P. and Adams, M. (1995) Cosmetic or drug. *The Regulatory Affairs Journal* 6(3) 197–201.
50. *FDA Investigations Operations Manual* (1994) US Department of Health and Human Services, 265 and 270 (October).
51. O'Reilly, J.T. (1993) *Food and Drug Administration,* second edition, chapters 17–25 and 26.

12 Quality

W.E. DUPUY

12.1 Introduction to quality

It is likely that readers of this book would expect a chapter on quality to discuss various analytical techniques (biological, chemical, mathematical, microbiological, physical, etc.) that are in common use in laboratories today. Twenty years ago that would probably have been acceptable. However, in the last two decades, ideas about quality have changed so much that it is more prudent to advise the reader of the latest philosophy, and of how it developed.

In the 1990s quality should touch every department in the company and should not be the sole responsibility of those of scientific specialisation. Quality as applied to manufacturing industry relates not only to that certain, customer-winning property of attractive goods, but first and foremost to the state of mind of those who produce them.

Quality is not just a certain level to be achieved once and for all, but means constant striving for an ideal goal, an infinity. As far as manufacturers are concerned, the quality of goods constitutes their signature and image. As for consumers, quality must provide motivation for the first purchase and, ultimately, complete satisfaction during use, thus providing the impetus for repeat purchases.

Another very important aspect of quality is that of safety. This seems obvious but, for that reason, is often not given sufficient attention. Even before providing attraction and satisfaction, the quality of a product should be a pledge and guarantee of its safety during use.

12.2 Definition of quality

The industrial meaning of the word quality has changed dramatically over the years. Modern quality methods are no longer the domain in manufacturing industry but extend to service industries, government institutions and health authorities. The word 'quality' comes from the latin '*qualis*' meaning 'of what kind'. Thus, in its original meaning quality was concerned with the composition or properties of an object without any implications of value. Today, however, quality is usually invested with a distinctly positive ideal.

Among the many definitions of 'quality' published in the technical literature, the one that is most concise and to the point is probably that of Juran [1]—"quality is fitness for use". Juran, however, gives no indication of how 'fitness for use' could be expressed in concrete terms or even be measured. The European Organisation for Quality Control [2] gave a more comprehensive definition: "Quality is the totality of features and characteristics of a product or service that bear on its ability to satisfy a given need!" Thus, quality is seen not as a single factor, but as a concept made up of several criteria.

In the US, the Pharmaceutical Manufacturers Association (PMA) has published its own quality definition: "Quality is the sum of all factors which contribute directly or indirectly to the safety, effectiveness and acceptability of the product!" This is a good definition, but would it fit cosmetics or toiletries? Providing that the weighting of the judges parameters is correctly apportioned (acceptability, for example, will require more evaluation criteria for a drug than for a cosmetic), the answer is yes.

As time has progressed, definitions of quality have become more and more complex. However, Crosby [3, 4] successfully introduced a relatively simple definition: "Quality means conformance to requirements". In other words, the definition of quality is very dependent on the intended usage. Quality department personnel cannot be concerned with the various mystical aspects of quality—beauty, elegance, value for money, or even fitness for purpose (unless this means conformance to a defined purpose). The emphasis on conformance throws the responsibility on to marketing and technical departments to define the requirement and the product, such that a product specification is produced. This specification should define the environmental and reliability characteristics that the product must have, the workmanship standards that are applicable, and so on. The quality department are responsible for checking conformance to this specification.

12.3 Inspection

During its development, quality has moved through three distinct phases. Initially the concept of quality control—'have we done it right?'—was developed. Quality control is not wrong, but its basis is inspection and is therefore questionable. Inspection is an activity that:

(i) verifies that an output is acceptable;
(ii) sorts good from bad; and
(iii) takes place after the output has been produced.

Inspection, although potentially stopping errors from reaching the customer, cannot be the basis for continuous improvement since:

(i) it does not prevent errors being produced in the first place;

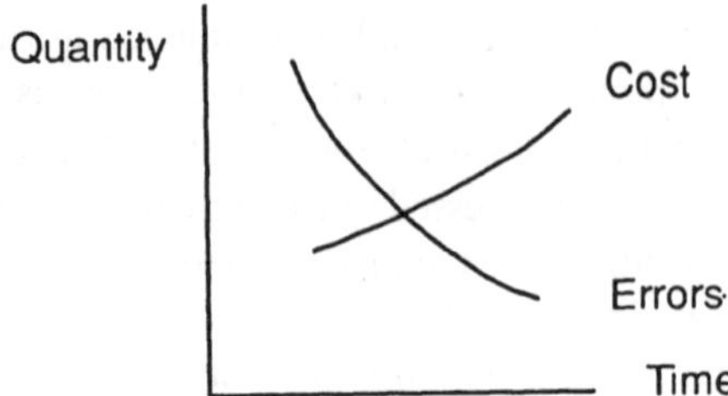

Figure 12.1 Inspection costs.

(ii) visual inspection is not 100% effective, therefore mistakes or errors will still reach the customer; and
(iii) it adds cost (Figure 12.1).

Inspection is, however, of use in four situations:

(i) to monitor and audit the process in order to provide necessary feedback to maintain control of the process;
(ii) to gather data about causes of mistakes or errors in order to take corrective action;
(iii) as a short-term measure until a permanent solution is implemented—this should only be done against an agreed time scale to ensure that inspection does not become the permanent solution; and
(iv) compliance with legal requirements.

12.4 Prevention

The second stage in the development of quality was the concept of quality assurance— are we doing it right? (see section 12.5.2). This was a strategy for stopping mistakes or errors from reaching the customer(s). Prevention is an activity that:

(i) stops mistakes or errors from being generated; and
(ii) takes place before or during the process.

Prevention is the only strategy for continuous improvement since it:

(i) stops mistakes or errors from reaching the customer:
(ii) adds value (Figure 12.2) by preventing defects from being produced in the first place and
(iii) will achieve customer satisfaction.

A study by IBM has shown that an increase in prevention costs is far outweighed by the corresponding decrease in inspection and failure costs (Figure 12.3). IBM found that if it costs £1 to correct a mistake at the design

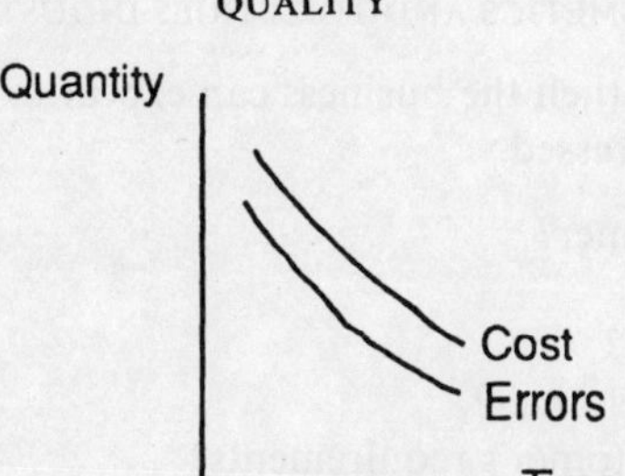

Figure 12.2 Prevention costs.

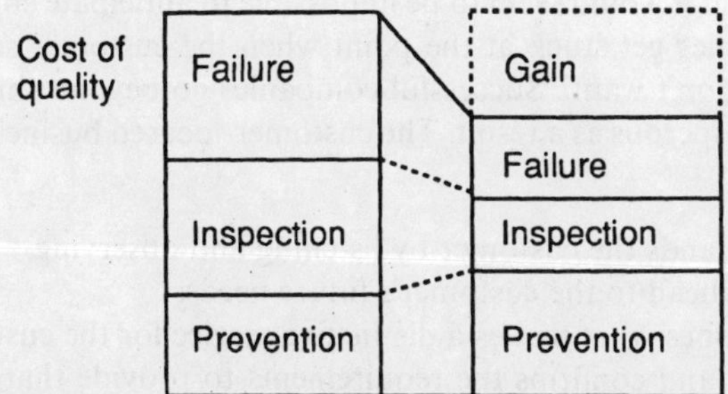

Figure 12.3 Quality costs. Prevention = investment cost to prevent mistakes or errors from occurring—what it costs to stop things going wrong; inspection = cost of checking the quality that has been achieved—what it costs to find the errors before the customer does; failure = cost of wasted resources due to mistakes or errors that reach the customer—what it costs to fix errors.

stage, then it costs £10 to stop it getting to the customer, and £100 to correct it after the customer has received it.

12.5 Total quality

In recent years the concept of total quality has developed the idea of assurance much further. Quality can be applied anywhere in an organisation. If everyone in the organisation is striving to improve his or her quality and that contribution is captured effectively, then the organisation is in a better position to compete for the customer's attention. This is total quality.

Customers are the most important part of a business, yet not all companies give them exactly what they want. In other words, many companies are not truly driven by customer needs. In such situations, a thriving organisation cannot exist. However, if customer requirements are used to drive all the activities of a business then the business can expect to prosper. A number of questions must be addressed.

- Who is the customer?
  - —the consumer?
  - —the retail trade?
  - —the internal customer?
- What are the customer's requirements?
- When were the customer's requirements last checked?
- How well are the customer's requirements being met?

The principles of satisfying customer requirements are equally relevant to both internal and external customers, with all the ensuing business gains.

As with all aspects of quality, it is far from easy to work out customer preferences, and it would seem to be impossible to anticipate emerging trends. Many companies get stuck at the point when the customer says in effect 'I know what I don't want'. Successful companies go beyond this and become even more prosperous as a result. The customer-focused business, department or manager:

(i) understands the customer by listening and observing;
(ii) looks ahead to the customer's future needs;
(iii) determines what makes a distinctive service for the customer;
(iv) defines and confirms the requirements to provide that service;
(v) focuses hard on precisely delivering the service;
(vi) sets benchmark standards to 'beat the best' of the competition; and
(vii) informs the customer of progress.

Total quality then is about conformance to the customer's requirements, and the total service given to the customer. It recognises that in the manufacturing industry, quality cannot be limited to product design and quality assurance, but encompasses every aspect of a company's dealings with its customers: (i) from sales literature through to after-sales service; and (ii) from response time to customer enquiries through to delivery performance. Total quality leads inevitably to a process of continuous improvement, even when the customer's requirements are being satisfied, and achieves on-going reductions in manufacturing costs. For this reason, it is frequently described as being about 'the elimination of waste'. Examples might include the following.

- Scrap
- Rectification and rework
- Downgrading of products
- Production hold-ups due to quality problems
- Warranty claims
- Product recalls

- Product liability claims
- Lost sales due to poor quality and service
- Redundant stock

Total quality is not a prepackaged formula, a collection of techniques and systems. No two companies will apply identical total quality, although most total quality programmes have many common aspects. The approach must reflect:

(i) the company's internal culture;
(ii) the expectations of the company's customers;
(iii) the company's manufacturing technology; and
(iv) how advanced the company is in terms of quality.

The adoption of total quality is increasingly being seen as the route to achieving competitive edge. What makes companies distinctive in the market-place is the difference between customer expectation and supplier achievement —if positive this defines the competitive edge. Research has indicated conclusively the importance of quality in determining a company's profitability. Profit impact of market strategy (PIMS), which correlates statistics about many companies worldwide, has concluded that relative product quality is the single most important factor that affects a company's long-term financial performance. Focusing on quality creates a cycle of improvement from which reduced costs become inevitable. Benefits include:

(i) reducing operating costs, through doing things right first time;
(ii) lowering inventory levels, from improved predictability;
(iii) improving staff motivation and commitment;
(iv) increasing sales, from improved customer satisfaction; and
(v) improving relationships with suppliers.

Total quality aims to improve customer service and reduce costs by harnessing everyone's commitment.

As stated previously, total quality is not a prepacked formula. However, certain aspects of quality apply to most manufacturing companies. These are: (i) product design; (ii) quality assurance; (iii) manufacturing processes and technology; (iv) cost of quality; and (v) human and organisational aspects.

12.5.1 *Product design*

There are four important design related aspects of quality: (i) market research; (ii) quality of design; (iii) design for manufacture; and (iv) product approvals.

12.5.1.1 *Market research* In many industries the customer's present and future requirements are established by market research. For consumer products, attitude and perception surveys are playing an increasingly important role in identifying the customer's subconscious or latent attitudes—as well as their more obvious requirements. Continuous improvement requires:

(i) that the customer defines improvements as well as value;
(ii) that customer requirements are continually changing and customers are continually demanding higher standards;
(iii) the need to be in a position to participate/influence tomorrow's customer requirements.

This means anticipating and not reacting to the customer's requirements. Anticipating the customer's requirements can only be achieved by having a close relationship with the customer.

12.5.1.2 *Quality of design* However well products are manufactured their quality cannot be improved beyond the inherent quality of design. In many industries up to 60% of quality problems can be traced back to design. In the cosmetics and toiletries industry this is where the most important aspect, safety, has to be built in. The following questions must be addressed.

- Are the raw materials legally allowed?
- Are they safe on the skin?
- Are they safe in combination with the other ingredients?
- Is the formulation well designed from a microbiological point of view?
- Will the preservation system be effective during the intended life of the product?
- Is the product stable in its intended packaging?
- Is the product and its packaging environmentally friendly?

The regulatory matters governing the production and sale of cosmetics will be discussed later in this chapter (see section 12.7 and chapter 11).

12.5.1.3 *Design for manufacture* Quality cannot be inspected into a product, but must be designed for manufacture. This includes:

(i) design with tolerances that can be held with existing machinery, or formulations that can be handled by available process plant; and
(ii) design so that the product can only be assembled in the correct and most cost-effective manner.

In many companies the designers do not involve production staff at an early stage of a product's development.

12.5.1.4 *Product approvals* An increasing number of industries and markets require product approvals, which must be taken into account at the design stage. Techniques available to improve the quality of design include the following.

- Value analysis
- Failure mode and effect analysis
- Taguchi
- Competitive benchmarking
- Quality function deployment
- Reliability testing

12.5.2 *Quality assurance*

As described earlier, quality assurance is the development of systems, procedures and disciplines that cover the complete manufacturing process—from the design stage through to final testing and commissioning to ensure that the products are produced to the correct quality standards. It includes:

(i) metrology—the science of measurement; and
(ii) statistical process control (SPC)—a powerful technique for monitoring quality during manufacture.

SPC can be used in both process and batch production industries. It has the added advantage that it can, in many cases, be carried out by the operators—a very effective way of involving them in the search for quality.

12.5.3 *Manufacturing processes and technology*

Manufacturing processes clearly affect quality and, together with technology, they need to be continually developed and improved to ensure that they are capable of consistent production of the required quality. Analytical techniques and solutions to problems are numerous. Some examples are listed below.

(i) Problem solving
- Pareto analysis
- Histograms
- Variation research
- Cause and effect analysis
- Scatter diagrams
- Regression analysis
- Taguchi
- Process mapping
- Brainstorming
- Control charts
- Project groups/task forces

(ii) Solutions
- Design modifications
- Operator training/retraining

- De-skilling
- Improved equipment
- Automation
- Better tolerancing
- Improved material specification
- Introducing or extending SPC
- Tighter process control
- Improved materials
- Preparation
- Vendor assessment

12.5.4 *Cost of quality*

The establishment of quality costs is an essential part of any quality improvement programme. Various methods are available, and should cover the following items.

- Type (i.e. prevention, appraisal, failure)
- Product line
- Manufacturing process or department
- Type of cost (material, labour)
- Causes (e.g. design, material, operator, training, etc.)
- Scope for reduction, with timescale and paybacks

These costs should identify where the maximum benefits of the quality improvement programme can be obtained from. It is important to continually monitor these costs once the programme is underway, to enable success to be quantified and to identify what still needs to be done.

12.5.5 *Human and organisational aspects*

12.5.5.1 *Top management commitment* The most important contributor to total quality is top management commitment. For many companies this requires a wholesale change of attitude—a complete cultural change within the company. Top management control relies on the top management being prepared to stop the production line and solve the problems, rather than letting the production line run for subsequent sorting. In this respect, the top management must lead by example—quality cannot be delegated. The same principle must apply at every level throughout the organisation. Various studies of successful companies show that, without exception, management is obsessed by quality.

12.5.5.2 *Education* To extend commitment to quality throughout each level of the organisation requires a major educational programme, from board level down to shop-floor or office worker. Although the content of these educational courses will also depend on the management level, most will cover the following aspects.

- The importance of quality to the company in achieving its corporate objectives
- How the quality performance of a company is perceived in the marketplace
- The company's own cost of quality
- Training in the quality techniques such as brainstorming, fishbone charts, pareto analysis, histograms and quality circles
- Ways in which those being trained can contribute to improving quality.

The courses should be supported with exercises and case studies that give customer focus, evaluate processes, anticipate working for continuous improvement and look to a leadership role in quality.

12.5.5.3 *Participation* Participation is an essential part of total quality, involving everyone within an organisation. The greater the involvement, the greater the rewards. Management needs to be far more open than they have been historically.

12.5.5.4 *Communication* The 'initial customer' concept is a very effective way to break down barriers and improve communications. Each department (or division) treats all other departments with which it deals as either customers or suppliers and:

(i) defines the products or services that it gives or receives;
(ii) agrees performance and quality measures; and
(iii) monitors performance against targets.

The lessons learnt from this approach can be salutary, and the results impressive. It can do more to transform attitudes than any other single change in a total quality programme. Internal customer/supplier agreements can be instituted.

As part of the continuing educational process, employees need to know what the company is doing and why it is doing it, on an on-going basis. Briefing groups, house magazines and notice boards can all play their part in heightening awareness of the importance of quality, and in reporting successes.

12.5.5.5 *Visible quality* People must be able to see how good quality is, and what the first time pass rate (or other measure) is. New measures may need to be introduced. Where possible, pass rates rather than reject rates should be stressed; in other words positive policing should be applied.

12.5.5.6 *Quality circles* Quality circles are voluntary meetings of peer groups to identify and resolve problems in their immediate working area. They can be a very effective way of involving the workforce in the search for better quality. Quality circles tap latent creativity of employees and involve them, where possible, in implementing the suggested solutions. The right environment, a commitment from the top and a willingness to listen by top management are all required.

12.5.5.7 *Self-inspection* Self-inspection is an integral part of the total quality philosophy, giving the operators the responsibility for the quality of their own work—once their attitudes are right. Most operators want to take over their own inspection, and often set themselves higher standards than inspectors—so-called, 'pride in the job'.

12.5.5.8 *Supplier involvement* A company's contacts with its suppliers are often between the buyers and the sales man. Companies should get much closer to their suppliers at senior management level, and involve them in their quality problems. Suppliers can be a very useful source of information and assistance. Most companies adopting total quality end up dramatically reducing the number of their suppliers, single sourcing wherever practical and offering long-term contracts in return for good performance. Supplier involvement is much more than vendor assessment. New relationships with employees and new partnerships with suppliers, should be developed.

Some companies have sensibly used common training material and common training sessions with their suppliers. One or two companies have even taken it upon themselves to train their key suppliers. The benefits of this are a common language and common understanding. It can provide detailed insight into processes and invites continuous improvement from which both parties can benefit. Developing in parallel with these initiatives have been electronic data interchange (EDI) where customers' and suppliers' computers interchange requirements directly and vendor managed inventories (VMI) whereby a customer's stock movements can be tracked using data interchange by the supplier's computer which will then schedule replenishment production. These initiatives have greatly improved accuracy and substantially reduced inventories to the benefit of both.

12.5.5.9 *Quality improvement projects* Quality improvement projects are undertaken by cross-functional and often multi-disciplinary project teams working on a problem or series of related problems. Members usually require training in effective group working and in problem-solving techniques.

12.5.5.10 *Organisation* As stated previously (section 12.5.5.1), the responsibility for quality must be with top management. In a manufacturing unit, this means that product quality must be the responsibility of the production department. It cannot be delegated to the quality assurance department—they should be facilitators.

Many companies have reorganised themselves around processes rather than stay with functional hierarchies. Processes not functions drive the business. A process view of an organisation illustrates:

- The interdependency of organisational departments,
- The primacy of the customer (consumer),
- The impact of customer requirements and response (e.g. consumer research),
- Continuous improvement, based on customer requirements and response,
- The importance of suppliers,
- The network of internal customer / supplier relationships.

This process view can have significant benefits allowing decisions to be made in the most appropriate place by the best qualified individuals. It also results in delayering, the removal of unnecessary levels of supervision and management.

12.5.5.11 *The quality plan* A quality plan should form part of each company's corporate plan and its objectives should be included amongst the organization's strategic business aims. The plan will need to be updated regularly—probably annually—and should include detailed information on quantifiable quality improvements and financial benefits, together with the resources required to realise them.

12.6 The new thinking

A simple way of illustrating the differences between the old and new concepts of quality is shown in Table 12.1.

Table 12.1 Differences between the old and new concepts of quality

Old	New
Quality is about products	Quality is about organisations
Quality is technical	Quality is strategic
Quality is for inspectors	Quality is for everyone
Quality is led by experts	Quality is led by management
Good quality is high grade	Quality is at the appropriate grade
Quality is about control	Quality is about improvement

12.7 Quality standards and guides

As one might expect, there are many quality standards and guides and much legislation (see section 12.8). All total quality approaches rely on some kind of underpinning quality system, and a good quality system should cover common ground no matter what the product or industry. The international standard for quality systems—ISO 9000—developed from the haphazard approach described earlier. It laid down internationally agreed criteria, with inspection from independent assessors (third party assessment) to ensure that the standard was met. Such systems were engineering-led but are fast gathering favour within the cosmetic and toiletry industry.

A number of cosmetic manufacturers are owned by pharmaceutical conglomerates, and their route to a quality system has been through the manufacture and fill of 'over-the-counter' (OTC) pharmaceutical products. Companies holding medicinal manufacturing licences are independently audited by inspectors to maintain minimum standards. In the UK, these companies must refer to the so-called 'Orange Guide' [5], which, although it has no statutory force, recommends steps that should be taken as necessary and appropriate by manufacturers of medicinal products, with the object of ensuring that their products are of the nature and quality intended. In the US, OTC review panels set product standards for certain cosmetic products.

Both ISO 9000 and the Orange Guide tend to focus on the manufacturing or operational side of the business, thus missing an opportunity to exercise some control over an industry frequently dominated by marketing, sales or finance. This is where total quality can be very effective—given leadership from the top, it touches the whole company. Wise companies now adopt a dual approach, following total quality whilst trying for ISO 9000 accreditation and/or complying with the requirements of inspectors.

12.7.1 *Links between ISO 9000 and total quality*

Put simply, ISO 9000 is a recognised and established example of a quality system, and all total quality approaches rely on an underlying quality system of some sort. At the same time, total quality pundits say that it is difficult for an organisation to successfully install ISO 9000 without first creating a management-led total quality approach. Consequently, it can be seen that both approaches are independent and the distinction between the two is blurred. The links between the two can be demonstrated by comparing the features of both alongside each other (Table 12.2). This brings out their strengths when linked together.

Companies advocating total quality often demand quality processes in advance of ISO 9000 requirements from their suppliers, although they still see ISO 9000 as a positive indicator. However, it is not entirely true to say that ISO 9000 is short-sighted or short-term focused, as it does require a quality policy and objectives, and organisation and management review, in addition to quality systems.

ISO 9000, BS 5750 (the UK quality standard) and the European standard EN29000 all had similar origins. Recognising the blurring referred to above,

Table 12.2 Comparison of ISO 9000 with total quality

ISO 9000	Total quality
Standards to ensure that things are done correctly	Focus on doing things correctly and on doing the right things (strategic)
Primarily product/service focused	Company-wide, covering all departments
A system	Philosophy/management approach
No requirements for employee involvement in improvement	Emphasis on total employee involvement/commitment/attitude change
Goal—meet the standard pass the audit	Goal—continuous improvement
Low visibility	Company-wide visibility

together with a wish for uniformity, the three standards have recently been revised and merged to become BS EN ISO 9000. Such standards are regularly reviewed and this was the case in 1994. In fact it is a requirement of the International Standards Organisation (ISO) that all of its standards are reviewed and either revised or revalidated every five years. This latest revision is part of a broader programme resulting from a long term strategy adopted by the international standards committee responsible for the ISO Series. This strategy has been published as a document called Vision 2000 [6].

Vision 2000 calls for a two phase programme that addresses the following issues

Phase one published in 1994:

(i) the need to correct inconsistencies and errors,

(ii) the need to improve the wording so that the standard is now more applicable to its wider usage, particularly sectors such as service industries,

and Phase two due to be published in 1998:

(iii) the need to make more significant changes to take account of the move towards the principles of total quality management.

The 1994 version has added emphasis and clarification. The new standard is not only updated but improved. Of major importance is the requirement placed on companies to operate as they see fit. The introduction of the standard states clearly that 'It is not the purpose to enforce uniformity of quality systems' and 'The design and implementation of a quality system has necessarily to be influenced by the varying needs of an organisation'

Virtually all clauses have a word change of some sort with varying degrees of impact. Significant changes are made in the following areas:

- *Policy.* The quality policy now has to 'be relevant to the supplier's organisational goals and the expectations and needs of its customers', i.e. it allows a customised approach to quality management. It will require companies to implement an appropriate level of quality management and not do anything less than is required to meet the needs of the customer. It also introduces the concept of management with executive responsibility for quality. This concept is used several times and emphasises that top management should be promoting quality and giving the lead. The greatest challenge remains in being able to demonstrate that the policy is being implemented and understood by all employees. Policies are one thing but actions are something else.

- *Planning.* The 1994 version has extended clause 4.2 to three sub-clauses:
  - (i) General—the standard now requires a quality manual (previously this was not a requirement although most companies did produce one),
  - (ii) Quality system procedure—this adds 'the degree of documentation required ... shall be dependent upon the methods used, skills needed and training required by personnel', i.e. only procedures that are actually required should be written.
  - (iii) Quality planning—this includes seven guidance notes amongst its eight requirements. Additional effort will undoubtedly be necessary to prove compliance.
- *Preventive action.* The new standard requires that action be taken before problems occur to prevent potential non-conformities.

  A reassuring clarification within the general paragraph is the comment 'actions taken shall be to a degree appropriate to the magnitude of the problem and commensurate to the risk encountered'. Companies may well find it advantageous to develop a risk analysis procedure to demonstrate why, or why not, data is analysed and action taken.
- *Purchasing.* Documented procedures are now required for purchasing. Those procedures need to evaluate the sub-contractor before selection, and the supplier shall define the type and extent of control exercised over the sub-contractor. There is no absolute requirement to conduct quality audits or obtain a questionnaire from sub-contractors, but there must be evidence of some form of assessment of capability.

12.7.1.1 *Bureaucracy versus the paperless approach!* Assessment bodies will frequently cite document and data control as the biggest cause of non-conformance to ISO 9000. The effort and resource often put in place to control this can be a burden in itself, without taking into account the associated record keeping and the remaining company information. Computers can help with the creation of documentation but, with networks, documents can be rapidly proliferated throughout the organisation, working against those trying to control it. In addition documents created by word processors and stored on disk are generally invisible unless printed.

However. software now available is specifically designed to address these issues, designed to act as a framework for building and running the documented system and managing the associated records. Document control issues are removed such that the time of the quality manager can be better spent being proactive towards quality matters, rather than being burdened with administration.

12.7.1.2 *The future* One of the great gurus of quality, Dr. Juran, made some interesting predictions at a conference in late 1994 [7]. It seems appropriate to close on these predictions:

- Quality competition will intensify with multi-nationals and common markets.
- There will be intense demands on suppliers.
- ISO 9000 will sweep across the world.
- Awards, e.g. Baldrige, will supply intense stimulus and there will be a growth in awards worldwide.

References

1. Juran, J.M. (1974) *Quality Control Handbook*, 3rd edn. McGraw-Hill, New York.
2. European Organisation for Quality Control (1971) *Glossary of Terms used in Q.C.* Rotterdam.
3. Crosby, P.B. (1979) *Quality is Free.* Mantor, New American Library, New York.
4. Crosby, P.B. (1984) *Quality without Tears.* McGraw-Hill, New York.
5. HMSO (1983) *Guide to Good Pharmaceutical Manufacturing Practice.* HMSO, London.
6. Vision 2000. BSI Standards, ref PD6538, BSI, 2 Park Street, London W1A 2BS.
7. Juran, J.M. (1994) *The Last Word.* 12th Annual Conference on Managing for Total Quality. Published in *Quality World*, January 1995, 30.

Further reading

CTFA (1991) *CTFA International Resource Manual.* 3rd edn. CTFA Inc. Washington, D.C.
Imai, M. (1986) *Kaizen* Random House, New York.
Oakland, J.S. (1989) *Total Quality Management* Heinemann Professional Publishing, Oxford.
Smith, W.S. (1989) *The Delighted Customer* Vol. 1—*The Quality Revolution* Quest Quality Consulting, London. Unilever Internal Publication.

13 Environmental issues

D.F. WILLIAMS

13.1 Introduction

Environmental regulation is widespread and virtually all embracing throughout the world, particularly in Europe and the USA. Much of the legislation in Europe is in a state of flux and still to be resolved into clear regulations but in the USA the environmental restrictions on industry have been in place for a long time. The Environmental Protection Agency (EPA) in the USA is long standing and environmental protection agencies have now been set up in the UK and in the European Union.

By and large the more draconian measures being put into effect in Europe and the UK under the Integrated Pollution Prevention and Control (IPPC) and the Integrated Pollution Control (IPC) bodies do not apply directly to the formulation or the manufacture of cosmetics and toiletries. The prescribed processes under these regulations do not include this industry and are more concerned for the moment at any rate with the heavier industry polluters of air, water and the land. Of much more concern to the formulator and manufacturer of cosmetic and toiletry products are the regulations and proposed tax levies associated with waste and with packaging.

Marketing attention is more and more becoming focused on environmental concerns as the pressure from consumers increases. The result of pressure from customers and retail outlets on the use of animals and animal-derived materials as well as ozone-depleting gases is now well known and established. The attention of the regulators is now turned to packaging waste and of the consumer to the origin of the raw materials used in the products. Attention to the use of energy is a significant concern also and formulations and methods of manufacture will need to take this into account. The use of water will be important as will all emissions from processing.

Concern is building up over the use and emission of volatile organic compounds (VOCs). VOCs are precursors in the formation of ground ozone (photochemical smog) and proposals for regulatory Directives are under consideration in the European Union (EU). Regulations which directly influence the formulation of cosmetic and toiletry products are already in place in a number of the States in the USA.

There are of course many other factors which may affect the environmental policies of companies and which may well influence the customer in the choice of products but in this chapter only those aspects concerned directly with the formulation, development and manufacture of the product are considered.

13.2 Raw materials

When considering the development of a new product it is common now to concentrate on materials that have a long history and which have not been tested on animals for many years. In the EU the 6th amendment to the cosmetic directive prohibits the use of animals for the testing of cosmetic products and raw materials from 1998. It is hoped that by the time this regulation comes into force there will be adequate alternatives to animal testing that will provide protection for the consumer and the manufacturers of the products from adverse effects. A problem still exists and will continue to exist in the regulations governing the notification of new substances. The law at the moment in the EU requires evidence of toxicological data from animal testing and until this is changed the cosmetic and toiletry industry will face a dilemma when considering the use of newer materials.

In the USA the state of California has set limits on the level of contaminants in raw materials which has resulted in the need for manufacturers to examine more stringently the levels of contaminants in their products. Contamination levels above the limits could result in the requirement to label products with compromising toxicological statements. No doubt the possible effects of contaminants on the environment will be considered in due course.

The effect on the environment of the raw materials used and the final product is of increasing importance and regulations requiring the labelling of chemical substances with an environmental declaration are spreading through the European Union. The main driving forces are the notification of new and existing chemicals and the regulations governing the control, packaging and labelling of hazardous substances both for supply and for transportation.

13.3 Energy

The conservation of energy is of course a prime concern but may not be the most significant factor in the manufacture of cosmetic and toiletry items. The use of heat is normally low and the mixing process does not often require large energy input. There are opportunities however for reducing these energy requirements by the careful selection of ingredients for a formula which may enable cold processing or lower temperatures for the preparation. In addition the energy for mixing can be minimised by similar considerations.

More efficient emulsifiers, lower melting waxes and fatty materials and more easily solubilised ingredients offer an approach to these objectives.

13.4 Water

The use of water as an ingredient offers little opportunity for savings since the formula requirement is often specific. Water for processing does however present a challenge. If hot processes can be eliminated then obviously the need for cooling water is removed. Lower temperatures and quicker mixing time will result in the use of less water for cooling. If water is required for cooling it may be possible to recycle it or to use the effluent from the cooling jackets or heat exchangers to heat some other material process. These points are obvious but can perhaps be built on with ingenuity and perhaps can be used to advantage in furthering the company's customer image.

13.5 Waste

As far as the cosmetic and toiletries industry is concerned the significant waste factors can be separated into two major areas; that produced as a result of the formulation and manufacture of the product and that produced due to the packaging of the product. Waste produced in the factory has of course been a traditional area of control from the cost point of view but it has taken on a greater significance in recent years due to environmental factors. Waste product from rejected batches of products must be disposed under strict control. Hazardous materials may be involved. Rejected raw materials fall into the same category. In Europe disposal must be carried out by registered waste carriers and of course paid for. The same principles apply to reject packaging materials and faulty components on line. It is important to ensure good quality componentry to avoid rejects and down time on the lines but now the disposal factor assumes greater significance in preventing cost increases. Preventative quality assurance measures and built in fail safe considerations at the design stage thus become even more important.

13.6 Packaging waste

It will be necessary as time passes to pay great attention to the way cosmetics are packaged for sale due to the regulations now being discussed in the EU and other countries. These regulations will in fact impose a levy on the packaging industry and allied industries designed specifically to discourage the use of excessive packaging. The imposition of tax on waste destined for landfill will also affect the design of future packs. Point of sale and merchandising units will have to be looked at with environmental considerations in mind.

Packaging waste provides a challenge and an opportunity particularly in the product development area. The prime target of most governments concerned with environmental problems is to minimise waste. The retail packs of consumer products therefore provide an avenue of approach towards achieving this aim offering the consumer an environmentally friendly product. In the USA, California and Oregon are imposing reductions in weight or minimum levels of recycled material in packs above 8 oz or 240 ml.

Design of packs that will use the minimum of material whilst still adequately protecting the product will become increasingly more important. Reusable packs and packs that can be recycled will offer further opportunities and failing all else packs that can be disposed of with no environmentally deleterious effect or indeed which may provide an environmental benefit must be considered.

Use of materials from sustainable sources and those which cause little or no environmental problems or do not use large amounts of energy in their preparation will increase in significance.

As ever the development of a product must ensure that the product and the pack are designed together and that all members of the team are aware of the environmental consequences.

13.7 Eco labelling

Consumer demand for information on the environmental acceptability of the products they buy has led to the spread of 'Eco labelling' schemes throughout the world. Individual countries and the EU are devising their own definitions and specifications required for the award of an 'Eco' label, an important factor in the development of products for the future.

13.8 Volatile organic compounds

The present situation on the use of VOCs is far more stringent in the USA where state legislation limits considerably the use of volatile materials in cosmetic and toiletry formulations. This drastically affects the formulation of hairsprays and fixatives, mousses and conditioners as well as nail polish remover, antiperspirants, deodorants and fragrances. These limitations will lead to new technology since at the moment solutions offered to date have not gained great consumer support. Elimination of the aerosol as we know it today seems to be the main target of the regulatory bodies.

In Europe at the moment the situation is still developing and is not yet clear. No overt regulations are aimed specifically at cosmetic and toiletry products. There is more concern over the sale and storage of petroleum, the coating, printing and paint industries where great changes have already taken place in formulations and processes. The forward looking cosmetic and toiletry formulator will be well advised to watch developments in the USA and use innovation to lessen the dependence on the use of volatiles.

13.9 Environmental management systems

It is worthy of note that public concern about the environmental performance of manufacturers of consumer products is growing. This concern may well influence the choice of products the consumer may make in the highly competitive marketplace. Any company that is able to convince the consumer of its concern and care about the effects of its activities on the environment may therefore well steal an edge on its competitors. It is also a distinct possibility that the retailers who place a great deal of emphasis on their environmental image may well force the pace and insist that their suppliers show evidence of their commitment to protecting the environment.

Throughout the world therefore a number of schemes are being developed that will enable companies to demonstrate such commitment through standards and certified environmental management schemes. The British Standards Institute have introduced BS 7750 and on the international front an ISO Scheme (ISO 14000) is being finalised. These schemes set out the requirements for a management system and structure which if adopted and accepted by independent accredited assessors enable companies to claim certified status. EMAS (Eco Management and Audit System) is a regulation introduced into the EU (Regulation 1836/93) which set up a voluntary scheme and was launched as a final scheme framework in April 1995. The above standards compliment this regulation. Other standards relevant to the Eco Management systems are BS ISO 14001 and 14040 (Life Cycle Analysis). In January 1995 these were all in draft form and others were under preparation covering aspects of life cycle inventory, impact assessment and improvement assessment. It is essential therefore that workers in the field of consumer product maintain awareness of the developments in this important area.

13.10 Sources of information

Environmental protection agencies in the USA, the Environment Protection Agency (EPA), the EU, the European Environment Agency (EEA) and the UK, Environment Agency (EA) have been set up to collect and provide objective and reliable environmental information and to assist governmental bodies in the USA, UK and European Union in formulating sound policies to protect the environment. The EU EEA (European Environment Agency) will publish a newsletter periodically and other countries throughout the world will no doubt set up environmental protection bodies in time.

Trade Associations provide regular summaries of the individual state's regulations and the development of international regulations and an extremely important source of information on rapidly changing environmental issues.

The *Official Journal of the European Community* (OJ) provides detailed information on all Regulations, Directives and proposed Directives operating within the European Union. Some of the more important and relevant environmental legislation is listed below.

13.10.1 *European Community*

- First Environment Action Programme (OJ C112 20-12-73).
- Fifth Environment Action Programme. Towards Sustainability (OJ C138 17-5-93).
- Proposed Directive for The Integrated Pollution Prevention and Control, a Framework Directive (OJ C311 17-11-93).
- The Directive on Packaging & Packaging Waste 94/62/EC (OJ L365 31-12-94).
- The Directive on Waste (Framework) 75/442/EEC amended by 91/156/EEC (OJ L194 12-2-75 & OJ L78 26-3-91).
- The Directive on Toxic and Dangerous Waste 78/319/EEC (OJ L84 31-3-78).
- The Directive on Hazardous Waste 91/689/EEC (OJ L377 31-12-91).
- The Draft Directive on the Landfill of Waste (OJ C212 5-8-93).
- Regulation on the Evaluation and Control of the Risks of Existing Substances EC Regulation 793/93 (OJ L84 5-4-93).
- EEC Council Regulation implementing the Montreal Protocol on Substances that deplete the Ozone Layer (OJ L67 14-3-91).
- The Directive on Air Pollution by Ozone 92/72/EEC (OJ L297 13-10-92).
- The Directive on the Approximation of the Laws, Regulations and Administrative Provisions relating to the Classification, Packaging and Labelling of Dangerous Substances. 67/548/EEC (OJ L196 16-8-67).
- The above amended and adapted by many Subsequent Directives up to 93/101/EEC (OJ L13 15-1-94).
- Decision establishing a list of wastes pursuant to Directive 75/442/EEC (OJ L5 7-1-94).
- Regulation on the establishment of a European Environment Agency (OJ L120 11-5-90).

- EEC Council Regulation on a Community Eco- Label Award Scheme 92/880/EEC (OJ L99 11-4-92).
- Working Document for a Draft Marking for Packaging Directive. 1-4-95 (Commission of the European Communities).

13.10.2 *USA*

- State Regulation of the Volatile Organic Compound content of Personal Care Products. April 1995 (CTPA).
- The Federal Clean Air Act (CAA) Amendments 1990.
- The State of California Proposition 65 (Contaminants in Raw Materials).

There are many other regulations governing waste solid and water quality operating in the various states and details can be obtained from the EPA.

13.10.3 *UK*

- The Environmental Protection Act 1990.

Appendices

Appendix I: List of suppliers

United States

Akzo Chemicals Inc., 300 South Riverside Plaza, Chicago, Illinois 50505

Allied-Signal Inc., A-C Performance Additives, PO Box 2332R, Morristown, New Jersey 07962–2332

Alzo Inc., 6 Gulfstream Blvd., Matawan, New Jersey 07747

Amerchol Corporation, 136 Talmadge Road, PO Box 4051, Edison, New Jersey 10818–4051

Amoco Chemical Company, Mail code 4101, 200 East Randolph Drive, Chicago, Illinois 50501–7125, 800-621-4567

Aqualon Company, 2711 Centerville Road, PO Box 15417, Wilmington, Delaware 19850–5417

BASF Corporation, 140 New Dutch Lane, Fairfield, New Jersey 07004

Bernel Chemical Company Inc., 30 Park Place, Englewood, New Jersey 07631

B.F. Goodrich Chemicals, 6100 Oak Tree Boulevard, Cleveland, Ohio 44131

B.F. Goodrich, Specialty Polymers & Chemicals Division, 9911 Brecksville Road, Cleveland, Ohio 44141–3247, 800-331-1144

Cascade Chemical Company, PO Box 438, Park Ridge, New Jersey 07656–0438

Croda Inc., 183 Madison Avenue, New York, New York 10016

Dow Corning Corporation, 2200 W. Salzburg Road, Midland, Michigan 48686

EM Industries Inc., Fine Chemicals Division, 5 Skyline Drive, Hawthorne, New York 10532

Finetex, PO Box 216, Elmwood Park, New Jersey 07407

GAF Chemicals Corporation, 1361 Alps Road, Wayne, New Jersey 07470

Goldschmidt Chemical Corporation, 914 E. Randolph Road, PO Box 1299, Hopewell, VA 23860, 800-446-1809

W.R. Grace & Co, Co.-Conn., Davison Chemical Division, PO Box 2117, Baltimore, Maryland 21203–2117

Henkel Corporation, Emery Group, 300 Brookside Avenue, Ambler, PA 19002

Heterene Chemical, 295 Vreeland Avenue, Paterson, New Jersey 07513, 201-278-2000

Hoechst Celanese, Colorants & Surfactants Div., Route 202-206 North, Somerville, New Jersey 08876

ICI Americas Inc., ICI Specialty Chemicals Wilmington, Delaware 19897, 800-456-3669

Inolex Chemical Company, Jackson and Swanson Streets, Philadelphia, Pennsylvania 19148-3497

Kelco, Division of Merck & Co., Inc., 75 Terminal Avenue, Clark, New Jersey 07066

Lonza, Inc., Specialty Chemicals Division, Corporate Headquaters, 17-17 Route 208, Fair Lawn, New Jersey 07410

National Starch & Chemical Corporation, Resins & Specialty Chemicals, Finderne Avenue, Bridgewater, New Jersey 08807

Reheis Inc., 235 Synder Avenue, Berkley Heights, New Jersey 07922

Rhone-Poulenc Inc., Surfactants & Specialties, 1099 Winterson Road, Linthicum, Maryland 21090

Rohm & Haas, Independence Mall West, Philadelphia, Pennsylvania 19105

Sandoz Chemicals, 4000 Monroe Road, Charlotte, North Carolina 28205

Scher Chemicals, Inc., Industrial West, Clifton, New Jersey 07012

Sherex Chemical Company, Inc., 5777 Frantz Road, PO Box 646, Dublin, Ohio 43017

Stepan Company, 22 Frontage Road, Northfield, Illinois 60093

Sutton Laboratories, Inc., 116 Summit Avenue, Chatham, New Jersey 07938

Tri-K Industries, Inc., 466 Old Hook Road, PO Box 312, Emerson, New Jersey 07630

Trivent Chemical Company, Inc., 45 Ridge Road, PO Box 597, South River, New Jersey 08882

Union Carbide Corporation, Specialty Chemicals Division, 39 Old Ridgebury Road, Danbury, Connecticut 06817

Van Dyk, Mallinckrodt Specialty Chemicals, Co., Drug & Cosmetic Chemical Division, 11 William Street, Belleville, New Jersey 07109

Wickhen Products Inc., Big Pond Road, Huguenot, New York 12746–0247

Europe/United Kingdom

Akzo Chemie UK Ltd, 1–5 Queens Road, Herstam, Walton-on-Thames KT12 5NL

Albright & Wilson, White Haven, Cumbria CA28 9QQ

BASF (UK) Ltd, PO Box 4, Earl Road, Cheadle Hulme, Cheshire SK8 6QB

B.F. Goodrich, Hounslow, Middlesex

BP Chemicals Ltd, Belgrave House, 76 Buckingham Palace Road, London SW1W OSU

Croda Chemicals Ltd, Cowick Hall, Snaith, Goole, North Humberside DN14 9AA

Dow Corning Ltd, Avco House, Castle Street, Reading RG1 7DZ

Durham Chemicals Ltd, Birtley, Chester-Le-Street, Durham DH3 1QX

GAF Europe, Rythe House, 2 Littleworth Street, Esher, Surrey KT10 9PD

Goldschmidt AG, Goldschmidtstrasse 100, D 4300 Essen 1, Germany

Henkel Cospha KGaA, Henkelstrasse 67, Postfach 1100, D4000, Dusseldorf 1, Germany

Hercules Ltd, 20 Red Lion Street, London WC1R 4PB

Hüls AG, Referat 1122, D 4370 Mari, Germany

Kelco AIL Ltd, 22 Henrietta Street, London WC2E 8NB

Unichema International, PO Box 2, 2800 AA Gouda, The Netherlands

Union Carbide UK Ltd, Union Carbide House, High Street, Rickmansworth, Hertfordshire WD3, 1RB

Wickhen Products Inc., Cumberland House, Greenside Lane, Bradford BD8 9TO

Appendix II: Useful addresses

ACOPER, Igancio Chiappe Lemos, Calle 37, No. 7–43, Bogata, Colombia

ASCOPA, Case postale 230, CH-1211 Geneva 3, Switzerland

Associacao Brasileria da Industria de Produtos Limpezae Afins, 1570-8 Andar, San Paulo, Brazil

Associacion de Fabricantes de Articulos de Tocador, Los Laureles 365, San Isidro, Lima, Peru

Associazione Nazionale dell'Industria Chimica (ASCHIMICI), Via Fatebenefratelli 10, I-20121 Milano, Italy

Camara Argentina de la Industria de Productos de Hygiene y Tocador, Paraguary 1857, Capital Federal, Buenos Aires (1121), Argentina

Camara de Industria Cosmetica, San Antonio 427, Santiago, Chile

Camara Nacional de la Industria de Perfumeria y Cosmetica, Talsan No. 54 Despacho 6, Mexico 1, D.F., Mexico

Camara Venezolana de la Industria de Cosmeticos y Afines (CAVEINCA), Edificio 'IASA' Plaza la Castellana, Oficiana 106 Apartado 3577, Caracas, Venezuela

Canadian Cosmetic, Toiletry and Fragrance Assn. (CCTFA), 24 Merton Street, Toronto, Ontario M4S 1A1, Canada

Chamber of Cosmetics Industry of the Philippines, PO Box 4541, Manila, Philippines

COLIPA, Rue de la Loi, 223 (Bte 2), B-1040 Brussels, Belgium

Cosmetic, Toiletry and Fragrance Assn. of Australia, 60 York Street, Sydney, New South Wales 2001, Australia

Cosmetic, Toiletry and Perfumery Assn. Ltd. (CTPA), 35 Dover Street, London WIX 3RA, England

CUPCAT, Adva. Agraciada 1670 Piso 1, Montevideo, Uruguay

Fachverband der Chemischen Industrie Osterreichs, Gruppe Korperpflegemittclindustrie, Schliessfach No. 69, A-1011 Wien, Austria

Federation Francaiso de l'Industrie des Produits de Parfumerie, de Beaute et de Toilette, 8, Place du General Catroux, F-75017 Paris, France

Indian Soap and Toiletries Makers Assn., P-11, Mission Row Extension, Calcutta 1, India

Industrieverband Körperpflege und Waschmittel e.V. (IKW), Karlstrasse 21, D- 6000 Frankfurt/M 1, Germany

International Cosmetic Regulations, Allured Publishing Corporation, PO Box 318, Wheaton, Illinois 60187, USA

International Federation of Society of Cosmetic Chemists, Delaport House, 57, Guildford Street, Luton, Bedfordshire LU1 2NL, UK

Japan Cosmetic Industry Assn., 4th Floor Hatsumei Bldg., 9-14, 2-chome Toranomon, Minato-Ku, Tokyo 105, Japan

Kemisk-tekniska Leverantorforbundet (KTF), Box 1542, S–111, 85 Stockholm, Sweden

Korea Cosmetic Industry Assn., 45-1, Pil-Dong, Jung-Ku, Seoul, Korea

Kosmetikkleverandorenes Forening, Boks 6780 St. Olavs Pl., Oslo 1, Norway

Malaysian Pharmaceutical Trade and Manufacturers Assn., 3rd Floor, Jaya Supermarket, Jalan Semangat, Petaling Jaya, Malaysia

Nederlandse Cosmetica Vereniging, Gebouw Trinderborch, Catharijnesingel 53, 3511 GC-Utrecht, Netherlands

New Zealand Cosmetic and Toiletry Manufacturers Federation, PO Box 9130, Wellington, New Zealand

PERKOSMI, C/O P.T. Kalbe Farma, J1 Jendral a Yani, Jakarta, Indonesia

Saebo Parfumeri Toilet & Komisktekniske Artikler (SPT), Ostergade 22, 1100 Copenhagen, Denmark

Savez Farmaceutiskih Drustave Jugoslavije Sekcija Za Kozmetologiju, Askercerva 9, 6100 Ljubljana, Yugoslavia

Sindicado da Industria de Perfumaria e Artigos de Toucador, Avenida Calogeras, 15, 4° andar, CEP 20030, Rio de Janeiro RJ, Brazil

Sindicado da Industria de Produtos de Perfumarias e Artigos de Toucador, Av. Paulista, 1319-9 Andar, San Paulo, Brazil

S.T.A.N.P.A., San Bernardo, 23-2, Madrid 8, Spain

Teknokemian Yhdistrys r. yu., Fabianinjatu 7B, SF-00130 Helsinki 13, Finland

The Cosmetic Manufacturers Assn., 292/37 Lan luang Road, Siyek Mahamark, Khet Dusit, Bangkok, Thailand

The Cosmetic, Toiletry and Fragrance Association Inc., Suite 800, 1110 Vermont Avenue N.W., Washington, D.C. 20005

The Proprietary Association of South Africa, PO Box 933, Pretoria 0001, South Africa

Unione Della Profumeria e Della Cosmesi (UNIPRO), Via Buonarroti, 38- Milano, Italy

Unione Panhellenique des Industriels et Agents de Produits Cosmetiques et de Parfumerie, 28 rue Academias, Athens, Greece

Verband der Kosmetik-Industrie, Breitingerstrasse 35, 8002 Zurich, Switzerland

Index